The Complete Home Carpenter

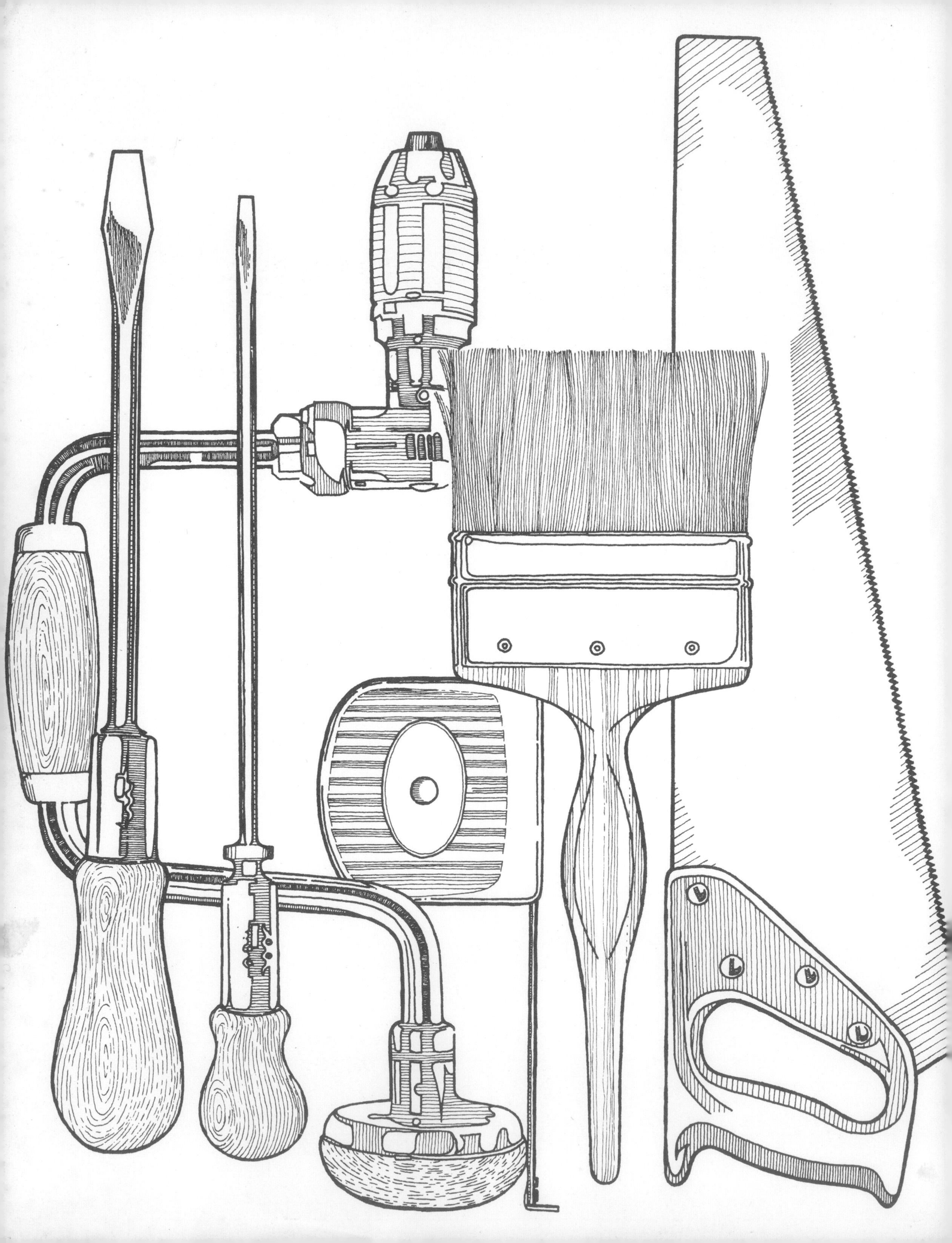

The Complete Home Carpenter

Editor: George Daniels

Bobbs-Merrill
Indianapolis / New York

Published by The Bobbs-Merrill Company, Inc.
Indianapolis New York
Published in Great Britain by Marshall Cavendish Ltd.
ISBN 0-672-52066-4
Library of Congress Catalog Card Number: 74-17659
Manufactured in the United States of America

ISBN 0-672-52067-2 Pbk.
First U.S. paperbound printing, 1978

About this book

Carpentry is one of the few do-it-yourself skills that offers something for everyone. Even the inexperienced carpenter can make attractive and functional furniture and carry out repairs to timber in the home. And carrying out a simple job gives you the skills and techniques to tackle more ambitious projects.

This book covers all aspects of carpentry. There are full instructions for making originally designed furniture for every room in the home — superb kitchen units, a beautiful gatelegged dining table and a Welsh dresser, modern bed platforms, a cocktail bar and built-in units for the living room, and a compact "mini" office for the study. If storage is a problem in your home there's a military chest to make, space saving wall units, and shelves and cupboards to make use of those awkward alcoves.

There are projects for the outside — Dutch doors for a farmhouse look, all kinds of fences — plus many involving structural carpentry. For example, if your family is outgrowing your home, you can build a mansard roof that will give you a valuable extra "floor". The book also tells you how to repair household timbers — the simple and the more complex jobs that can both be expensive if done by a tradesman.

There is carpentry to decorate your home too, with full instructions on timber panel walls, room dividers, paneled doors and applying veneers. The sparkling color photographs show just how good these features will look in your home.

Step by step photographs show you how to undertake the projects in the book and with the full directions and easy to follow diagrams you can't go wrong. There are a host of useful hints too that will simplify your work.

Data sheets tell you exactly what tools to use for a job, what the different types of screws, nails and bolts are used for and the variety of carpentry joints. In addition, two chapters tell you about the power tools that will speed up your work.

This book is a collection of information and ideas that you are unlikely to find elsewhere in one place and as such it is invaluable to both the beginner and the advanced carpenter.

Contents

The Complete Home Carpenter

CHAPTER 1

Storage box on wheels: 1

Storage boxes can make very attractive items of furniture, in addition to being extremely practical. The modern tendency toward smaller houses and rooms means that storage planning becomes increasingly important. When articles are inconvenient to store in one place, the problem can often be overcome by using storage boxes. This basic unit affords scope for personal design and it is easy and inexpensive to make.

This unit, because of its simple structure, is a good introduction to carpentry. The work involved provides practice in the basic technique of glued and nailed butt joints, and the use of some important tools. Basically all furniture is of either frame or box construction. As the storage box uses both methods, the experience of making it will assist the handyman in a wide variety of projects later on.

If the box is required for some purposes its interior may have to be divided. Three different uses are illustrated here, and many more variations could be added. Handles can be attached, and the surfaces can be painted, polished, or covered with material or wallpaper.

Buying tools

In general, buy the best tools you can afford. A reputable dealer or a respected brand can serve as good guides to quality. If you buy a high grade hammer, you can select a type with the head permanently bonded to the handle, so you need not worry about it loosening. A troublefree hammer is important, as you'll use it in most of your work. Good hand saws are also important because they ease your work by cutting with a minimum of effort. Poor quality saws are often slow cutting. While you are shopping for your hand tools, also look at the power tools you may buy later on, and gather any available folders concerning them. Then, when you are ready to buy them you'll be in a better position to make a choice.

Tools required

The tools needed for making the wheeled storage box are:

Folding rules and *steel tape rules.* For carpentry these are essential tools. For small work a 1 ft. or 2 ft. boxwood folding rule is handy. For large jobs a wooden "Zig-Zag" type of folding rule up to 6 or 8 ft. is stiff enough to extend across fairly wide openings like stairwells or wide doorways. For greater length with compactness, choose a steel tape rule, up to 12 ft., which rolls up into a case only about 2 inches square.

Try square with a fixed blade. This is essential for seeing that lumber is square and for marking lines at right angles to the edges. Eight inches or 9 in. is the most useful size; it is wide enough to cover most shelves, for example, but not too unwieldy to use on a narrow molding.

Marking knife for scoring lines across the grain of the wood. It marks more accurately than a pencil and, by severing the fibers on the wood surface, helps ensure a clean cut with saw or chisel. A handyman's knife, such as the Stanley model with replaceable blades, will serve both as a marking knife and for several other jobs.

Panel saw about 22 in. long with ten points (teeth) to the inch. It is designed for finishing work, such as cutting moldings, but will serve in a pinch for almost anything—from ripping (down the grain) or cross-cutting (across the grain) in heavy lumber down to quite fine bench work.

Back saw about 12 in. long with 14 points to the inch. The back saw has a stiffening rib along the back of the blade to help you cut a perfect straight line. Used for fine cutting, it is essential if your ambitions lie in furniture making.

Bench hook to hold lumber steady when using a back saw. It is usually homemade (see instructions below).

Plane. There are many different types of planes, but the handyman's best "all-purpose" is a smoothing plane about 9¾ in. long with a 2 in. cutting edge. The Stanley No. 4 is one. A 6 in. block plane is also handy.

Claw hammer for driving and removing nails. Hammers come in several patterns and weights, and should be chosen to suit the individual. The 14 and 16 oz. weights are popular.

Awl for starting holes for screws, in the absence of a drill.

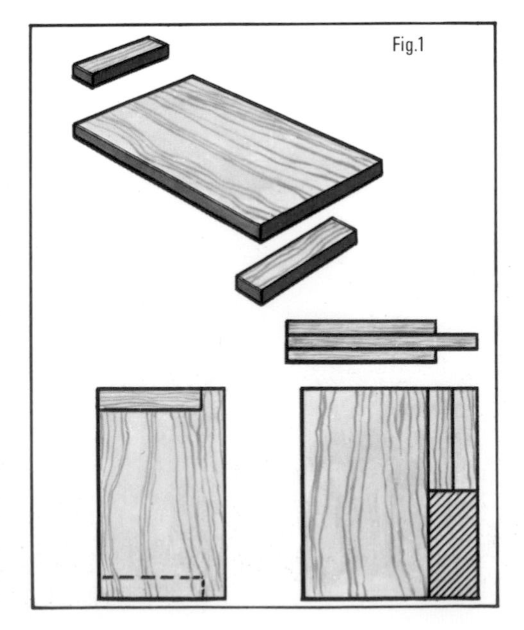

Screwdrivers with ¼ in. and 5⁄16 in. blade widths to fit Nos. 6 to 12 screws.

Nail set to drive finishing nail heads below the surface of the wood. When nail set holes are filled and smoothed down, a better surface is obtained than one with exposed nail heads.

Nail sets are made in several sizes, so buy a size that matches the head size of the finishing nails on which you will use it. It is also wise to buy a center punch at the same time. This looks like a nail set, but has a pointed tip instead of a flat one, as on a nail set. You use it to make small indentations in which to start drill bits without risk of slipping.

The bench hook

Carpenters use the bench hook (Fig. 1 is an exploded diagram) to hold wood firmly while cutting, planing, or chiseling it. The length of batten or planking is held firmly against the upper lip, or edge, while the lower lip prevents the hook from slipping across the table or workbench. Since the upper lip is cut at an exact right angle, the experienced carpenter uses it as a guide while he cuts a straight line "by eye."

Although bench hooks can be purchased, they are so simple to make that it is usual to see the homemade variety in a workshop. And if you are starting off at carpentry, making a bench hook will show you how to measure, mark, cut and fasten lumber without a disaster—if you make a bad job, you can afford to throw it away and start again.

For materials, you need a rectangle of wood about 12 in. x 10 in. x 1 in. thick—"nominal" 1 in. pine is good—and two pieces of nominal 1 x 2 pine about 8 in. long. Nominal sizes are usually used in buying lumber and are smaller than full size, due to seasoning and smoothing.

Scrap wood is good enough, but make sure that none is warped.

Measuring and marking

"Measure twice, cut once" is the first principle of carpentry. It saves a lot of waste!

When measuring with a folding rule (or any other measuring stick of considerable thickness), always stand the rule on edge so that the lines on the rule actually touch the timber at the points where you want to mark it (Fig. 2). This will avoid sighting errors of the kind shown.

Lumber is almost never exactly straight. So you will find, if you try to mark around a board in a continuous line before cutting it, that you usually finish at a different point from where you started; the slight error caused by the curve in the board has "compounded." On a piece of good "dressed" lumber the variation will probably be negligible. But on rough-sawn

Right. *Burlap for the "bar," paint for the toy box, felt for the sewing box—each box is finished to suit its intended use.*

Fig. 1 *(left). An exploded view of the bench hook. This can be cut from scraps of any reasonably good lumber.*

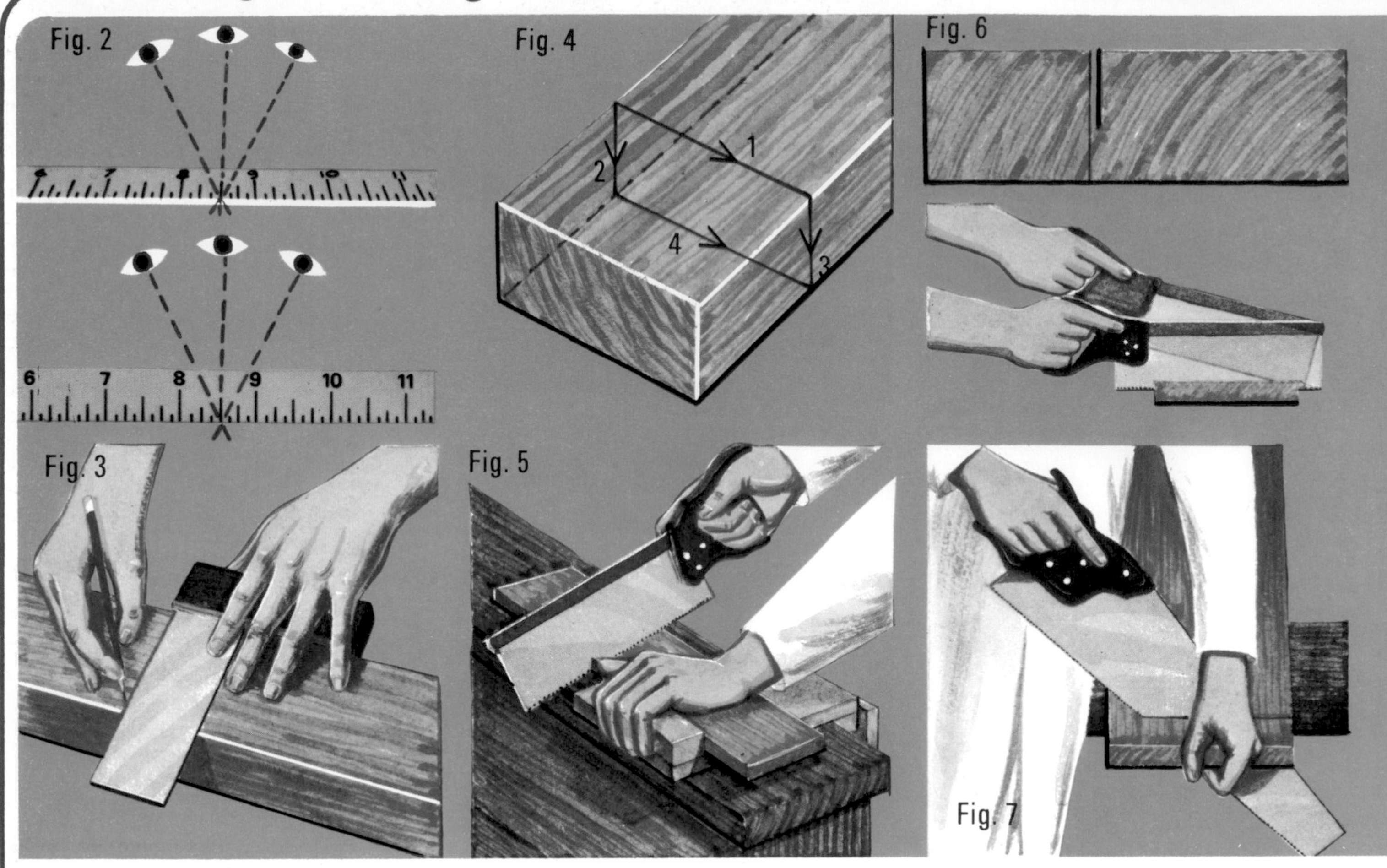

Fig. 2. *If the rule markings are not touching the surface of the wood, inaccuracies will result. A folding rule, because of its thickness, is more likely to create this problem.*
Fig. 3. *The correct method of using the try square and marking knife or pencil.*

Fig. 4. *How to mark a sawing line around a piece of timber. Note that lines 1 and 4, and 2 and 3, run in the same direction. This helps ensure a perfect square cut.*
Fig. 5. *Using the back saw and bench hook. Start by steadying the blade with your left thumb while you draw it back a few times.*

Fig. 6. *Start the saw cut with the blade at an angle of 45°. As you proceed, try to keep the saw at about that same angle.*
Fig. 7. *Using a panel or cross-cut saw. Steady the work with your left hand until the offcut is nearly severed, then hold the offcut to prevent it splintering away.*

lumber sometimes used for economy or antique effects you could easily find yourself ⅛ in. or even ¼ in. out of line. The problem then becomes, "Where *do* I cut?"

So it is good practice to get into the habit of always marking in the correct order (Fig.4). The first stage of making the bench hook, then, is to mark a "good" end on each of the battens. Use the try square as shown in Fig.3, pushing it against the timber with the thumb of your left hand and holding the blade flat with your fingers. Hold the pencil or the marking knife against the blade of the try square, and mark a line toward you across the timber. Working in the order given above (you will have to swap hands at one stage), score lines on the other three surfaces.

Using the back saw

The next stage is to use the back saw to cut the batten. Since you have no bench hook—yet—you will need a substitute like scrap wood nailed temporarily to the work bench and a good firm grip to hold the work steady. What you try to do is to saw on the waste "throw-away" side of the marked line, so that the line is just barely visible after the cut is made. In fact, you should *always* cut on the waste side of the line, not down the middle; this avoids the possibility of cutting "short."

Hold the back saw with its teeth almost parallel with the surface of the wood (Fig.5). Use your left thumb to guide the blade while you make your starting cut by drawing the saw backwards two or three times. Now try to saw smoothly and easily, using as much length of saw as possible for each stroke and letting the weight of the saw do the work. (Short, jerky strokes and heavy pressure will simply carry the saw off line.) Keep your line of vision over the saw to help you cut straight. When you reach the bottom of the cut, use three or four extra strokes to make sure you do not leave a fringe of fibers protruding.

If at your first attempt you have cut a squared line, the marking line will still be faintly visible all round. If not, call this the "bad" end of the batten—and start again!

Once you have two battens with true ends, use the same marking and cutting process to square the ends of the 12 in. x 10 in. board. This time, use the panel saw, holding it at a slightly steeper angle.

Nailing

If you "can't drive a nail straight," do not worry—carpenters seldom do, either. A nail driven straight does not always have great holding power. So, in many jobs, alternate nails are driven at opposing angles for greater strength. For the bench hook, you should use the technique known as "dovetail nailing" (further details of nailing techniques are in the next chapter.) Since the hook will have to withstand considerable pressure in use, you will need six 1¼ in. finishing nails, plus wood adhesive, for each batten.

Start by placing one batten along the top left-hand corner (as you look down at it) of the hook board, and drive one nail at each end until it is just catching the board below. Use light taps of the hammer to start the nail while you are holding it. Then, when it is in far enough to stand by itself (and your fingers are out of the way) use full strokes of the hammer. When the nail head is almost even with the

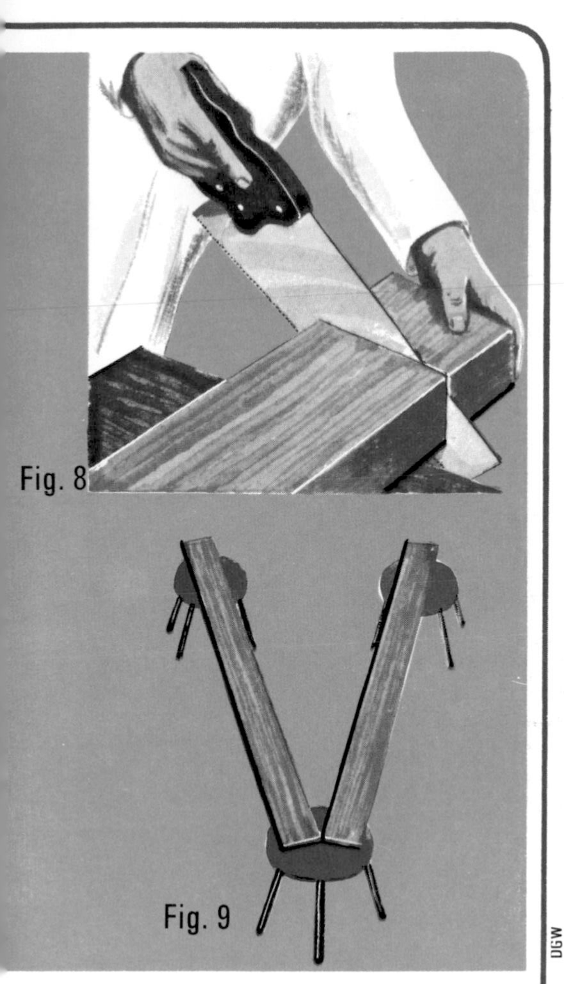

Fig. 8. *When sawing heavy lumber, take a firm grasp of the offcut when you are near the end of the sawing line.*
Fig. 9. *A suitable temporary support for sawing a large panel. There are many other ways in which you could improvise an adequate support.*

wood surface, finish driving it with a nail set. This eliminates the chance of the hammer striking and denting the wood.

With your first two nails part-driven, lift the batten off the base and spread some white glue all the way along it. Replace the batten, check that it is properly aligned—the part-driven nails will stop it skidding about on the glue—and drive home the two nails. Now drive in the four intermediate nails.

The opposite batten is fastened on in the same way as the first. Note that it, too, goes on the left-hand side and not diagonally opposite the first one.

Materials for the storage box

The storage box has dimensions of 2 ft. 2 in. x 1 ft. 6 in. x 1 ft. 4 in. The complete exterior can conveniently be cut from a piece of 6 ft. x 3 ft. plywood, although a partitioned box would need a larger sheet. For a box this size, plywood ½ in. thick is required. If cutting the panels is too difficult a proposition, ask your lumber dealer to cut them, using the dimensions in Fig.10 as a guide.

In addition to the panels, you will need two 7 ft. 6 in. lengths of 1 in. by 2 in. wood batten; several dozen 1¼ in. finishing nails; four caster wheels with screws; one squeeze bottle of white glue.

Cutting the panels

The panels are first marked out on the sheet plywood, then cut, then planed down slightly to the marked size.

If you intend cutting the panels from a single sheet, measure the outlines according to Fig.10. Use the try square and marking knife for marking at right angles to the edges. Then extend the lines with the rule and marking knife. You must allow enough space between the panels for both the saw cut and for planing—⅛ in. should be adequate. If you have no other straight-edge long enough to cover the longer dimensions, use the "manufactured" edge of the first panel you cut out as a check that your other lines are straight.

The fine-toothed panel saw listed above is well suited for cutting plywood. A coarser saw, with fewer teeth or "points" to the inch, would produce bad splintering along each side of the cut. Before beginning to cut, be sure that the plywood is placed on a suitably smooth surface that will not rock or slip, and that the line the saw blade has to travel is clear of obstructions.

Accurate sawing requires a good working position. Your feet should be slightly apart and your body balanced so that your sawing arm is free to move without hitting your body. If your arm hits your body, the saw will wobble and work away from the marked line. Like the back saw, the panel saw demands an easy, flowing movement. Your job is to guide, withdraw and steady the saw; the cutting action comes from the saw's own weight on the forward stroke.

As you near the end of the saw cut, you will need someone to hold the panel you are cutting off so that it does not wrench away from the sheet and leave great splinters. You cannot solve this problem by laying the sheet across two planks and sawing down the middle; the sheet will merely sag inward into a slight V-shape and jam the saw.

Cut the two sides, two ends, and transverse partition (matching the ends) if wanted, but do not cut the top or bottom yet. Leave these until the box is assembled, when you can adjust their size to match any slight variation in your outer "walls."

Fig. 10. *The plywood sheet: a laid-out view showing the panels in position of assembly. Label each panel with a pencil before starting work. Note that the end panels must be shorter than the side panels by the thickness of the lid. Check this before cutting, otherwise the finished lid will not sit flush with the side panels, and you will have unnecessary planing to do.*

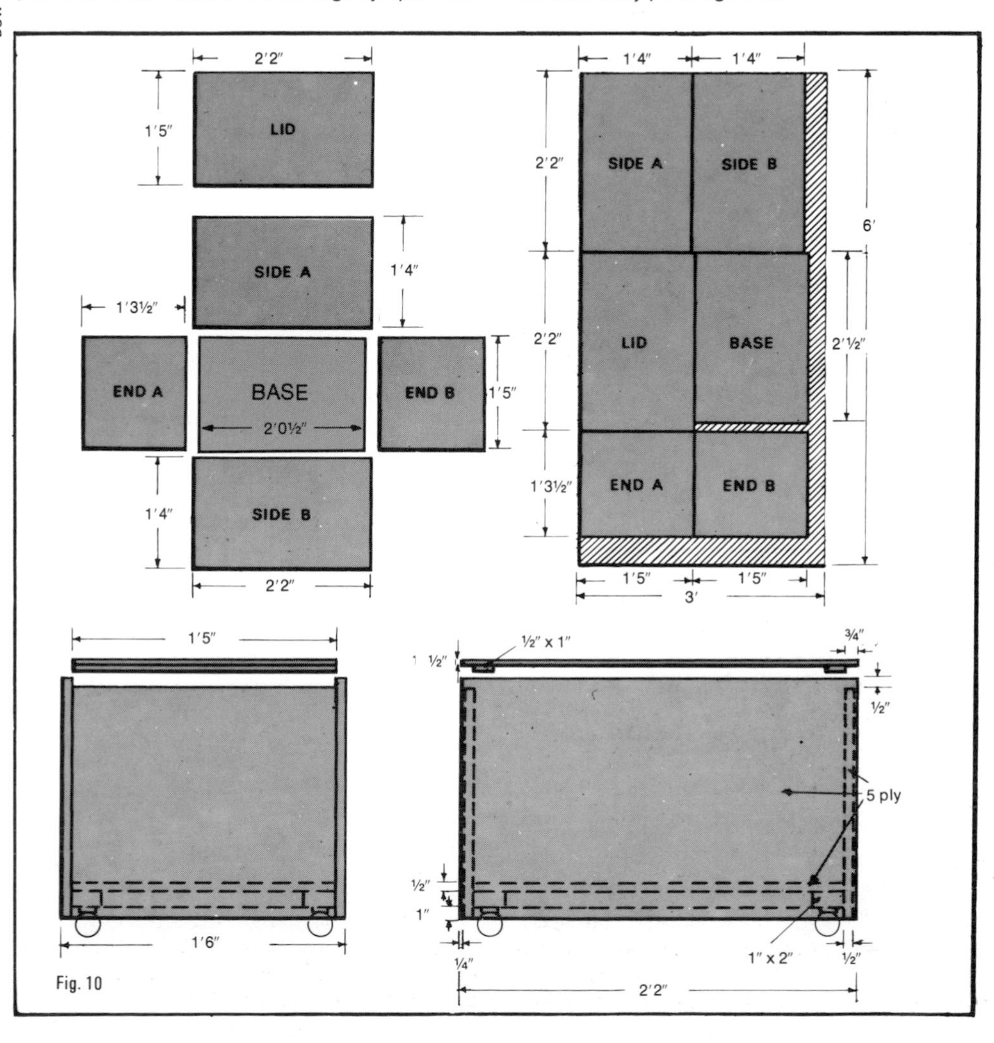

MENDELSSOHN
ZINO FRANCESC
TCHAIK
OF THE ANIMALS

CHAPTER 2

Storage box on wheels: 2

Planing is one of the most important arts in carpentry. It isn't just done to remove surplus wood; most really superb woodwork fitting depends on the plane for essential preparatory work. The second stage of building the wheeled storage box introduces one of the most basic techniques in planing–the edge cutting and trimming of plywood sheets. It also introduces joining techniques for sheet materials.

Planing the panels

Using the plane, smooth the edges of all four panels down to your knife-marked lines. A sharp plane blade is essential. Blunt cutting tears the surface of the wood, making it unsightly and sometimes unusable. Set the plane blade to "fine" (with very little blade showing). When planing, take a firm stance behind the work, making sure the panel is secure.

In planing plywood edges, you must always work from one end of the edge toward the center, then from the other end toward the center. This way, the plane never passes outward over a corner, which is likely to cause splitting outward of the plywood layers. The final stage of the work consists of planing the middle portion of the edge to make the entire length straight and even.

It is extremely important to keep a constant watch on the marked line to which you are planing, so that the end result will be a smooth, planed edge exactly to the line. A long straight-edge, such as a 4 ft. length of steel or aluminum strip is useful in marking and checking the planed edge for straightness. Use medium grit sandpaper to slightly round off the sharp corners of the lid before applying the finish. Or, you can do it with a single light pass of the plane along the corner.

As you plane, make sure your edges are kept at right angles by frequent checks with the try square. To do this, fit the square snugly into the corner of each panel; the inside of the handle and the inside edge of the blade should line up exactly with the edge on the panel. Make sure that each pair of panels is identical in size.

Assembling the box

Next, you must mark the panels for assembly. The two end panels will be set in (recessed) by ¼ in., so a nailing line must be drawn along the outside of each side panel opposite where the end panel butts. Measure a point ½ in. inward from the edge of each panel, then line the ruler along the points and draw a line through. With the awl, make starting holes along the line for the nails at about 2 in. intervals, then drive each nail through the panel so that the tip is just protruding through the opposite side; this will help locate the side panel to the end panel.

Nailing is much easier when the wood to be nailed is resting on a solid surface. For this job, the easiest way is to make a kind of table out of the two end panels, with the side panel resting on top of them. If you place the farthest end panel against a wall (Fig. 2), it is easy to steady the side panel while you nail it.

Use dovetail nailing. Spread a thin film of white glue over both surfaces. Now locate an end panel, make sure that there is an even ¼ in. overlap by measuring from the end, and hammer home the nails.

When you have nailed on both sides, use the hammer and a nail set to drive the nail heads just below the surface of the wood. Do this before the adhesive starts to set as the punching action helps bring the surfaces closer together, ensuring a stronger joint.

Once the box is assembled, before leaving it for the adhesive to set, measure the diagonals to ensure that the box is square. Run the ruler across from corner to corner; each diagonal should measure the same. If not, slight hand pressure will adjust the corners. The assembly must be left on a level surface to set, otherwise twisting will occur.

Fitting the bottom frame

Battens are fitted as a frame around the underside of the base panel to stiffen the box and provide a mounting for the wheels. Do not measure the lengths of batten with a ruler; instead, use the "direct" method. Hold the batten along the inside edge of the box and mark the inside length on the batten with a pencil or the marking knife (Fig. 3). This eliminates mistakes in transferring measurements from a ruler. Cut the four lengths of batten with the back saw, using the bench hook to steady the pieces. Remember to cut on the "waste" side of the line. Check that the four battens fit together snugly inside the box.

Now use the same "direct" method, by placing the box on your sheet of plywood, to measure and mark the bottom panel. Cut it out with the panel saw. Make sure that it, too, fits snugly, paring it down with the plane if necessary.

The next step is to fasten the battens and bottom panel together so they can be fitted into the box as one unit. Mark your nailing line 1 in. inward from the edge all around the top face of the base panel and make starting holes for nails at about 2 in. intervals along the line. Drive the nails in so that the tips are fractionally protruding through the other side. Spread

Opposite. *Another use for the box-on-wheels. This is a storage unit for records, avoiding the necessity of carrying piles of records to and from the record player. The range of alternative uses is limitless.*

Fig. 1 *(top). Checking a frame for squareness. Each diagonal is measured from corner to corner, and both sets of measurements should be identical. If they are not, the "long" corners must be pressed inward. A smaller frame or unit could be checked with the try square to ensure that the edges are at right angles.*

Fig. 2 *(middle). Three panels can be formed as a table for nailing the butt joints. One upright panel and the far end of the top panel should be placed against a wall and the whole assembly kept rigid by the pressure of your hand pushing it against the near end of the top panel. This keeps the panels firm against the wall.*

Fig. 3 *(bottom). A batten is marked "direct" over the box into which it will be fitted. Direct marking eliminates errors in transferring measurements from the rule to the working surface of the wood. Because of the greater accuracy achieved, this method should be used whenever possible in preference to the rule transfer method.*

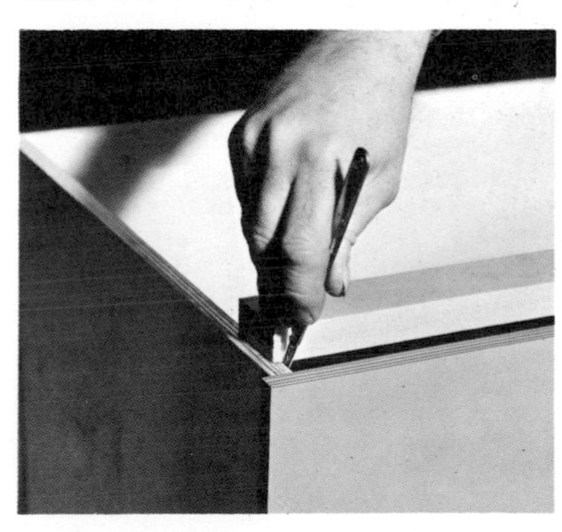

adhesive on both panel and batten surfaces to be joined, place battens in position and nail into place. (The nails are driven in through the ply panel first because this ensures that the nail grips the full thickness of the panel, while still leaving an adequate length through the batten. It is a general rule in nailing that the thinner piece is nailed to the thicker piece whenever possible.)

When the adhesive on the base panel "sub-assembly" has set, fit it in the bottom of the box, planing down any edges that make fitting difficult. The base is inset or recessed by 1 in. Check that this recess is the same depth all around by making a mark on the inside of each side or end panel about halfway along. Then see that the battening of the base panel is level with these marks all around.

Before nailing the base in position, make sure that the recess is not too deep for the caster wheels. If it is, and the casters cannot move around freely, the depth of the recess will have to be reduced.

Now nail the base in place through the outside of the side and end panels. To make a neat job, mark a line around the outside of the box at a depth equal to the depth of the recess, plus half the depth of the batten. In this case the line would be drawn 1½ in. deep.

Making the lid

(If you intend covering the box with wallpaper or fabric, leave the lid until the covering has been completed, otherwise the thickness of the material may make the lid too tight.)

The lid fits *between* the side panels and *over* the end panels and is, in effect, countersunk. This is a removable lid—without hinges—and battens will have to be joined to the bottom of the lid, and recessed, so that the lid will not slip off in use. As the lid fits between the two side panels, it needs only two battens fitting inside the two end panels to make it secure, although you may like to fit battening all around for a neater finish.

Cut and plane the lid to shape, planing it to a loose fit (about 1⁄16 in. undersize) to leave room for the paint. Then carefully mark lines on the underside of the lid just inside each end panel; this can be done by measuring in ¾ in. from the lid ends. Cut two (or four) lengths of batten to fit, not too tightly, inside this line. Spread adhesive on the surfaces to be joined and place in position on the lid. Using only one nail per batten, lightly nail in position. You can now place the lid in position to check whether it fits. If it does not, it is easy to re-place the batten(s). When the fit is perfect, nail the battens securely, as before, through the ply lid.

Finishing off

Every nail head must be driven in. Carefully fill all the holes left by the recessed nail heads with a wood filler such as Plastic wood and, when it is dry, sand smooth with fine grit sandpaper. If you intend to paint it, the storage box should be finished by the same method used for other furniture. Follow the instructions on the can of paint used. If it is to be wallpapered, or covered in fabric, thinly coat the whole surface of the box with thin shellac and leave it to dry. This prevents stains coming through from the wood, and provides a good base for applying the covering or wallpaper paste.

The casters are screwed in position after the finishing and decorating work has been done.

Making the partitions

The partitioning has been omitted from the general construction details because in a general purpose box none might be needed, while boxes for specific purposes would require differing numbers of partitions.

Constructing a partition to divide the box in half (across its width, for example) is quite easy. Measure and cut the partition and check that it fits. On the outside of the side panels, mark the exact middle at the top and bottom edges. Draw a nailing line between the two marks. Drive a finishing nail through about 2 in. inward from each end of the line, so that the points are just protruding on the inside. The partition panel can now be located on the nails, and the nails driven home. This method ensures that the nails are driven into the middle of the partition, and avoids splitting the edges. A further two nails can now be placed at equal distances inside the first two.

Any number of parallel partitions can be added, using the same procedure. For T-shaped, H-shaped or more elaborate forms of partitioning, make up the partitions as a complete sub-assembly before dropping it into the box and nailing from the outside.

Storage tray

A useful accessory, especially for a toolbox or sewing box, is a tray that fits inside the top of the box so that small items can be stored there, leaving room for larger items underneath. This is begun by fitting lengths of 1 in. x 1 in. batten below the inside top edge of the side panels. The battens will provide a lip, or rail, to hold a panel of ¼ in. plywood on which the tray is built up.

To fit the rail batten, mark a line along the inside of the side panels. Remember to allow enough depth to accommodate the thickness of the plywood main lid, the battening underneath the lid, the thickness of the tray base itself, and the depth of the partitions or receptacles fixed to its surface. The *top* of the rail batten must be on the *lower* side of the line.

Some kind of support will be needed while the batten is glued and nailed from the inside—otherwise you risk knocking out the side panel. The best idea is to lay the box on its side on a table. Hammering in a confined space is difficult. Two tips to make it easier: **1.** Drive the nails through the batten before you put it in the box; **2.** Use the nail punch to drive the nails home, hitting with short strokes of the hammer.

Wood or plywood tray partitions are measured (to dimensions suiting the items you will store), marked, cut and fitted to the tray base by the same methods you used in making the wheeled box itself. Since the joints are carrying little weight or stress, white glue is quite strong enough to hold them together, provided the edges are straight. So you need use only brads to fasten the joints while the glue dries.

Alternatively, you can fit a plastic sewing box to the tray base with small screws or bolts. To punch the necessary holes in the plastic, first mark the hole positions on the bottom of the box, keeping them well clear of the partitions. Now hold a nail or skewer in a pair of pliers while you heat it red-hot—a gas stove flame will do the job in seconds—and quickly, before it can cool, drive the nail through the plastic. Be careful that the heated end of the pliers does not rest on the plastic, or the hole may be somewhat over-sized!

If the storage tray covers only half the inside area of the box, it can slide across the rail in the manner of a sliding door, providing access to the contents underneath without your having to remove the tray. But if it is to cover the whole inside area, handles may have to be attached at either end so that the tray can be lifted out.

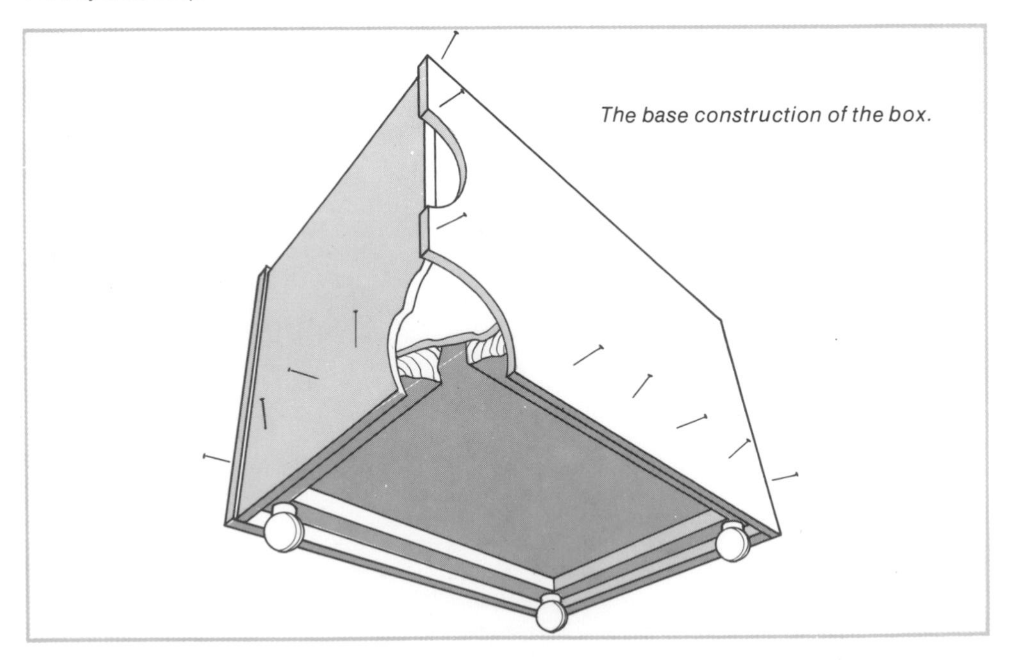

The base construction of the box.

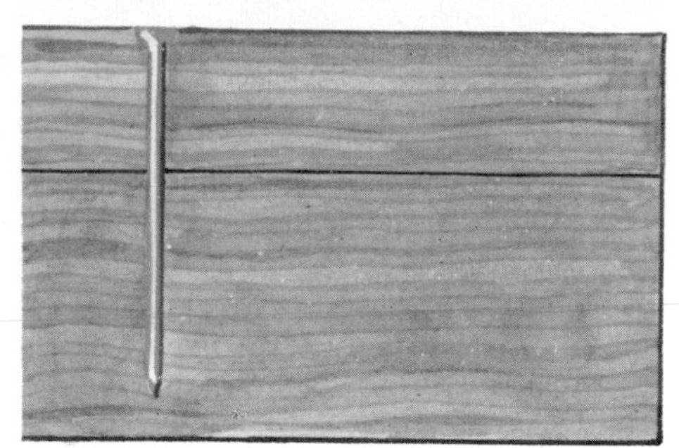

1: Always nail the thinner piece of timber to the thicker.

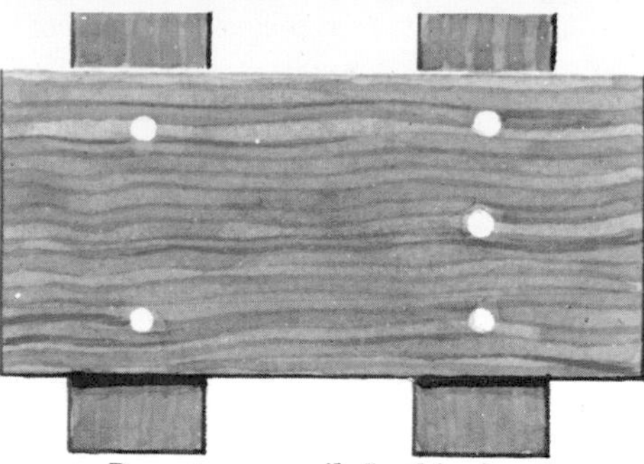

2: Do not over-nail. In this situation, two nails hold as well as three.

3: Nailing overhead, lean backwards and use normal 'forehand' nailing stroke.

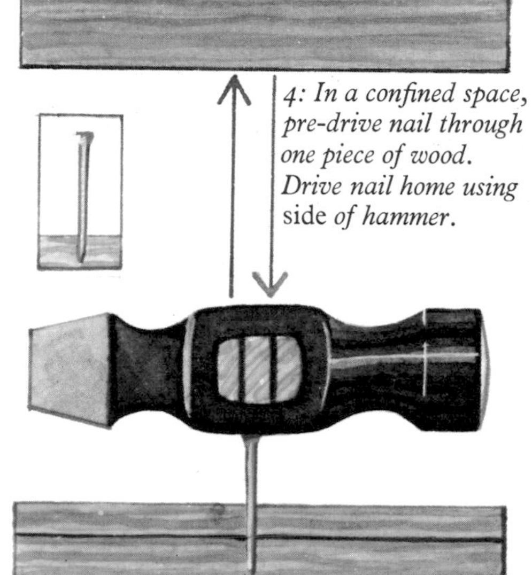

4: In a confined space, pre-drive nail through one piece of wood. Drive nail home using side of hammer.

5: Blunt nail points help avoid splitting the wood.

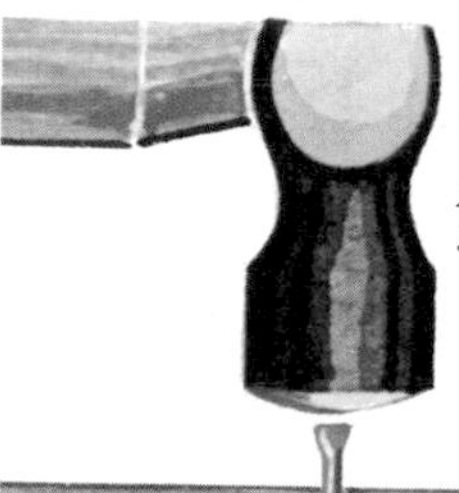

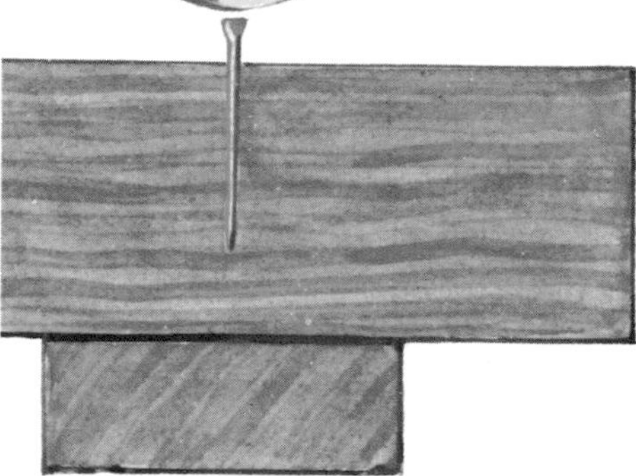

6: Provide adequate support under nailing point to avoid nails skidding or bending.

7: Block of wood under hammer avoids bruises while extracting nails. For big nails, extend hammer handle with length of pipe for extra leverage.

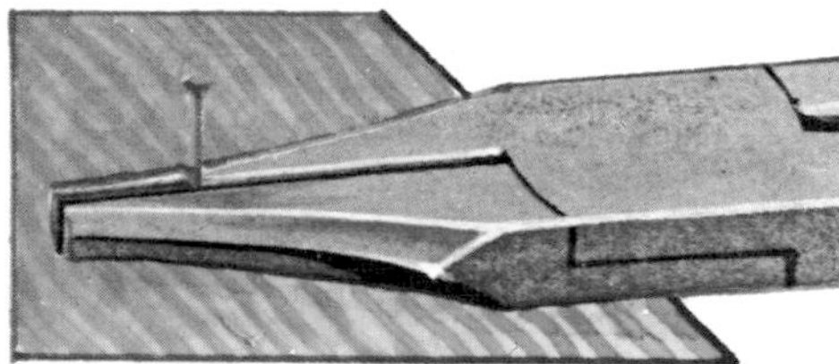

8: Pincers or pliers hold nails too small to be gripped by finger and thumb.

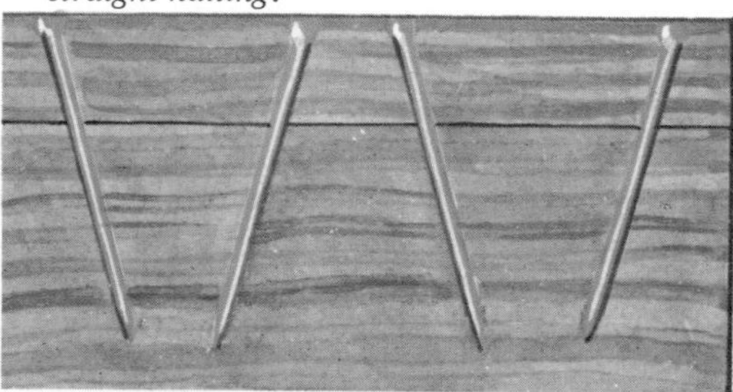

9: Dovetail nailing — nails at opposing angles — makes stronger joint than straight nailing.

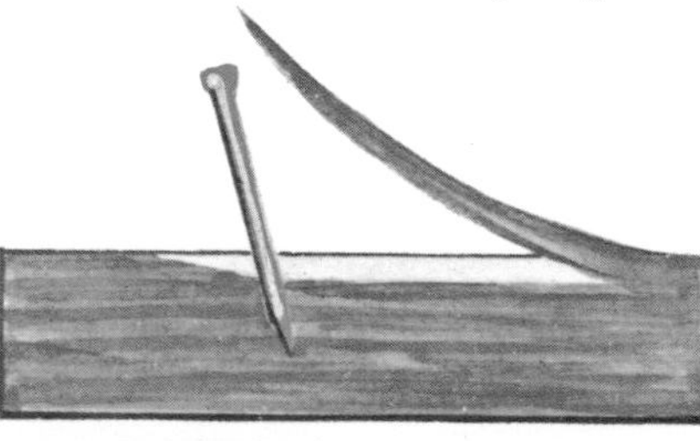

10: For 'secret' nailing, lift surface with chisel and drive nail under the paring.

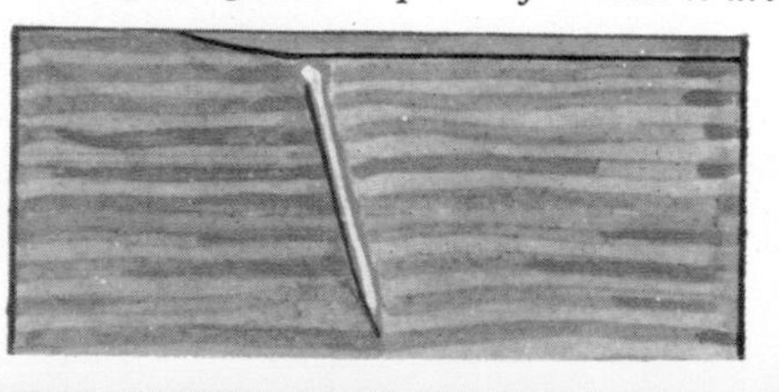

Glue paring back in place after nail is driven.

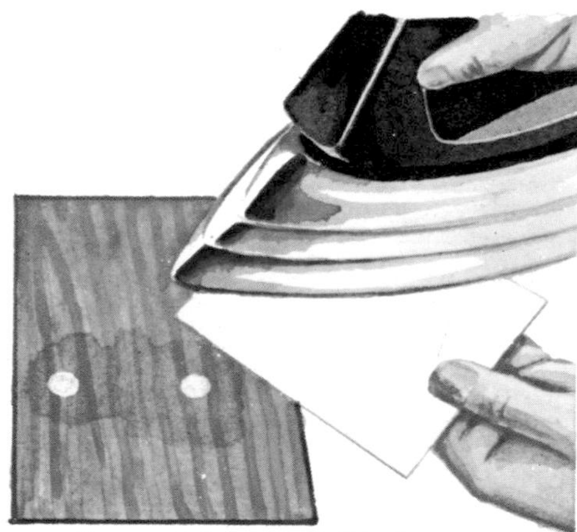

11: Iron and damp cloth lifts hammer bruises from new woodwork.

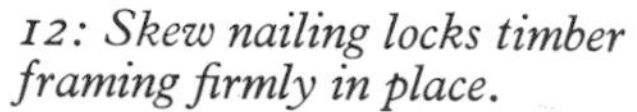

12: Skew nailing locks timber framing firmly in place.

CHARLES·I
Mary QUEEN OF SCOTS
The Art of the Renaissance
The Decline of Imperial Russia

CHAPTER 3

Build a roomy bookcase

This bookcase is practical, attractive, and easy to make. The particleboard, available at lumberyards and building suppliers, enables you to keep costs down when you do your own cabinetwork.

The bookcase is made in 9 in. by 5/8 in. particleboard with the rear panel in 1/4 in. AC grade fir plywood. (The 9 in. widths may be sawed from a large panel if necessary. And 3/4 in. thickness may be used.)

Making it will introduce you to two new tools, the plow plane and the paring chisel, and to the techniques for making a rabbet joint and a stopped dado joint.

About particleboard

Particleboard is made by bonding wood particles or chips together with synthetic resin under heat and pressure. The resulting material can be used like wood for many purposes but costs less. It is usually sold in its natural form, though it's sometimes available with various surface materials bonded to it, including wood veneer and plastic laminates.

Panel sizes vary with the source. The 4 ft. x 8 ft. panel size is one of the commonest. Some suppliers also stock counter widths, such as 2 ft. x 8 ft., and sometimes narrower widths like 10 in. or 12 in. for use as shelves. (You can cut these sizes from a 4 ft. x 8 ft. panel.) The bookcase illustrated was built from veneered material 5/8 in. thick, but could be built just as well from 3/4 in. material without veneer if that is the type stocked by your supplier. If you are working with particleboard in its natural form, a paint finish is the easiest to apply, but you can veneer the surfaces if you have the skill and the facilities.

Working in particleboard

Particleboard, although easy to work, is easily bruised. Even small wood chippings on a bench cause indentations if some types of board are placed over them and pressed down. Always provide a clean, smooth surface to work on.

The veneer may tend to flake or chip during sawing—this is more apparent on the underside of the cut—and the coarser the teeth of the saw, the worse the flaking will be. So always use the finest saw blade possible.

Flaking can be eliminated if the veneer is scored with the marking knife before you start sawing. When marking out any lines that will be sawed or chiseled, cut right through the veneer, using several light strokes rather than one heavy one. Then, if you follow the correct procedure and cut on the waste side of the line, the veneer will not chip at all.

Left. *The finished bookcase, an attractive addition to any home. The unit could also be used as a china cabinet.*

Tools required

Some of the tools have been introduced in earlier chapters—a claw hammer, fine tooth panel saw, 12 in. back saw, try square and marking knife. New tools for this project include a marking gauge, a 3/4 in. bevel edge chisel, and a fine nail set. In addition you will require:

Plow plane. This plane plows a groove through wood. The width of the groove is set by the size of the plane blade. The plow is guided along the groove by a fence or runner, which can be adjusted. (A dado plane could also be used.)

Paring chisel, 1/2 in. for forming and cleaning out dado joints.

Materials

The quantity of particleboard can easily be estimated from the drawing (Fig. 1). Allowing about an inch of wastage for each section of board, you require six 31-in. lengths. Other materials:

A 22 in. x 31 in. sheet of AC fir plywood, 1/4 in. thick (cut from 24 in. x 48 in. "handy panel").

A small roll of veneer strip, for covering exposed edges of the particleboard.

Finishing nails, 1 1/2 in.

White glue.

Proprietary wood filler to match the veneer. If you cannot get a matching color, buy two cans of filler, one darker and one lighter, and blend to match; or use ordinary putty, tinted. (If you make a practice of keeping on hand three tubes of tint for varnish—one dark brown, one red and one yellow—you will be able to match the common wood stains fairly well.

Sliding glass doors, if required, have to be made up by your glass dealer. Do not order until the case is finished and the runners inserted. Then measure carefully.

The final finish will require turpentine substitute; clear matt polyurethane; fine sandpaper; grade 0 steel wool, and oil stain.

Marking out

All the lines that are to be sawed or chiseled out must be cut with the marking knife. Any lines used for location, and which will not be cut, are marked with the pencil—the lines can then be rubbed when no longer required.

Carefully measure each length of particleboard, double-checking each measurement. For each length, score a line heavily across the veneer using the try square and marking knife. Repeat this until you have cut straight through the veneer. Follow this line right around the board, using the method which is given in CHAPTER 1, "Measuring and Marking" to ensure that the lines on both sides of the board correspond with each other.

You will now have a line which, if sawed properly along the waste side, will allow the board to fall neatly apart without damage to the veneer.

It is best at this stage to cut all the lengths you need—two sides, top, two shelves, and plinth (the panel below the bottom shelf). This allows you to lay out the panels in rough form to check that all dimensions are correct.

The rabbet joint

Joints in carpentry have two quite separate objectives—strength and appearance. Many are designed to conceal the end-grain of lumber, which is less attractive than side grain and often more difficult to finish smoothly.

The rabbet—sometimes spelled "rebate" or even "rabbit"—is a stepped joint. It is often used to hide the sawed edges of paneling, as in paneled doors and the backs of furniture. It is also used, as in fitting the top of this bookcase, to provide a joint which is both more rigid than a butt joint (because there are three separate touching surfaces) and neater in appearance.

In this project, you start by marking out the rabbet joints for the top. These are located at the top inside edges of the two side panels, and under the ends of the top panel.

Set the marking gauge to half the depth of the particleboard. Holding it firmly against the top inside face of one of the end panels, lightly scribe a line along the veneer. Do not press too hard or the veneer will chip. Repeat on the opposite side panel. With the gauge set to the same depth, run a line along the top end of each side panel. Now lay the blade of the try square along the scribed lines, and sever the veneer with the marking knife.

The marks for the same joint are now made along the under sides of the ends of the top panel, and along the extreme ends of the top panel. Fig. 7 shows how the joints will match after waste material has been cut away.

The same procedure is carried out along the edges of the side and top panels, and of the bottom shelf, where the back of the case will be located. This rabbet is for the plywood back.

The dado joint

Dado joints, like rabbets, are used both for extra strength and for better appearance. In a bookcase, they provide a solid support for the shelves, which otherwise would rest solely on the strength of nails driven through the sides. At the same time, the cut ends of the shelves are tucked neatly out of sight.

A "stopped" dado joint is one which, instead of running for the full width of the board, stops just short, giving a neater finish at the most conspicuous point—in this project, at the front edge.

To mark out for the stopped dado joints start by clamping the two side panels together. The lines marked for the rabbet along the inside faces should touch each other; this end of the pair of panels represents the top of the bookcase. Decide where the joint for the first shelf will be and, with the try square, mark two lines across the panel the thickness of the particleboard. Use the direct marking system for this by using another piece of board. These

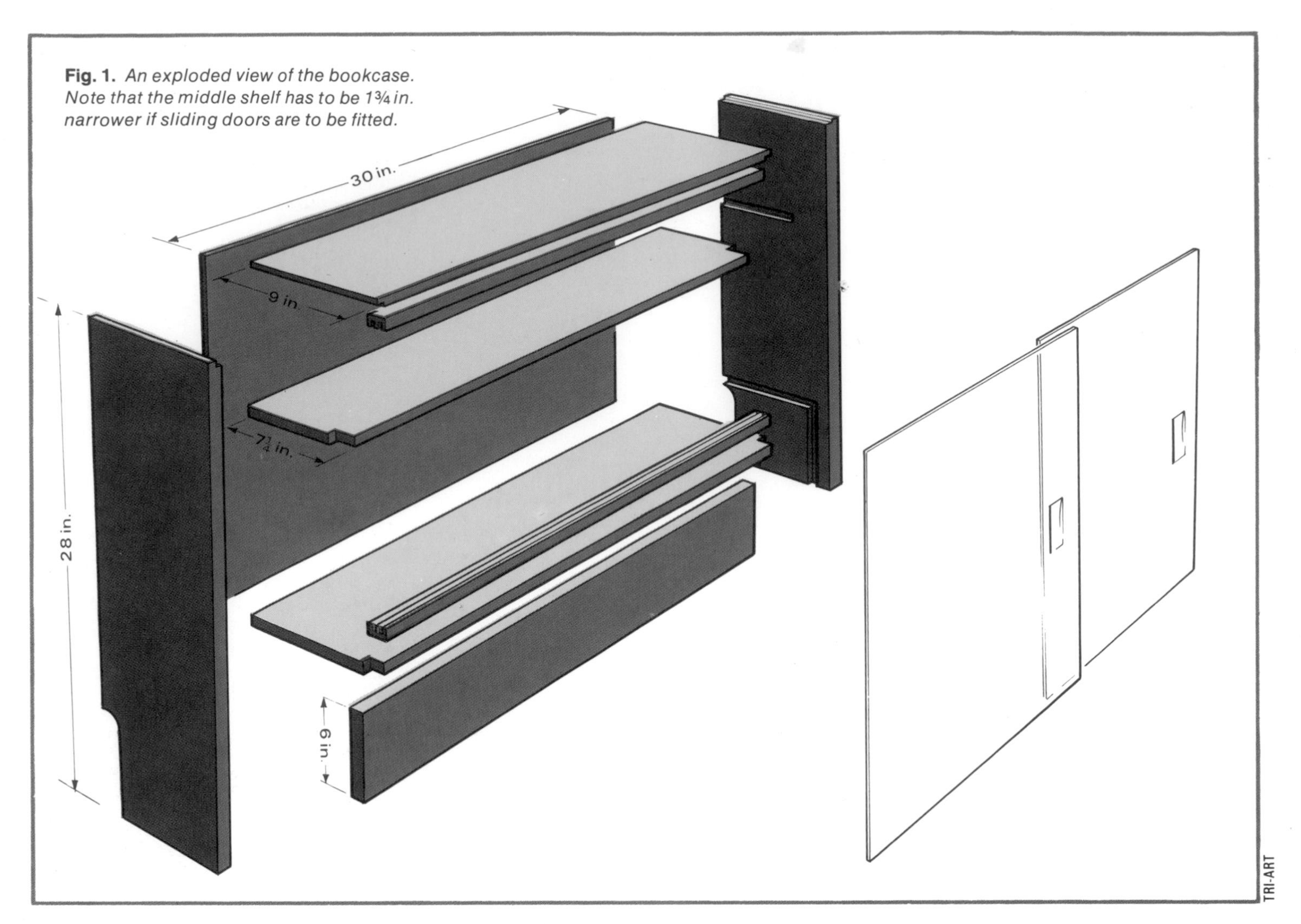

Fig. 1. *An exploded view of the bookcase. Note that the middle shelf has to be 1¾ in. narrower if sliding doors are to be fitted.*

marks run across what will be the back of the case.

Repeat the two marks at the point where the bottom shelf will be located.

Separate the side panels. Lay the try square blade along the inside of one panel, and draw a line, with the marking knife, from the topmost mark, along the face, to within 1½ in. of the opposite edge of the board. Repeat this with the next mark, and you will have twin parallel lines, ⅝ in. apart, running from the "back" of the panel, and stopping 1½ in. short of the opposite side. At this point run the marking knife between the two lines, joining them. You now have a clearly defined outline that will have to be chiseled out to house the ends of the top shelf.

The same procedure is carried out for the bottom shelf. But in this case, when the outline has been marked out, an additional pair of lines are marked, beginning at the point where the dado joint "stops," and running down to the bottom of the panel at right angles to the joint (Fig. 2).

Making the joints

Start with the stopped dado joint for the top shelf. Place one of the side panels on a clean bench, with the markings upward. With the ¾ in. chisel held vertically, lightly tap along the marking lines at the stopped end. The bevel edge of the chisel should be inside the joint. The idea is to remove the veneer for about 1½ in. along the joint (Fig. 2). Once you have done this, the particleboard beneath must be removed down to half the thickness of the board. Do this by starting in the middle between the marked lines, chipping gently with the chisel and working outward. It will chip quite easily. Blow into the joint as you work, to keep the space clear. You now have a rectangle, ⅝ in. wide, about 1½ in. long, and 5/16 in. deep. Check the depth carefully—the whole joint depends on its being accurate.

The remaining two lines will be worked along with the back saw, the recess you have just made providing a clear area for the end of the saw blade.

Extend the width of the remaining marking lines with the chisel, so that you have a channel wide enough to take the back saw blade. This is done by running the chisel along the marked lines slightly inside the joint (Fig. 3), removing a sliver of veneer.

With the back saw, carefully saw along the lines (Fig. 4) down to half the depth of the board.

Then, with the paring chisel, starting at the "open" end of the joint, carefully chip a groove right along the joint (Fig. 5). Start with the chisel "upside down"—that is, with the beveled end facing downward—and turn it over only to pare out the last few slivers of wood. (The reason for doing this is that, held "right way up," the blade tends to dig in deeper and deeper, rather than to cut straight, because of the wedging action of the bevel.)

You have now made a stopped dado joint. Check with the edge of the try square that it is flat along the bottom. Repeat this for the lower shelf. The only difference here is that once you have completed the stopped section of the joint, work out the section to take the plinth (Fig. 2).

The rabbet joints—for fitting the top panel to the sides, and the plywood sheet to the back—are slightly simpler to make. The plow plane cuts a rabbet out of the edge of a piece of wood by cutting a deep narrow groove, to the depth required, along one side (Fig. 6). This is repeated along the other marking line, which is at right angles to the first one. The two grooves, when they meet, remove a square of wood, forming a step-shaped recess, or rabbet (Fig. 7). (Instructions should come with the plane.)

The most important point in cutting the rabbet is to set the plane fence or guide to the correct width (Fig. 6). When using the plane, start the cutting near the *far* end of the line. For each short stroke you push the plane away from you, but the starting point of each successive stroke is a little nearer to you than the one before. In short, you work in a series of overlapping movements.

Assembly

Once you have made all the joints, you will have all panels ready to fit together like a jigsaw. Before you go any further, trial-assemble them, without glue, to make sure that they fit.

At this stage it is best to finish all the surfaces that will be on the inside of the bookcase. The method for this finish is given below.

To assemble, start by placing one of the side panels on the bench, "inside" upward. Spread adhesive along the dado joints where the shelf ends will be located, and stand each shelf end in the joint. Spread adhesive on the exposed ends of the shelves, and place the opposite side panel over the ends. You now have the shelves dadoed in the side panels.

While the assembly is in this position, drive finishing nails through the outside of the side panel, which is resting on top of the upright shelves into the ends of the shelves. Place the nails at approximately 1½ in. intervals. With the nail set, drive all heads below the surface.

Carefully lay the assembly down, and turn the whole unit over so that the opposite side panel can also be nailed to the shelves. With a damp cloth, remove immediately all traces of any glue that has been squeezed from the joints.

Next, stand the assembly upright. This will enable you to glue, place and nail the top panel to the side panels.

Now lay the unit down again, spread adhesive in the joints to take the plinth, and slide or tap this into place. Again be sure to wipe off any excess glue immediately.

Check the frame for squareness by measuring diagonally from corner to corner. If the two diagonals are not identical, adjust by pressing the longer side.

Leave the frame for the adhesive to set.

The back of the case frame has a rabbet groove running around the inside edges. With the panel saw, cut the plywood sheet to fit snugly into this. As with the side panels, the plywood is fixed with adhesive and finishing nails, but in this case the nails can be spaced more widely—say, at 3 in. intervals. Or brads may be used.

Sliding doors

If sliding doors, in glass or wood, are to be fitted, these should not be cut or ordered until the runners have been attached. There are many types of brand-name runner, the most usual being plastic channels that are nailed to the top panel and bottom shelf. The top runner is deeper than the bottom one, so that doors, when fitted, can be inserted into the top runner first, then dropped into the bottom runner and still leave enough room to hold the top of the door.

For this reason, vertical door measurements are taken by measuring the distance between the base of each channel, and deducting the depth of the bottom channel. The upper shelf must be narrower by the width of the channels.

When fitting plastic channeling especially, remember that it is easily bent out of line, and that jammed doors can result. So check carefully for straightness as you go.

Applying veneer

At certain points—the skirting recess, both ends of the top panel, and the bottoms of the side panels—the veneer will have been removed, exposing the particleboard. This is replaced from the roll of veneer strip.

Veneer strip is made with either an adhesive backing or a plain backing requiring glue. Another type has a dry adhesive that melts when heated; this is ironed on.

If any exposed areas are narrower than the adhesive strip, cut the veneer back with the marking knife until it fits. Then cut the strip to size and attach it according to the method advised by the manufacturer.

Fig. 2. *The end of a stopped dado joint, chiseled out ready for the next stage. This is the shelf-plinth junction.*

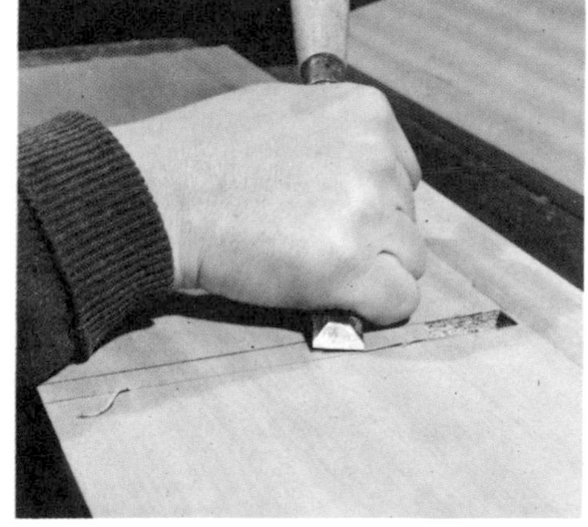

Fig. 3. *When the end of the stopped dado has been chiseled out, a groove for the saw blade is cut with a chisel.*

Fig. 4. *With the back saw, carefully cut along the lines, down to half the depth of the board.*

Fig. 5. *The stopped dado joint is completed by chiseling out the material between the lines cut by the back saw.*

Fig. 6. *The plow plane cutting the first groove for the rabbet joint. Another groove is then cut at right angles.*

Fig. 7. *When the two rabbet grooves have been cut, they are fitted to form the joints at the top of the bookcase.*

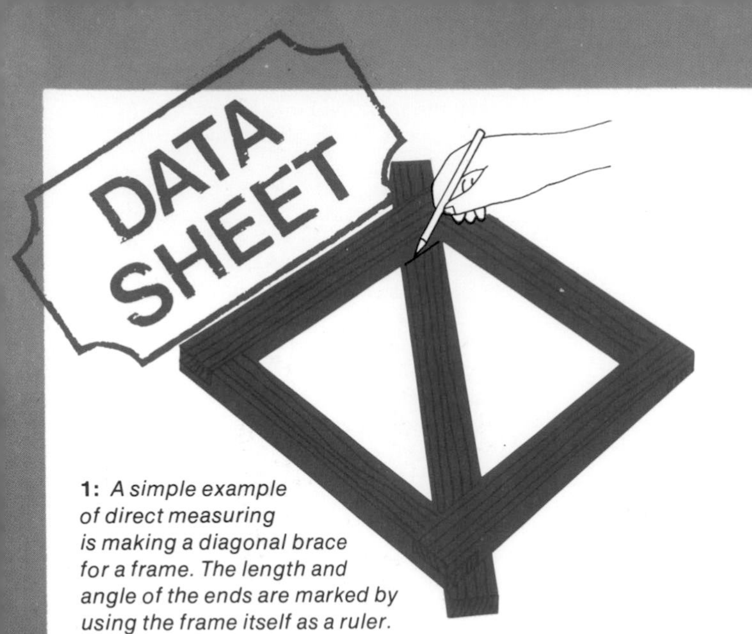

1: *A simple example of direct measuring is making a diagonal brace for a frame. The length and angle of the ends are marked by using the frame itself as a ruler.*

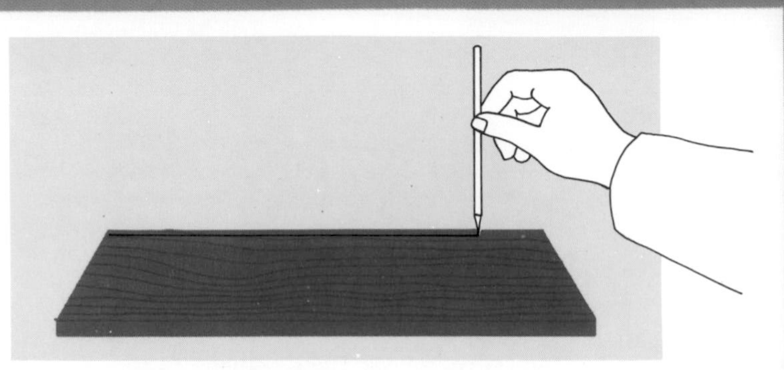

2: *To scribe the back of a shelf to an uneven wall, run a pencil along the wall to trace its shape on to the shelf. Cut along the pencil line.*

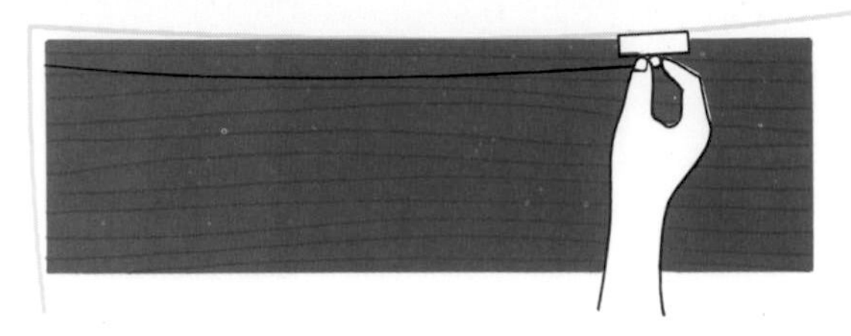

4: *Fitting a shelf to a corner: scribe it first to one wall, then to the other (below). Use a block behind the pencil on irregular walls.*

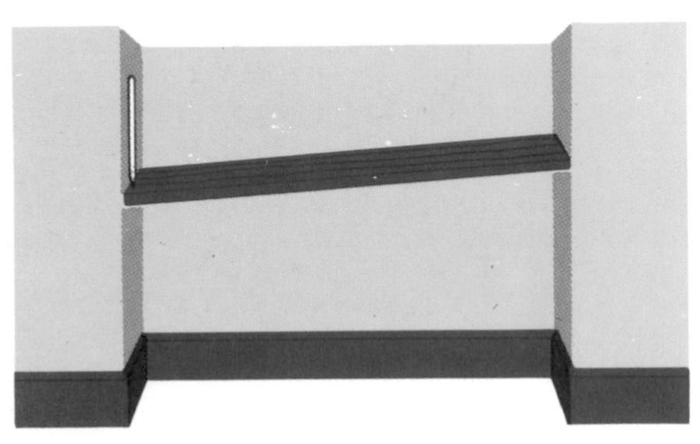

3: *Shelves should be scribed to alcoves by making them slightly too large and jamming them in diagonally at the right height.*

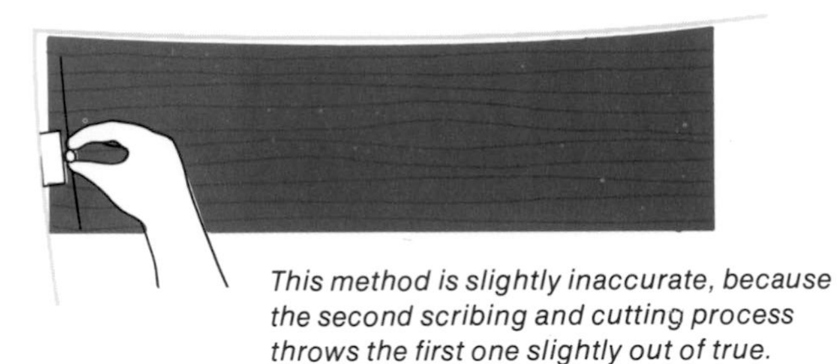

This method is slightly inaccurate, because the second scribing and cutting process throws the first one slightly out of true.

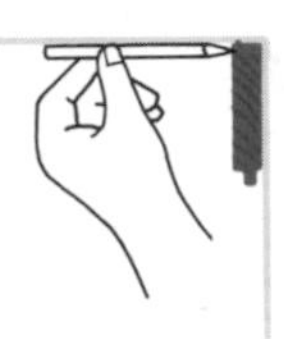

5: *Three stages in fitting T&G boarding to a chimney breast.* **Left.** *The first board is set upright with a level and scribed to the back wall.* **Center.** *The last board on the side wall is scribed flush with the corner, using the point of the pencil.*

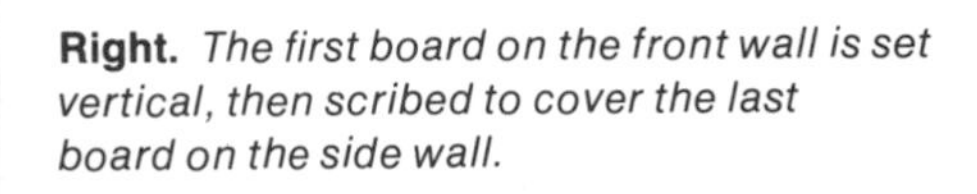

Right. *The first board on the front wall is set vertical, then scribed to cover the last board on the side wall.*

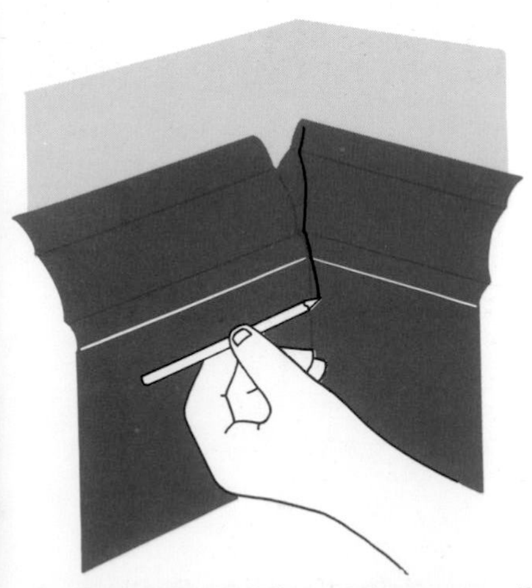

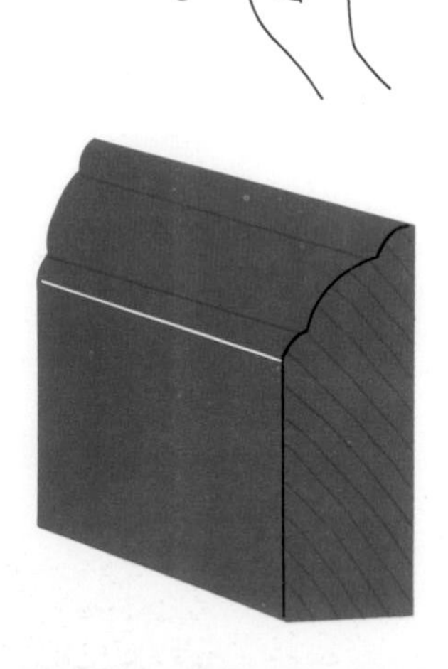

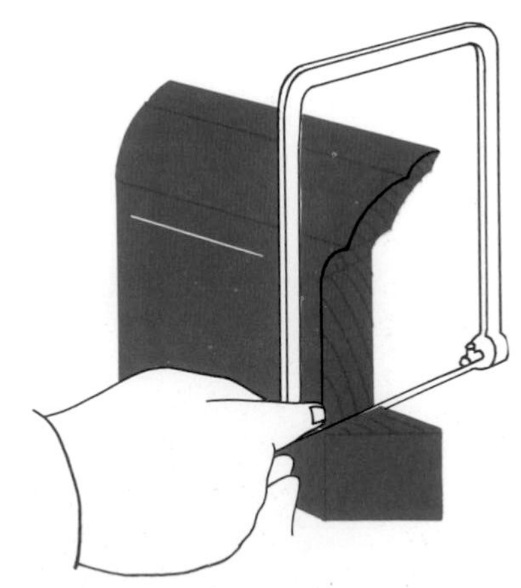

6: *Don't miter moldings at the corners; if they shrink, it leaves an ugly gap. Cut one piece off square and scribe the other to fit it.*

An alternative method: cut the end of one piece at exactly 45 degrees. Then use a coping saw to cut down the line the slanted cut makes with the front surface. This will produce the right profile automatically, without the need for scribing.

CHAPTER 4

Modern coffee table: 1

If you always wanted to make your own furniture, but were frightened off by the apparent complexity of the job, here is your chance. You can make this elegant, professional-looking coffee table simply in a few hours, with a minimum of fuss and bother.

This project introduces a *lap joint* and a *middle lap joint,* and the use of *screws and screw cups.* It begins by showing how to make a shooting board, a simple but necessary device that allows you to plane the end or edge of a piece of wood exactly at right angles to the faces. If you make the shooting board first you can use it when you build the coffee table—and for many other jobs.

Tools required

In addition to the tools listed in earlier chapters, the following will complete this project and set you up for most carpentry jobs you are likely to encounter.

Set of bevel-edge chisels. In this type of chisel, the sides of the blade are sloping and so will not stick when being used to make a narrow crevice. It is a good idea to buy a matched set of chisels in sizes from ¼ in. to 1½ in. You will need only the widest of these at the moment, but the others will be useful later. Buy chisels with smashproof plastic handles that can be hit with a mallet. These are more expensive than the wooden kind, but will last longer because they do not split or fray.

Hand drill (the type that looks like an egg-beater) and a range of *twist drills* (drill bits) which cut metal as well as wood, in a range of sizes between 1⁄16 in. and ¼ in. It is cheaper to buy each size separately and not to get a made-up set. Drills fit into the *chuck* (holder) of the hand drill and are fixed in place by turning the outer ring of the chuck clockwise.

Countersink bit, for making recesses in wood to take the heads of flathead screws.

Screw-Mate cutters, such as the Stanley, for making both the holes and recesses for flathead screws. These useful cutters are numbered in the same sizes as the screws they are for.

Edging clamps for holding the glued joints of the table together while they dry. C-clamps can be used, but edging clamps are more useful as they have an additional device to hold a third edge to the main parts being clamped. Buy large ones. You can use large clamps for small jobs, but not small clamps for large jobs.

Woodworker's vise. Buy a simple vise that clips on to the edge of any table or bench. If your vise comes with unpadded metal jaws, cut two wood blocks about ½ in. thick and fit them between the jaws to keep them from making marks on wood clamped in the vise. Glue or tape them lightly in place to stop them from falling out when the jaws are open.

Marking gauge. This simple wooden tool (mentioned earlier for the bookcase) scribes (scratches) a line parallel to the edge of a

Fig. 1 *(above, left). Chamfering the corner of the stop with a chisel. The tool should be held upright.*

Fig. 2 *(above, center). Using a center punch. The small dent it makes keeps the drill from slipping sideways.*

Fig. 3 *(above, right). The top plate held in place on its bearers by a clamp with jaws padded with scrap wood.*

Figs. 4–5 *(left). Two stages in marking the position of the bearers on the top plate. The line is squared first up the back edge of the plate, and then along the top face.*

board. It consists of a wooden shaft with an adjustable crosspiece sliding up and down it. At one end of the shaft there is a small metal spike. You set the crosspiece with a rule to the desired distance from the spike, and slide the tool up the edge of the piece of wood to be marked. The crosspiece keeps the spike a constant distance from the edge, and the spike scribes a straight line.

Making a shooting board

A shooting board is a simple wooden device that enables you to automatically plane the edges or ends of a board perfectly square with the faces. Essentially, it consists of an upper step, or plate, on which the work rests with its edges protruding slightly beyond the edge of the step so it can be planed. A wooden stop at the far end of the upper step prevents the work from sliding. The plane slides on its side along the parallel lower step. As the sides of the plane are at right angles to the bottom of the plane, it planes the edge of the work at right angles to the surfaces of the steps, hence, at right angles to the face of the work resting on the upper step.

You can make a shooting board out of any old wood you have around, provided that it is not warped. It does, however, have to be made accurately, or it will not plane at right angles. Size is not important. The dimensions given below should be taken as minimum sizes, except for the thickness of the top and its bearers, which should not be exceeded.

Plywood is the most suitable material for the two largest pieces, because it is both cheaper and less likely to warp than solid lumber. The large bottom plate measures 9 in. x 24 in. The smaller top plate is 7 in. x 24 in. The wood for making these pieces should be ¾ in. thick, unless you have a very narrow plane, when the top one should be ½ in. thick.

The three bearers that space the top and bottom plates apart are ½ in. thick in either case (though they can be less for very narrow planes), and measure 6½ in. x 2 in. The stop should be made of a piece of good quality ¾ in. hardwood or plywood 7 in. x 2 in. The grip on the bottom, which holds the device steady in use, should measure roughly 9 in. x 2 in. x 1 in. and can be made out of any kind of wood. You will also need some adhesive and screws: nine 1 in. No. 8s and four 1½ in. No. 8s.

Mark out the pieces, and cut them out with the panel saw. Make sure, by looking along each long edge, that none of them is in the least warped. Test the two large pieces for *winding* (twisting) by laying them down flat, putting a straight batten across each end and checking that the battens are parallel.

Planing the pieces

The next stage is to plane an absolutely straight, accurate front edge along the top plate. The edge must be at right angles to the flat faces of the plate.

Always hold the plane parallel with the line you are cutting along. The grain of wood never runs exactly straight. Plane with the slope of the grain (see Fig. 9) and not against it, or the plane blade will catch on the ends of the fibers and produce a rough result.

When planing the end grain of a piece of wood, do not run the plane over the far edge of the surface, or it will split the wood. Plane from one edge to the middle and stop there (see Fig. 10). Then turn the wood around and plane the other edge to the middle to finish the piece off. This will keep the end of the wood straight and symmetrical, though it will not be as well-finished as if you had used a shooting board.

Check the angle of the planed surface with a try square, and make sure of its straightness by looking along it from one end. Any curve in the wood is magnified when seen from a low angle.

Plane one of the narrow sides of the stop with particular accuracy. Then use a chisel to *chamfer* one corner of the opposite side (Fig. 1). This means cutting the corner off at an angle of 45°. You will not need to hammer the chisel to cut the wood; just scrape the corner off bit by bit until you have removed about ¼ in. from each side of the corner.

Marking up and assembly

Next, use a try square, ruler and marking knife to draw two lines on the base plate at right angles to its accurately planed edge and 2 in. from each end. These mark the position for the outside edges of the two outer bearers. Mark the inside edges another 2 in. farther in. Then draw another line halfway along the board and two more lines 1 in. either side of it. These mark the position of the third bearer.

Draw a line up the center of each bearer along its wide face and cross it at right angles with three lines, one in the middle and one 1 in. from each end. Then make a small x on the outer cross lines halfway between the center line and the edge of the wood on one side of the center line only, and mark the middle cross half way between the center line and the opposite edge (you are in effect making a triangle). Punch a small dent on each x with a center punch, nail set or old blunt nail, to give the drill a spot to start on (see Fig. 2). Put each bearer in turn on the baseplate accurately on its lines with its far end flush with the rear edge of the plate. Holding the bearer in place, drill a 7/64 in. hole through each x almost through the baseplate. Then remove the bearers and enlarge the holes in them (but not those in the baseplate) with the 11/64 in. drill. Use a countersink bit to open out the tops of the holes so that the screw heads will be level with the wood.

Spread a thin film of white glue on the bottom of each bearer and its place on the baseplate. Then put the bearers down in position and immediately screw them to the baseplate, using 1 in. No. 8 flathead screws.

Now lay the top plate on the bearers exactly in position and hold it there while you mark the position of the bearers on the plate's rear edge. Draw lines from that edge forward along the top to show where the bearers lie under the plate. Sketch in the position of the bearer's screws—freehand if you are confident enough (this is a precaution against putting the next

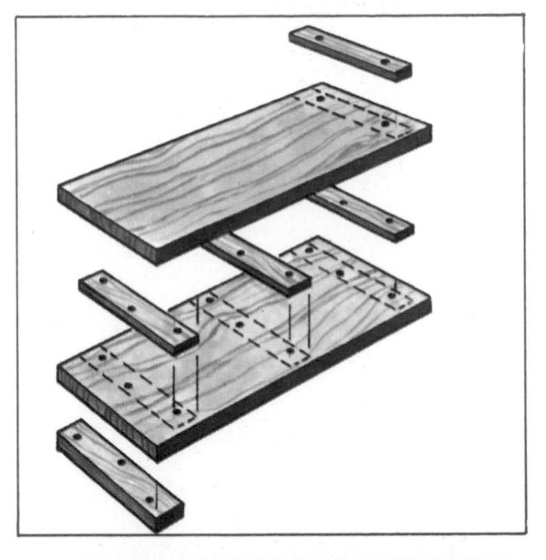

Fig. 6 *(top). The parts of the shooting board —from top to bottom, the stop, the top plate, the three bearers, the baseplate and the "hook" for fixing it to the bench.*

Fig. 7 *(middle). Ensure that the stop is on straight before you screw it down. The slightest inaccuracy will affect everything planed on the board.*

Fig. 8 *(bottom). Using the shooting board to plane a piece of lumber. The board keeps plane and wood exactly at right angles to each other.*

screws on top of the bearer screws). Then mark the center line of each bearer on the top plate. Make two x's on this line—one 1½ in. from the rear edge, and one 2 in. from the front edge.

Clamp the top plate in position with one of your C-clamps, padding the jaws on each side of the board with a small square of scrap wood. Then drill a 5/32 in. hole through each x through the bearer and deep into the base. Enlarge the holes with a 11/64 in. drill in the top two pieces of wood only. You can drill to the exact depth (1 in. is about right) if you stick a piece of tape to the drill bit that distance from the end and stop drilling when the tape reaches the wood. Countersink the holes and unclamp the top plate. Glue it to the bearers and screw it home with 1½ in. No. 8 flathead screws.

The stop should be mounted on the top board with its center just in from the center of the right-hand bearer (to keep the screws from hitting each other). If you are left-handed, put it at the other end. Its exact position is not important, but what is vital is that its inside edge, on which the wood rests, is *exactly* at right angles to the front edge of the top plate. Check this with the try square. The stop should project about 1/8 in. over the front edge. The chamfered corner should be at the front on the edge away from the work. Screw the stop down as you did the top plate, using the same size screws and holes. Do not glue it, because you will want to take it off sometimes for replacement.

The final stage of the assembly is to glue and screw a grip on to the bottom of the plate. You put it at the opposite end of the board from the stop, so that you can hook the board over the bench like a bench hook and plane away from you. If you prefer to plane from side to side, put the grip below the front of the board so that you can lock it in a vise sideways. Exact positioning is not important.

To finish the board, rub wax polish on the surfaces on which the plane slides to enable it to glide easily. The plane slides on its side on the baseplate, with its cutting edge not quite touching the near edge of the top plate. The work being planed extends just beyond the edge of the top plate. To plane end grain, place a parallel-sided piece of scrap wood between the work and the stop, with its outer end flush with that of the work. Then, when the plane blade shaves across the end of the work, the scrap wood is also shaved, but it prevents the corner of the work from splitting out.

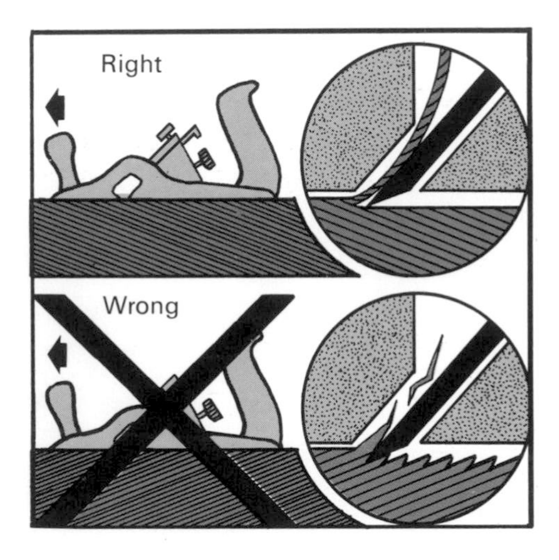

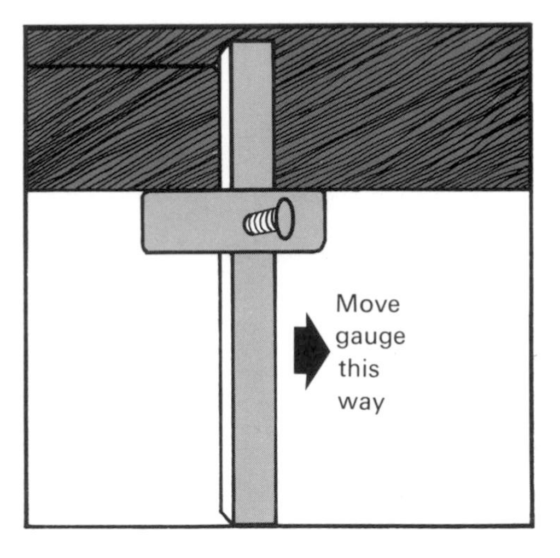

Fig. 9 *(above left). Always plane* with *the grain of wood, and never against it, so that the plane iron cuts cleanly.*

Fig. 10 *(center). When planing end grain without a shooting board, always plane from the outside edges to the middle, first from one edge, then from the other.*

Fig. 11 *(right). Slide a marking gauge* against *the grain, so that the slope of the grain holds it in place.*

The coffee table

The choice of wood for the table depends on you. The construction is suitable for both hardwood and softwood. A really elegant version for the living room can be made out of teak or birch, using solid wood for the frame and veneered plywood for the top. The wood can be given a natural oiled finish with a subtle sheen which will blend in with any type of furniture, old or new.

A less expensive version of the table can be made out of softwood and a manufactured board. The frame can be made of good-quality pine, which has a clearly marked grain that looks good when varnished, or it can be made of the cheapest type of softwood and painted. The top can be made from particleboard and covered top and bottom with plastic laminate such as Formica; cover the edges with a stick-on edging strip. Alternatively, you can use ordinary particleboard or plywood without a laminate. Use plywood with a good quality surface; or (if available) a wide pine planking. If the top is made of plain particleboard or plywood, it should be edged with a narrow wood strip to match the frame.

The lengths and sizes of the lumber you will need are: for the top, a piece 18 in. wide, 44 in. long, and 3/4 in. thick. For the frame, a 7 ft. length of nominal 1 x 3. Also for the frame, have your lumberyard "rip" (on the power saw) a 12 ft. length of nominal 1 x 3 to 2 in. width for the parts of that width shown on the drawing.

These lengths allow 10 percent extra for wastage, but you may find that the dimensions vary slightly from the measurements given—and not always by the same amount. Watch out for this and measure everything yourself.

You will also need seven 2½ in. No. 10 screws and screw cups, and four 1¾ in. No. 8 plated oval head screws. If you are using teak, the screws should be brass.

Starting work

Once you have bought the lumber for the table, the next stage is to cut it roughly to length. Following the drawing, cut each part of the frame, making it slightly longer than it will finally be, so the ends can be sanded smooth after sawing.

In buying the lumber for the table, keep in mind that you will be buying what lumber dealers call "nominal" sizes, which are smaller than full size. The explanation: the lumber was full size when it came from the saw at the mill. But some wood was removed in the smoothing process, and the wood shrank somewhat during its seasoning (drying). So "nominal" 1 in. thick lumber is really 3/4 in. thick. And nominal 2 in. lumber (not used in this project) is really 1½ in. thick. As to widths, nominal widths are ½ in. less than full width up to 7 in. Lumber of nominal 8 in. width and greater, is actually 3/4 in. less than full size. So a nominal 1 x 7 is really 6½ in. wide, but a nominal 1 x 8 is actually only 7¼ in. wide.

The grade of lumber usually used in this type of cabinetwork is classified as "selects and finish" lumber, though other terms are used for some species. Simply explain to your lumber dealer that you want an appearance grade. If you want to cut costs you can use one of the "board grades," which have various natural characteristics, such as knots and minor flaws. In the higher board grades, such as No. 1 and No. 2 "common," however, you are likely to find very attractive material, especially if you can examine it before you buy it. Often this is possible when the lumberyard isn't busy.

CHAPTER 5

Modern coffee table: 2

The lap joint is one of the most useful joints in carpentry. It is a means of joining the corners of a frame where moderate strength and a neat appearance are called for. The marking, sawing and chiseling techniques needed to make it are introduced at this stage.

Marking up the pieces

Once the pieces are cut a little over-size, mark them up accurately for final cutting. The pieces of the frame come in pairs, one at each end or side. Mark up each pair of pieces together, to make sure they come out the same length.

The marks made by a marking knife are quite deep and, of course, will not come out of the wood. For this reason, it is essential to mark only where you intend to cut. The lap joints used in this frame will be spoiled if marks are visible on the side of the wood opposite the cutout you make for the joint.

The four pieces that make up the rectangular central part of the table's frame are visible from both sides of the table. The face side (good side) of each piece should be the side that is cut out to make the joint, and the face edge the edge toward the inside of the rectangle. You can judge the best side by eye.

Lay each matching pair of pieces in the vise, face edges uppermost. Measure them for length, but mark the upper surfaces only. Now measure 2 in. in from each of the end marks, and rule another two lines across both pieces of wood. Take the wood out of the vise and use a try square to continue the end marks around the wood. Take the other pair of marks only part way—across the face side, and halfway down each edge. These lines mark the cutouts for the lap joints (see Fig. 2).

Using the marking gauge

To "halve" the wood, set your marking gauge as exactly as you can to ⅜ in., and use it to scribe a short test line on one of the edges, but somewhere less than 2 in. from the end, so that it will be covered up later. Then turn the gauge around and mark the same surface from the opposite side. If the two marks merge into one, you have set the gauge right. If they do not, adjust the gauge till they do.

Now scribe around the end of each piece, starting from one end of the cutout mark, continuing the 2 in. to the end of the wood, around it and back up to the other end of the mark. Keep a wary eye on the grain of the wood as you use the gauge, because the grain can carry the gauge point off line. If you show signs of getting stray marks on the "good" side (instead of "waste" side), reverse the direction in which you are gauging so the grain helps keep you *on* line.

Sawing for a lap joint

When sawing a section out of a piece of lumber, the "rip" cut (running in the same direc-

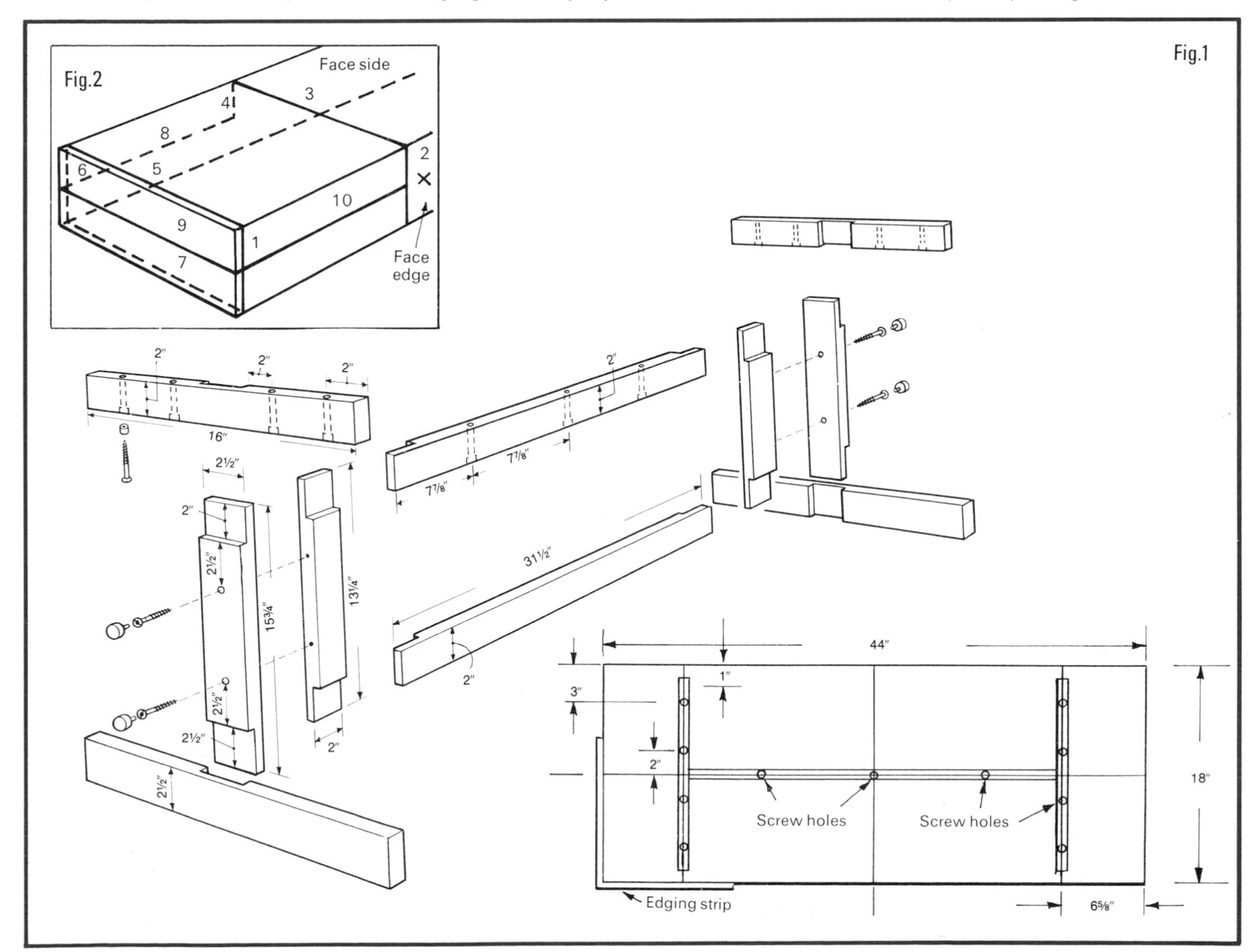

tion as the grain) is done before the "cross" cut (running across the grain). If you try to reverse the order, you will find that, before the rip cut is complete, the lumber will split down to the cross-cut mark, leaving you with an unnecessary clean-up job with the chisel.

Really fine sawing depends largely on how you position your body. You should not stand square-on to the work, because your right elbow would hit your body as you draw the saw back. Stand with the work at your left side and cross your right arm over your stomach to reach it, moving your arm from side to side to saw. This position sounds a bit twisted, but will soon become familiar.

Place the work in the vise so that the marked sawing line is visible from where you stand and so it will not be obscured by the saw blade. The rest is simply a matter of care. Saw on the waste side of the line, and don't rush.

To start your lap joint, set the wood upright in a vise with the face side away from you. Put your left thumbnail on the far end of the marked line. Lower the back saw until the blade just touches your thumbnail. Then make a tiny notch on the corner next to your thumbnail by drawing the saw blade back (see Fig. 5).

Continue to saw very gently, dropping the saw handle slowly until the blade is horizontal, and cut a groove about ⅛ in. deep along the end of the wood. The near side of this groove should just touch the marked line. (A common "first time" mistake is to forget that the cut itself has width, and thus take out too much wood.)

Now lower the handle of the saw until the blade is at a 45° angle to the end of the wood, as in Fig. 6. (You may prefer to tilt the wood instead of the saw, to keep your hand in a comfortable position.) Saw down until the deep end of the cut you are making almost touches the line drawn across the wood 2 in. from the end. Then turn the wood around in the vise and make a similar cut from the other side. Finally, straighten the saw (and the wood, if you have tilted it) and saw out the V-shaped piece of wood left inside the cut.

Now that you have a clean, straight rip cut, finish the cutting out with a cross cut. Start this cut the same way as when ripping, but when you have lowered the saw blade so that it is parallel with the surface of the wood, continue to saw with the blade level until you have reached the right depth. Keep checking the angle of the saw from both sides to avoid cutting too deeply.

When you have made an L-shaped cut in the end of each piece of wood, clean out the angle of the L with a chisel, used *across* the grain.

Finishing the pieces

Once all four pieces of the rectangular center frame are cut to shape, drill three screw holes in the top horizontal piece. They are to hold the tabletop on, so they should be drilled through the wood from edge to edge—a job that needs a vise for the wood and a steady hand with the drill. One hole should be in the middle of the piece, and the other two halfway between the middle and the end (see Fig. 1.). Mark the position of the holes lightly in pencil. Punch a small dent on each mark and drill the holes with an 11/64 in. drill for No. 8 screws. Sight along the length of the wood, rather than across it, so that if you drill slightly off line the drill point will not come out through the side of the wood. Do not countersink; instead, enlarge the end of the hole away from the tabletop with a drill of the same diameter as the unthreaded shank of the screws you'll use. Use a screw cup for a decorative effect on the surface.

Put the pieces together without glue to make sure they are a good fit. If they are not, you can sometimes improve the fit by switching the pieces around.

Finish the face edges of each piece of wood (which will be on the inside of the frame, and hard to get at later) before you put the frame together. This finishing is done in various ways, depending on how the wood is to be treated. All methods start by leveling any irregularities in the wood, usually with sandpaper (remove as little wood as possible). You lightly sand the wood with a piece of fine sandpaper wrapped around a block of scrap wood.

If you are using teak and sandpaper to give it an oiled finish, you need do nothing more at this stage. The same applies if the wood is to be painted. If, however, you are going to wax-polish the wood, give the face edges a light coat of sanding sealer or, failing that, clear shellac. When it is completely dry, re-sand the surface. Then apply a thin coat of wax polish to stop glue from sticking to the wood in places where it cannot be sanded off.

Gluing up the frame

Squeeze out a fairly generous amount of white glue on both surfaces to be joined, and spread it thinly with a piece of scrap cardboard. Fit the frame together and hold each glued corner in place with an edging clamp. All the clamps should be put on the same way, so that the frame will sit level on a level surface. To avoid marking the wood, pad the jaws of the clamps with pieces of scrap wood or hardboard, all of which must be of the same thickness to keep the frame level.

Before the glue sets, check that the joints are tight-fitting. Wipe away any excess glue with a damp cloth—even a spot of glue left on the surface will show through most finishes. Check also that the frame is square by measuring the diagonals between opposite corners; the two diagonals should be exactly the same. The measuring can be done with an ordinary steel tape rule, but a more accurate way is to sharpen the ends of two laths, each slightly longer than half the distance to be measured (see Fig. 8). Then put one pointed end in each inside corner and grip the laths firmly where they overlap. Lift the laths out of the frame, without disturbing their setting, and lay them across the other diagonal. The points should fit into the corners exactly. If the frame is not square, slacken off the clamps and adjust it.

Look at the frame from the side and end to check that it is not twisted. Lay it down on a flat level surface to dry. If it is twisted, weight it to hold it flat. Leave the frame to dry overnight.

Using the chisel

The next job is to make the two end frames. They are made in the same way as the center frame, except that the joints are in the middle of the two horizontal members of each frame instead of at the ends. This means that the chisel must be used to cut out for these joints.

The upright members have cutouts at the ends only. Make them first, remembering that the bottom cutout of each is 2½ in. deep instead of the usual 2 in.

To make the cutouts in the center of each horizontal piece, first measure the exact width of the upright members—it may vary from the standard 2½ in. Then divide the measurement in half and, using a pencil, mark a line that distance on either side of a center line drawn across the face edge of each piece. Check that your pencil lines are accurate by "direct measuring"—that is, laying the upright on the area where it is supposed to fit—before using the marking knife. Mark so that each upright will fit its own pair of cross-pieces; lumber components cannot always be cut so accurately that they are interchangeable.

Having established accurate widths for the cutouts, use the marking knife and gauge, still set to ⅜ in., to complete the outlines of the cutout.

Put each piece of wood separately in the vise in a horizontal position, with the marked surface uppermost. Saw along the lines running across each piece, cutting halfway through the wood and stopping at the marking gauge lines. Try to saw just inside the lines you have marked, to make the joint a tight fit. But do not go more than 1/32 in., or a "shaving thickness," inside the marked lines, since enlarging an under-sized cutout with a chisel is difficult to do accurately. Add a pair of saw cuts inside the first two, dividing the area roughly into thirds, and not quite as deep as the main cuts. These extra cuts will make it easier to chisel out the wood.

To chisel away the surplus wood, angle the chisel blade at 45°. Push the chisel across the grain with your hand so that you cut a slanted piece out of each end of the section to be removed (see Fig. 3). The lower (outer) side of the cut should reach almost to the full depth of the proposed cutout. Cut halfway across from each side, turn the wood around and cut from the other side. This will leave a roughly tapered piece of wood in the middle of the cutout. To remove this lump, turn the chisel blade upside down and knock out most of the wood by hitting the chisel with a mallet. (If you have no mallet, use your hammer lightly.) Finish off by turning the chisel right way up, holding it flat and (pushing by hand instead of using the mallet) working from both sides to the middle. Check the straightness of the cutout with a try square.

Finishing the end frames

Drill two 11/64 in. holes through each upright member, one 2½ in. from the shoulder (the

NELSON HARGREAVES

edge of the cutout) of each joint and both on the center line of each face. These holes are for fixing the end frames to the center frame, and should be countersunk. The oval head screws you are using require it.

Drill four 11/64 in. holes from edge to edge of each top horizontal member, two 2 in. from each end, and two 2 in. from each shoulder of the center cutout. These holes are for fixing the end frame to the tabletop, and should be drilled like the holes in the center frame.

Saw the pieces of the end frame very slightly longer than their marked length. Plane the outside ends of the horizontal members to their exact marked length, using the shooting board, but leave the uprights oversize for the time being.

Finish the inside surfaces of the pieces as you did those of the center frame. Include all edges that adjoin inside corners—that is, both edges of the uprights, the bottom edge of the top, and the top edge of the bottom. Finish also the back sides of all the pieces of the end frames. If you are using wax polish, keep it off the middle ¾ in. of the width of the upright member, which is going to be glued. You can do this by sticking a ¾ in. wide strip of cellulose tape down the middle.

Now fit the pieces of each end frame together. If your sawing has been accurate, they

Top left. *A teak frame and teak-veneered top give this table a touch of luxury.*

Top right. *In this version, a plain white frame sets off a rich rosewood-veneered top.*

Above left. *An ultra-modern table with glass top supported on small rubber pads.*

Above right. *A tough plastic laminate protects the top of this pine table.*

should be a tight fit. If the fit is too tight, you can take a bit out of the cutouts using a sanding block, but rule a line first to keep you straight, and take out only a little at a time. Glue and clamp the parts together, and measure them to ensure they are straight. Leave them to dry overnight, with both ends of the horizontal members propped up straight and level—not with the clamps resting on the floor, which might give the parts a twist as they dry.

When the glue on all three frames is properly set, sand, finish and polish the outside surfaces of all of them. The end grain showing at each joint will be protruding slightly beyond the wood on either side, because you have sawed slightly over-size. Sand off the excess, sliding the sanding block from the outside corner of the frame toward the inside. Finish and polish (if required) in the usual way.

Assembling the frames

The rectangular center frame is slightly shallower than the end frames because it is not meant to touch the floor. Turn all the frames upside down, hold the end frames one at a time to the center frame in the exact position they will occupy, and mark the position of the screw holes in the center frame by tapping a small nail through the holes already drilled in the end frame. Take the end frame away and check that the nail dents are exactly centered. If not, make new dents in the correct positions.

Drill a hole 1 in. deep and 7/64 in. wide through each dent. Then glue and screw all three frames together with your 1½ in. No. 8 oval head screws. The oval head screws, usually chrome plated, are decorative in themselves. If you like the effect, however, you may also use screw trim washers, sometimes called screw cups. These are chromed metal rims that you slip on to the screw to seat under the head. As they vary in form, if you use them measure to see if they hold the screw head above the wood surface. If so, it may be necessary to use a slightly longer screw in order to

Fig. 3. *Chiseling out wood for a middle lap joint. First a slanting slice of wood is removed with the angled blade.*

Fig. 4. *Then the wood between the slanting cuts is removed with the chisel blade held bevel edge downward.*

Fig. 5. *Starting the "rip" (along the grain) cut for a lap joint. A small notch is made in the far edge of the wood.*

Fig. 6. *The saw handle is lowered and the wood tilted in the vise to make a deep diagonal cut.*

Fig. 7. *Laying an end frame out to dry. The wood is resting on its ends and not on the two edging clamps. Note how the clamps are padded with scrap wood.*

Fig. 8. *Testing the center frame for squareness. The two sharpened laths are laid first across one diagonal, then across the other.*

ensure ample holding power. Trim of this type is made in sizes to match common screw sizes.

Check that the table frame is sitting straight and true in all directions. Small adjustments can be made by slackening off the screws a little, moving the pieces and tightening the screws up again.

While the frame is drying, prepare the top. Cut it to its finished size and trim its edges with the plane. Mark the position that the frame will occupy on the underneath of the top, using Fig. 1 as a guide. If you are gluing laminate to the top, glue on the backing layer (underneath of the board) before you mark the frame position. (See below for instructions.)

Drill eleven 7/64 in. holes for the mounting screws to a depth of ½ in., being careful not to drill through the top. A piece of tape round the drill ½ in. from the end is a useful guide to the depth of the hole. Measure for the screw length by pushing a stiff wire through the horizontal upper frame member into the hole drilled in the underside of the top.

The method used to finish the top depends on what it is made of.

If you are laminating a particleboard top yourself, it is best to laminate both sides to ensure that it does not warp. Cheap, thin laminate is used for the underside. Remember to buy enough good-quality top laminate to cover the edges as well as the top, unless you are covering them with wood strip.

More finishing techniques

If you are using a veneered particleboard top, the edges can be finished with wood strips. Do the ends first. These should be cut slightly over-size, put on with white glue, and held down with edging clamps while the glue sets. Tighten the clamps to the top of the table, padding the jaws with scrap wood. Then do up the clamp's extra screw (padded with wood) on to the strip to hold it tight. (If you have no edging clamps, tack the edging on with small brads, punch their heads below the surface and, after the first coat of paint or other finish, fill the holes with plastic wood.)

When the glue is dry, trim off the ends of each strip with a plane. Plane inward, so that the plane presses the strip down on to the table edge and thus prevents splitting. Now glue on the side strips so that they overlap the end strips. Finally, trim the sides to length and finish all the edging with fine sandpaper—held tightly around a sanding block.

If the frame is to be finished differently from the top, now is the time to paint, varnish or polish it. When the finish is dry, turn the top upside down and screw the frame to it—and the table is complete.

If the table is made of teak, it can be finished simply by rubbing in teak oil. A tougher finish that looks exactly like oiled teak can be achieved by painting the teak with a 50/50 mixture of clear polyurethane varnish and turpentine substitute. Allow it to dry, sand it down with fine sandpaper, and put on another coat. It is important to paint along the grain to avoid leaving brushmarks. Finish with Grade 0 (very fine) steel wool and a little teak oil.

CHAPTER 6

Timber paneled walls

Natural timber boarding can be used to decorate any room in a house—from a living room to a bathroom. The colors and patterns which can be achieved with different wood grains, finishes and board profiles are almost limitless, and since timber is long-wearing, redecorating problems will be minimized.

Most woods used for building purposes can be used for decorating interiors. The choice of lumber, therefore, should depend on cost, availability and appearance. Many softwoods are very pale in color, while hardwoods range from the creamy tones of ash, natural maple or birch to the deeper hues of mahogany, cherry and walnut.

When choosing a wood for timber paneling, it is important to consider the moisture content of the wood; timber which has a high water content will shrink and warp in a heated room.

Solid timber planks, or boarding, are cut to one of several standard profiles. Square-edged planks may be used, but tongue and groove—or "T and G"—planks are more commonly used for internal timber linings, as they allow for easy formation of interlocking joints. Different groove designs, such as rabbeted, V-jointed, squared and extended shiplaps, are available in lengths which will fit the height of an average room. On a long wall, however, horizontal boarding may need to be joined end-to-end.

Board thickness is usually ¾ in., widths 6 in. or more. The thickness determines both the rigidity of the boarding and the spacing of the fastenings. Allowance for movement of the planks is made when the tongues and grooves are originally machined. As shown in Fig. 5, the tongue is made not to extend the full depth of the groove.

Design considerations

So much depends on the use of the room, its size, shape and lighting, and on personal choice, that it is difficult to lay down rules for choosing internal linings. The basic factors which should be considered are: whether to line the walls and ceilings in wood (wall paneling methods may be applied to ceilings) or only one or the other; whether to cover all the walls with timber or to leave some in a contrasting finish; what kind of wood to use; the angle at which to install the boards—vertical, horizontal, or diagonal; and the extent of coverage that is desired—the full or partial height of the walls.

Another point to consider is that if furniture in a room is a jumble of pieces of different

Below. *A small, rather confining bedroom can be made to look rich and cozy with timber lining on all walls.*

MICHAEL BOYS

heights, horizontal boarding will tend, unfortunately, to accentuate this aspect by providing guidelines for the eye. On the other hand, rooms such as kitchens and laundry rooms, which usually have worktops and appliances all about 3 ft. high, will usually look better with horizontal boarding. It helps make their (usually small) walls look longer.

A final consideration is your own skill. A beginner will find it quite easy to board one plain wall to make a "feature wall." But covering a whole room involves a range of problems—replacing door and window trim, for example—and requires a bit more skill, or experience, or patience at any rate.

Preparatory work

Before fastening any boards to new walls, be sure that the walls are dry; newly plastered walls will take at least two months to dry and concrete walls will take at least four months to dry before boards can safely be attached to them.

The method of attaching internal timber linings will depend primarily upon the type of wall to which they are to be fastened.

Furring strips of nominal 1 x 2 lumber are usually required. On a typical wall, with wood framing behind the wall surface, these strips are simply nailed through the wallboard into the studs (internal wall posts). You can locate the studs by tapping the wall lightly with a small hammer, or by means of a "stud locator" sold by hardware stores. If the paneling is to run vertically, the furring strips must run horizontally. For horizontal paneling the furring strips run vertically, directly over the studs. For diagonal paneling the furring strips may run either way. They're usually spaced 16 in. apart, center to center.

On masonry walls, as in a basement, the strips (usually 2 x 4s) are fastened to the walls with masonry fasteners, such as Rawl plugs. You need a power drill (which you can rent) and a masonry bit to drill the holes for the fasteners. Ask for specific instructions for the type of fasteners you buy, if they are not printed on the container.

Re-wiring and re-plumbing

You may want to put in some new light fixtures or move existing ones or, if you intend to line a kitchen or bathroom, you may need to rearrange some pipework. Any work of this sort should be carried out at the "bare wall" stage rather than later, to avoid damaging the wall paneling. Also, it is advisable to mark the wall surface where any new wiring has been installed. This will help you to avoid accidentally severing or damaging cables when drilling or nailing the paneling.

Wall preparation

Moisture protection. In most rooms no special moisture problems are likely to be encountered. In bathrooms and kitchens, however, measures must be taken to protect the paneling from the effects of moisture, and to prevent the moisture from penetrating to the outside of the house wall, where it can cause paint peeling and other damage. If your house is a new one it may already have a moisture barrier in its walls. If you're not sure the barrier is there, you can add one simply by covering the old interior wall with polyethylene sheeting made for the purpose and sold by lumberyards. It is usually fastened to the wall with a staple gun. The paneling boards, if they are to be used in a kitchen or bathroom, should be protected by a moisture-proof finish. A good sealer should be applied to the back surface of each board and to the edges before the panel-

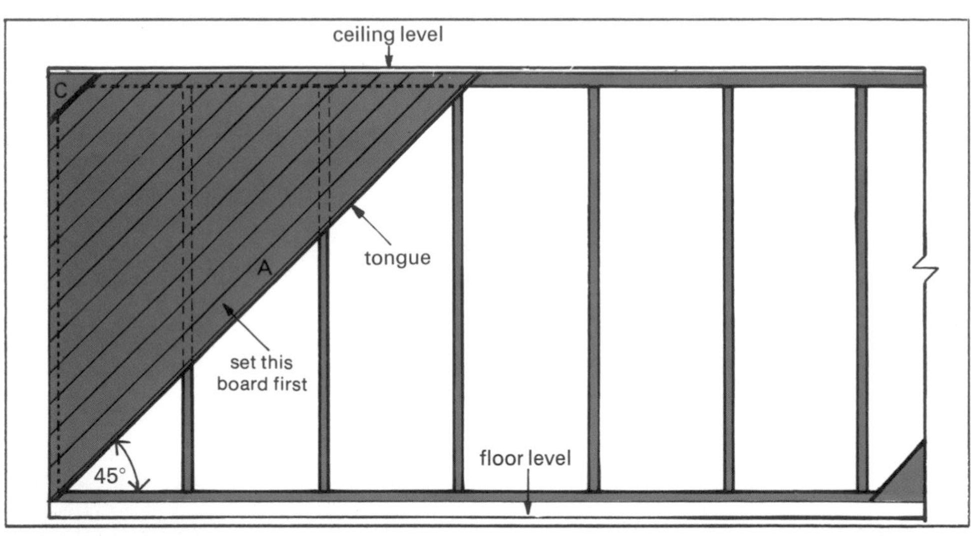

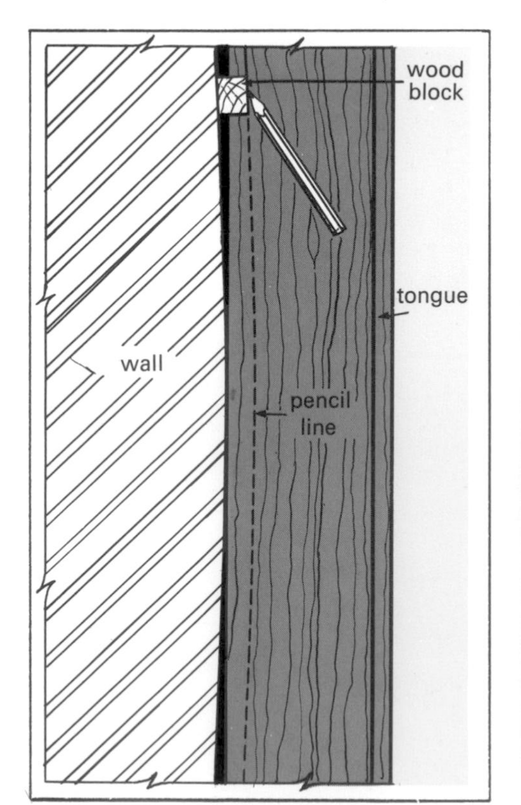

TRANSWORLD FEATURE SYNDICATE

ing work begins. A moisture-proof varnish should be used on the exposed surface. This need not have a high gloss. Polyurethane types are available with a satin finish. Let your paint dealer help in the selection of both products.

Insulation. You can add fiberglass or foam insulation between the furring strips described shortly, before the paneling goes on. The foam type, in some forms, also serves as a moisture barrier, in addition to saving heat.

Conditioning the timber

Ideally, keep all timber in a dry, damp-free place, preferably in the room in which it is to be used. Store it for at least a week before using it, to allow the wood to acclimatize to the room's general temperature and to permit excess moisture to evaporate from it.

Calculating quantities of furring

Estimating the amount of lumber needed for furring strips is simply a matter of measuring and adding. If the strips run vertically, you'll need one to match each stud. Don't space them farther apart, as this would either require a gap at least double that of the studs (too wide) or the placing of strips between studs, which would result in nailing merely into the wallboard. This provides poor holding power for the nails. Horizontal furring strips (for vertical paneling) may be spaced 2 ft. apart from center to center if you are using the usual ¾ in. thick paneling boards. Be sure, however, that

ELIZABETH WHITING

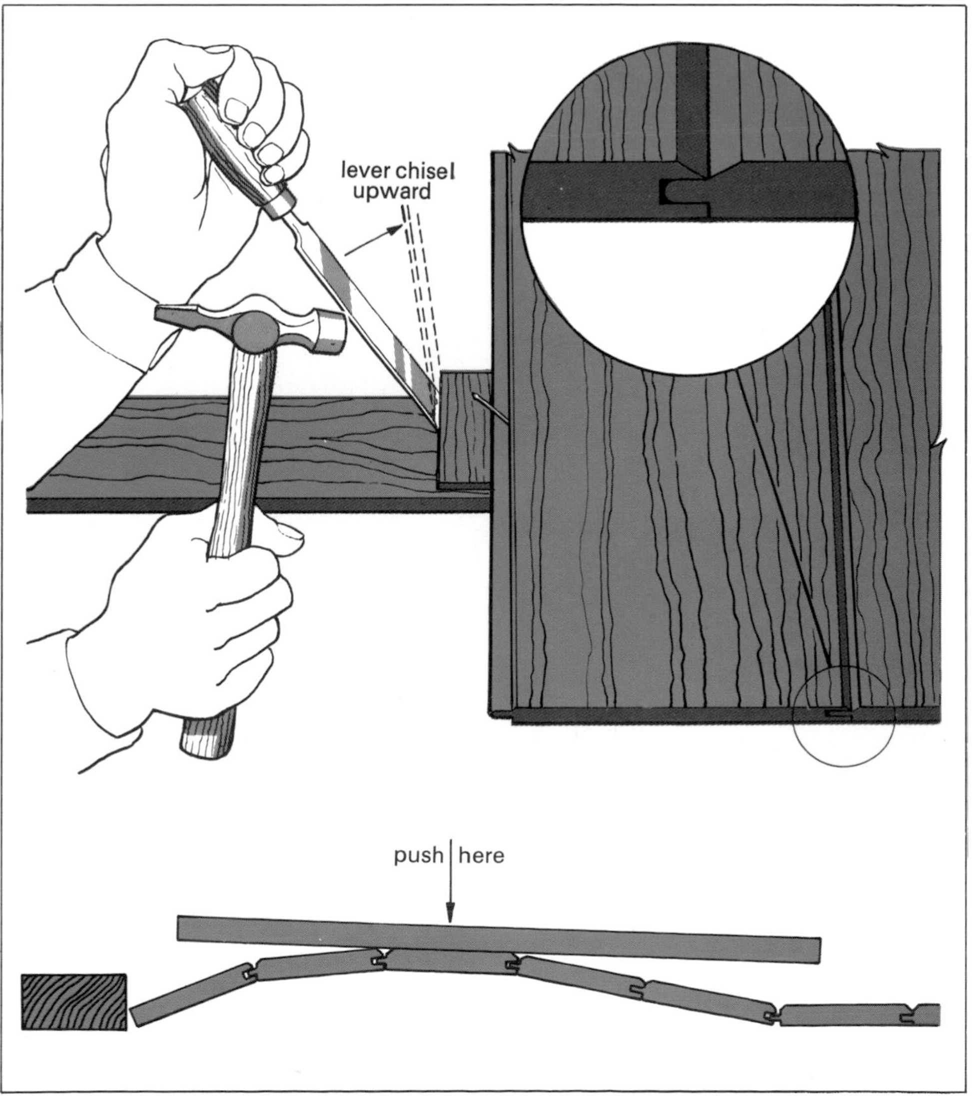

Fig. 1 *(far left). When an adjoining wall is "out of true" (shown here in exaggerated form) the first board must be scribed to fit the curve or slope. See that the board is vertical—it too may have a curve which you must straighten—as you nail it temporarily. Then trace the outline of the wall by running a wood block and pencil down the wall. Cut the board down the pencil line, and move it over into its correct position.*

Fig. 2 *(above left). Attaching diagonal boards to vertical battening. Note that you must also have top and bottom battens; these are to give you nailing for the ceiling molding and the skirting board, as well as for the ends of the T & G boards. Whichever way your boarding is to run, you always need to "frame" the wall with the outside battens.*

Fig. 3 *(left). Timber boarding used as in this kitchen, for walls, storage units and a dining shelf, will help disguise awkward features and unify the room's appearance.*

Fig. 4 *(top right). A plain room can be given a distinctive appearance by the use of T & G boards, in this case set diagonally.*

Fig. 5 *(center right). Clamping vertical boarding. Drive the chisel point into the batten and pull it upright to squeeze the board against its neighbor. This helps to eliminate ugly gaps as the timber dries.*

Fig. 6 *(bottom right). Since the last two or three boards cannot be clamped, fit them together into a bow shape and "spring" them into place with your fist.*

BAVARIA VERLAG

the strips are nailed into every stud along their length. No nailing between studs is necessary, as the nails would merely penetrate the wallboard and have poor holding power. Whichever way the furring strips run, it is essential to also have furring all the way around the perimeter of the wall—top, bottom, and both vertical ends.

Getting an even surface

If the wall on which the furring strips are to be mounted is uneven or irregular, there are a number of ways in which the furring strips can be fastened so as to present a flat, even surface for nailing the paneling. Keep in mind, too, that the paneling boards can flex somewhat without appearing distorted. So a reasonable amount of curvature over the wall surface is permissible without likelihood of a distorted appearance.

Above. *Natural grained furniture combined with timbered walls for a striking effect.*

If there are noticeable high spots on the wall surface, as from built-up plaster on a hastily made plaster repair, simply chip off the excess with a chisel. As the entire wall will be covered by the paneling, the resulting rough area won't be visible, anyway. To check for overall curvature of the wall, drive a small finishing nail into it close to one corner, and draw a cord taut from the nail to another nail driven close to the opposite corner of the same wall. Any curvature, concave or convex, will be made immediately apparent by the straight line of the string. Wooden shims can then be nailed to the wall to bring each furring strip out just enough to provide an even overall nailing surface for the paneling. Some of the furring strips, of course, will be in contact with

ELIZABETH WHITING

Above. *A narrow hall and archway gains warmth and richness from all-over timber.*

MICHAEL BOYS

Above. *One timbered wall, plus matching shelf unit, makes a pleasant kitchen.*

the wall. The same shimming process must, of course, be followed with the furring strips that rim the perimeter of the wall. In many instances, however, little or no shimming will be required, especially in newer houses.

About trim. The baseboard of the walls must be removed before the furring is applied, as furring at the base of the wall will be in its place. Window trim may be removed and replaced over the paneling later. Or, to save work, it may be left in place and the paneling fitted around it. Either way, furring must be used around all window and door frames. If the paneling is fitted around window and door trim, a quarter round molding may be used to conceal the juncture between the paneling and the window or door trim. Measure the existing trim before deciding on the method to use. If you are short of working time, it may be worthwhile to improvise as necessary to do a fitting job around existing trim.

Electrical outlets. The simplest way to treat electrical wall outlets is by merely leaving them where they are. Mount small wood frames around them, equal in thickness to the furring strips, and just large enough to permit removal of the outlet cover plates if it should become necessary. Cut the paneling boards to provide an opening matched to the inside of the wood frames, and round the edges of the panel opening with coarse, then finer sandpaper for neatness. The recessed outlets are not inconvenient, and cord plugs are less likely to be accidentally knocked out of the receptacles.

Estimating the amount of paneling. As board paneling requires trimming to length in almost all cases, you need to buy somewhat more than the square footage of the wall. Although the extra amount required depends on the individual job, a rule of thumb often applied calls for 1150 square feet to cover 1000 square feet of wall when the paneling is either vertical or horizontal, and 1350 square feet to cover 1000 if the paneling is diagonal. You can apply the same general proportions, of course, to smaller amounts.

Splicing. If the paneling boards you are using on a horizontal paneling job are less than the length of the wall, they may be joined either at the furring strip or by means of a "butt block." The latter is simply a length of wood as wide as the panel board and as thick as the furring. It should be about 1 ft. long. Nail it to the wall with its mid-point where the ends of the panel boards will meet. Then nail the panel boards to it, after carefully fitting their ends for appearance. Always stagger the joints so they do not come next to each other in adjacent boards. This is usually easy if planned in advance. If you prefer to join panel boards at a furring strip, 1 x 3 furring strips (instead of 1 x 2) will make it easier.

Selecting and matching paneling

All natural timbers vary in color, grain and figuring, so there may be some slight variation from plank to plank. To get the most pleasing effect, stand the boards against the wall and arrange them to suit your own taste before cutting and nailing them.

Fastening the boarding

There are three important points to note when paneling: **1.** Always begin vertical paneling from a corner to avoid wastage, and to establish a "true" board, so that following boards can be correctly joined on and do not "stagger" out of line. **2.** Always "square off" the first board with the adjoining wall and ceiling before nailing, as few ceilings and walls are ever perfectly straight. **3.** Always leave a ⅛ in. gap between the paneling and the floor and ceiling. This allows the panels to expand slightly in damp or humid atmospheres and can be concealed by small "quarter-round" molding.

Scribing

If the wall is out of plumb or is slightly curved, it may be necessary to trim your first board to fit exactly into a corner. To do this hold the board against the corner, flush to the butting wall and the ceiling; use the spirit level to make sure it is truly vertical. The exact contour of the wall is then traced off on to the board with a pencil (Fig. 1) and the board trimmed or cut accordingly. This process of tracing an angle or contour directly on to a board is called *scribing.* The advantage of scribing over measuring is that any unevenness in a surface will automatically be transferred in outline to the piece being traced.

For trimming such an edge of a board, you can usually flex the panel saw blade enough to follow the mild curves.

Fastening T & G boards

Vertical positioning

Begin at an internal corner of a room and scribe to fit the groove side of the first board into the corner, as described above. Remember to leave about a ⅛ in. gap at the ceiling and floor levels. Keeping this first plank vertically true, begin nailing about ¼ in. away from the corner edge. Skew nail the board at about a 45° angle into the support framework behind, using 1¼ in. finishing nails, a small hammer and a nail set to push the nailhead well below the surface of the board. Work down the plank from ceiling to floor, nailing at intervals to correspond with the furring behind. Any holes can be filled with a suitable wood filler.

On the other side of the plank, place nails along the tongue of the board and skew nail with a hammer and nail set (Fig. 5). Fit the groove of the next board over the tongue, thereby hiding the nail heads. Next, part drive the nails in this board, and *clamp* the board up close to the preceding one using a chisel (Fig. 5). Clamping should always be done on the tongue of a T and G board, as any marks will be covered by the groove of the next board. It is necessary to clamp tightly; otherwise gradual shrinkage of the timber will make your wall open up at the joints. When you come to the third board, clamp and nail from the bottom up instead of from the top down—clamping always in the same order will gradually "skew" your boards out of line.

When you reach the last two or three boards to be mounted along a wall and into a corner, do not clamp them. Instead, cut them very slightly oversize and fit them together in a slightly bowed shape (Fig. 6). Then bang or "spring" them into place with your fist. These last boards will need to be nailed through their surfaces, and the holes filled with wood filler.

DATA SHEET

All about joints

1. The butt joint is the simplest of all joints in carpentry. It may be made straight or right-angled, and needs nails, glue or screws to hold it together.
2. The doweled joint is basically a butt joint reinforced with dowels—lengths of wooden rod. Both halves of the joint are drilled at once, where possible, to make the holes line up.
3. The blind doweled joint is better-looking, because the ends of the dowels do not show. The two rows of holes are drilled separately, so great accuracy is essential.
4. The mitered joint has a very neat appearance, because no end grain is visible. Unfortunately, it is a very weak type of joint unless it is reinforced in some way.
5. The lap joint is used at the corners of a rectangular frame. It is simple to make, has a reasonably neat appearance, and is quite strong if glued together.
6. The middle lap joint is a variant of the usual L-shaped lap. It is used in conjunction with the previous type of lap in the construction of simple frameworks.
7. The cross lap joint is the third member of this versatile family. It is used where two pieces of timber have to cross without increasing the thickness of the frame.
8. The lap dovetail joint is an extra-strong type. Its angled sides make it impossible to pull apart in a straight line, though it still needs glue to hold it rigid.
9. The housed dado joint is used for supporting the ends of shelves, because it resists a downward pull very well. It, too, must be reinforced with glue or screws.
10. The stopped dado joint has a neater appearance, but is harder to make because of the difficulty of cutting out the bottom of the rectangular slot neatly.
11. Tongue-and-groove joints are most commonly found along the edge of ready-made boarding. But a right-angled version is also found, for example at the corners of cabinets.
12. The end rabbet has a recess cut in one side to hide most of the end grain. It is often found in cheap cabinet work, because it is easy to make with power tools.
13. The mortise-and-tenon joint is a very strong joint used to form T shapes in frames. The mortise is the slot on the left; the tenon is the tongue on the right.
14. The through mortise-and-tenon joint is stronger than the simple type. It is often locked with small hardwood wedges driven in beside, or into saw cuts in, the tenon.
15. The haunched mortise-and-tenon is used at the top of a frame. The top of the tenon is cut away so that the mortise can be closed at both ends, and so retain its strength.

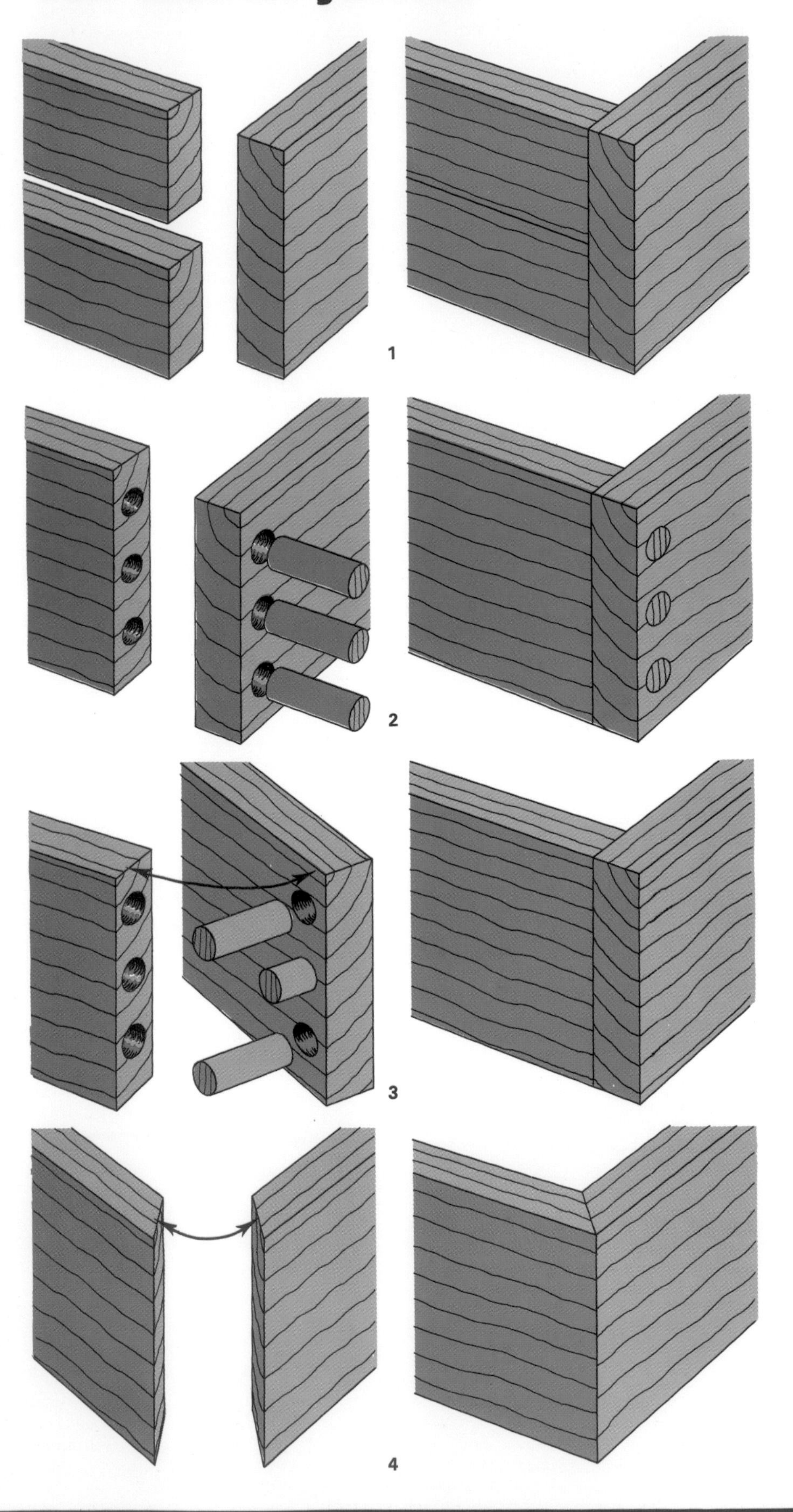

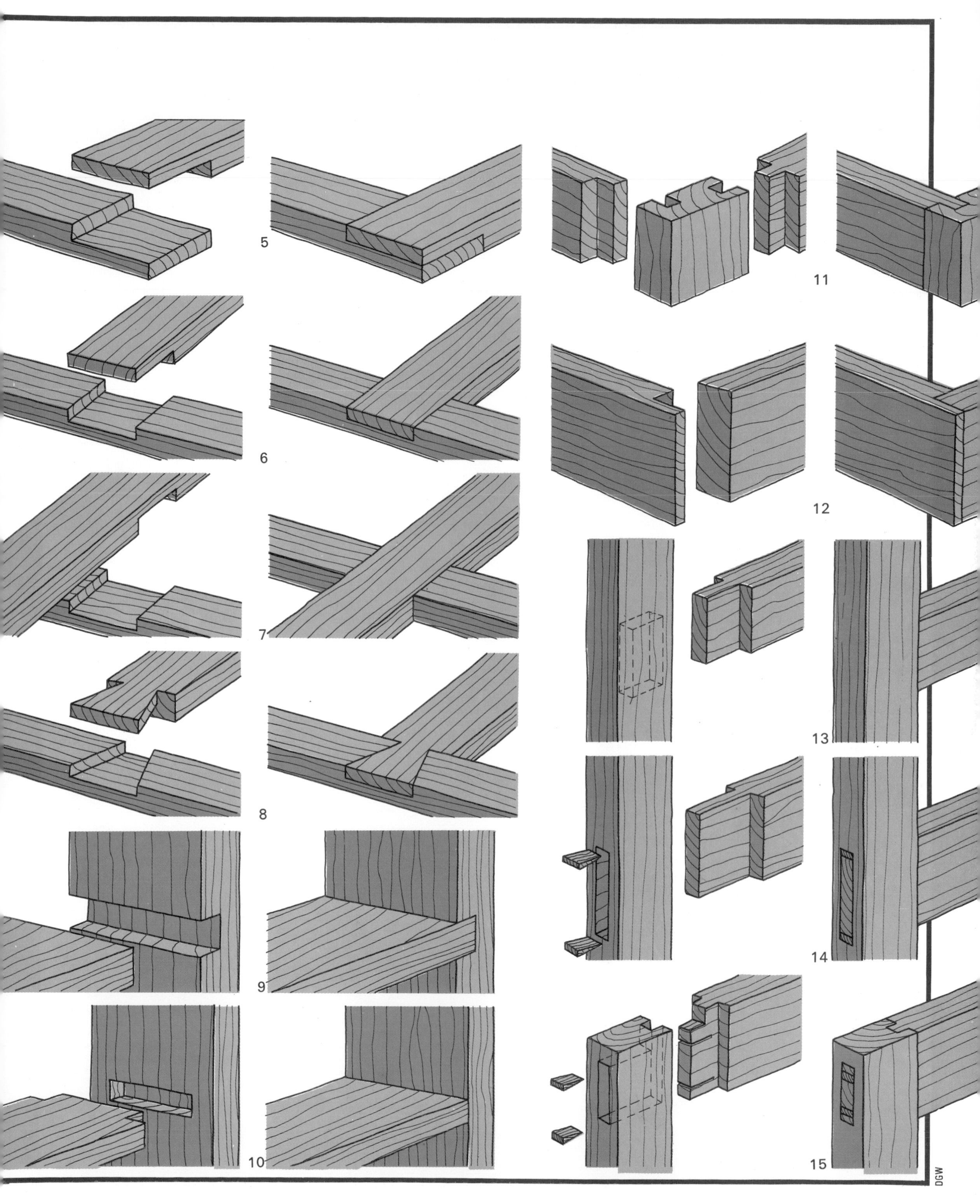
5
6
7
8
9
10
11
12
13
14
15
DGW

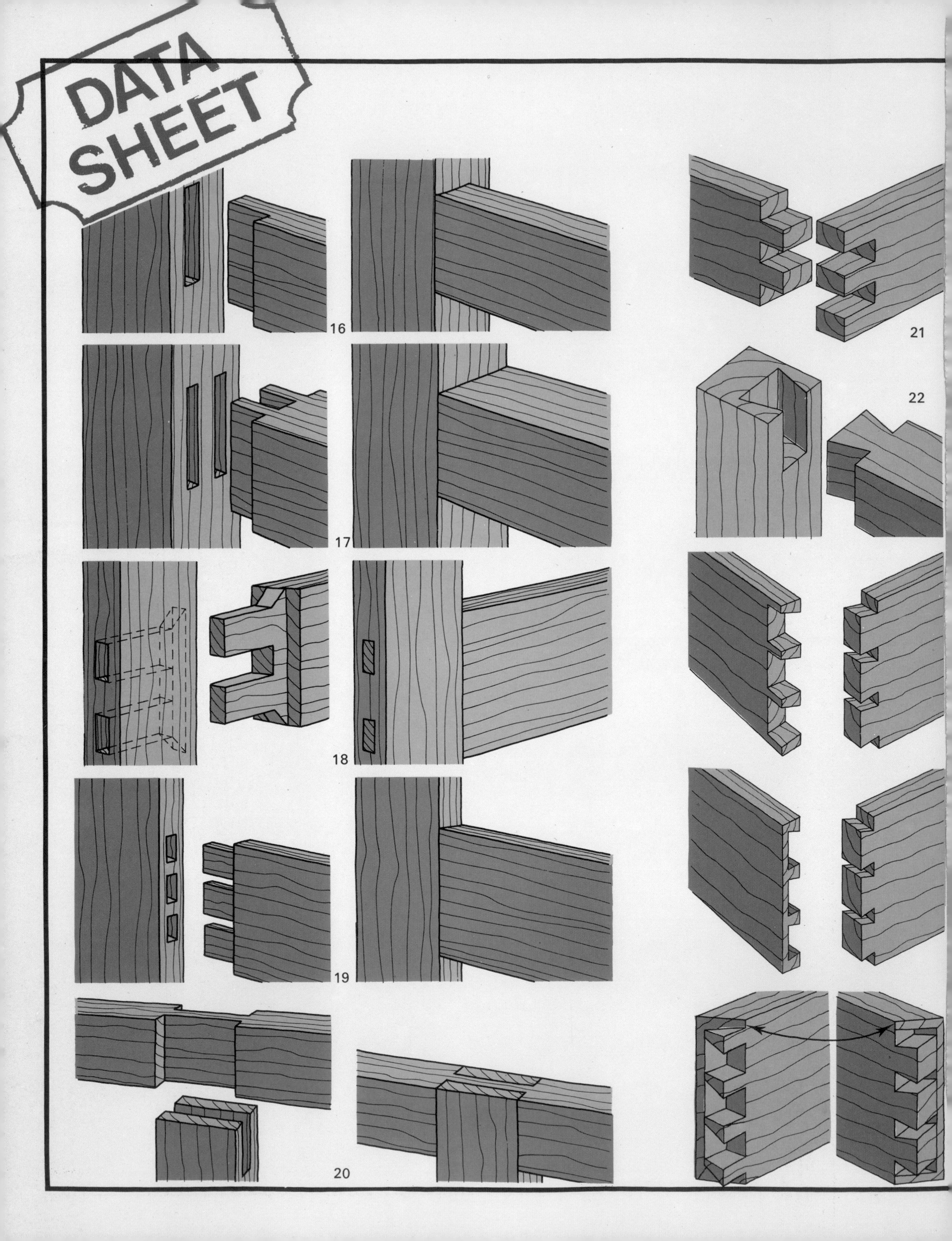
DATA SHEET
16
17
18
19
20
21
22

Four more kinds of mortise-and-tenon joint:

16. The bare-faced tenon is offset, with a "shoulder" on one side only. It is used for joining pieces of different thicknesses.

17. Twin tenons are used in very thick timber. They give the joint extra strength by doubling the area of the gluing surface.

18. Forked tenons add rigidity to a deep, narrow joint. The angled edge of the tenon is sometimes found in a haunched m-&-t joint.

19. Stub tenons are used on even deeper joints, but are weaker and less rigid than the forked tenons shown above.

20. The through bridle joint is used to join cross members along the run of another member in a framework.

21. The box joint is quite strong and has a decorative appearance. It is used for the corners of wide frames and boxes.

22. The single dovetail, like all dovetails, is extremely strong and hard to pull apart. It is used at the corners of heavy frames.

23. The through multiple dovetail is used at the corners of cabinetwork where great strength and a good appearance are required.

24. The lapped dovetail is nearly as strong, but also has one plain face. It is used in very high-quality cabinet work.

25. The mitered secret dovetail is also used in very high-quality work. It looks like a mitered joint, but grips like a dovetail.

26. The stopped lap dovetail looks like an end rabbet. It is slightly easier to make than a mitered secret dovetail.

27. The cogged key joint is like a dovetail with the tails subdivided into smaller tails. It is extremely strong and rigid.

28. The scarfed half lap joint is used for joining frame members end-to-end where only moderate strength is required.

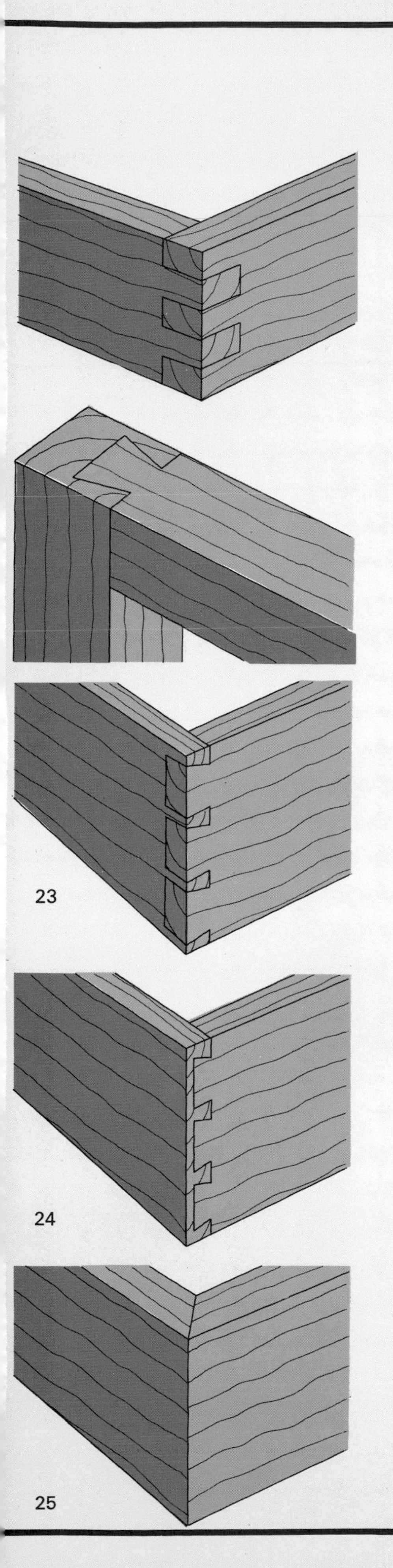

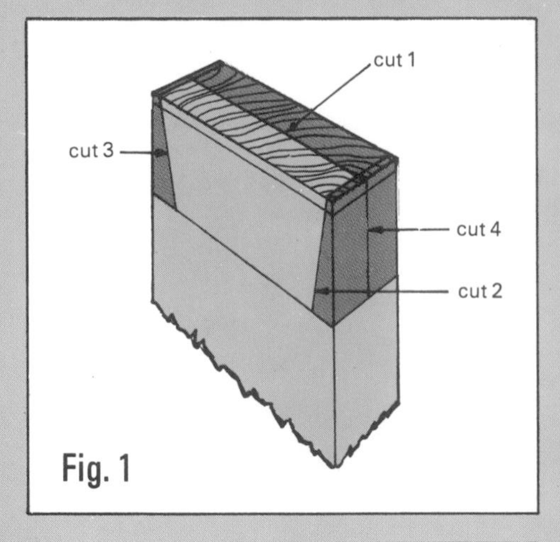

Fig. 1

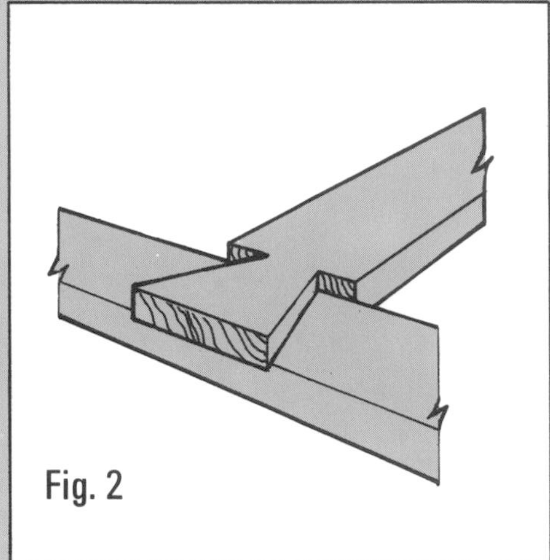
Fig. 2

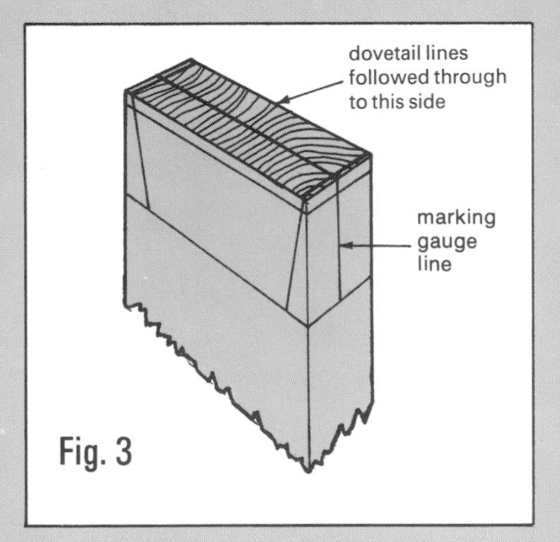

Fig. 3

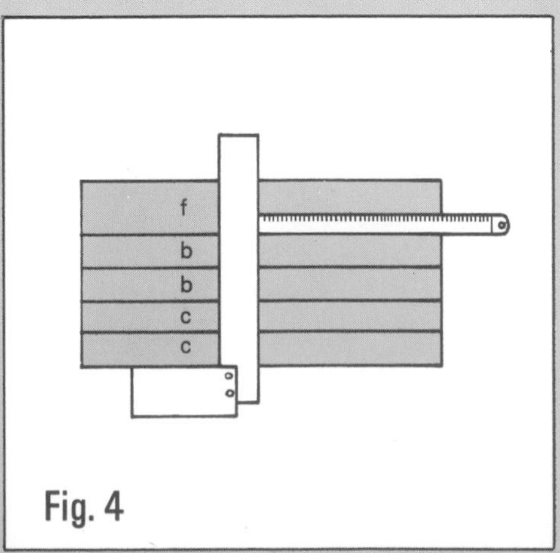

Fig. 4

TRI-ART

CHAPTER 7

Sawhorse-bench

This combination sawhorse and portable workbench is easy to make from stock size lumber. Although it is the usual sawhorse size, it can be modified to suit the user. If you are taller than average, you may want to make it a little higher if you plan to use it for tool work. If so, simply lengthen the legs.

If you look at Fig. 10 you will note that the legs D are joined to the top rails C with simple lap joints. But they are joined to the lower rails E with a lap dovetail, which is somewhat trickier to make, but stronger for that particular part of the project. It is also a good type of joint for many other uses, so this is an easy place to try your hand at making it. As the sawhorse is a workshop accessory, appearance isn't too important, and if your first attempt at making lap dovetail joints happens to look a bit crude it won't matter. And if you glue the joints with plastic resin glue there are simple tricks to make them very strong even if you bungle somewhat and end up with a loose-fitting joint.

The tricks that can save the day for you are simply a matter of filling the gaps in a loosely fitted joint with slivers of wood that have been sawed and planed to fit. Thus, if your bungled joint is so loosely fitted that it has a gap of 1/8 in. on each side, all you need do is cut a lengthwise sliver of wood a little over 1/8 in. thick and plane it down until it fits snugly in the gaps. Then cut a piece for each gap and push it in place with a liberal coating of glue. The joint will be almost as strong as if you had done it right. And once it is sanded smooth on its exposed surfaces, the fitted slivers will have a workmanlike appearance, as if you had intended to make the joint that way in the first place. Keep this trick in mind any time you do a poor fitting job. Until you become skilled at joint work it can save you trouble and material.

The top support

The top support, or backbone of the sawhorse is made up of two pieces of stock 2 x 3 and one piece of 2 x 6, all 30 in. long. If you are working with leftover lumber, cut the pieces a little over length, then use a try square to mark the ends square for cutting to exact length. If you are buying new lumber for the job you can have it cut to the length you want at some lumberyards. There may be a small charge for the cutting, but if it's done on a typical lumberyard power saw, the cuts will be perfectly square, saving you the work of hand sawing.

The legs

Stock 2 x 3 lumber is used for the legs, which are also 30 in. long. The upper and lower rails to which the legs are joined are also of stock 2 x 3 lumber. The upper rails are 30 in. long, the lower ones, 25½ in. To see how these parts fit together take a close look at the drawing, Fig. 10, and the large photograph. The details of joining them are also shown in the smaller drawings.

Putting it together

The first step in assembling the sawhorse is that of making the top support or backbone. The 2 x 3s are nailed to the sides of the 2 x 6 and also glued to it. Use plastic resin glue (available at most hardware stores) and apply it by brush to both contacting surfaces. This means that the inner surfaces of the 2 x 3s will be coated with glue along with the portion of the outer surfaces of the 2 x 6 that will be in contact with the 2 x 3s. Wipe off any squeezed-out glue.

On the side where the legs will be hinged, drive the nails through the 2 x 3 so they won't be in the way of the screws that will hold the hinges. On the other side (where the legs won't be hinged) drive the nails a little below the center line of the 2 x 3, as some of the upper portion of the 2 x 3 will be cut off later to create a level surface for the top platform.

Making the joints

To make the lap joints that join the legs D to the top rails C, you need your back saw and chisel, and your try square for marking. The outer edge (the edge nearest the end of the piece) of each of the lap cuts in the top rail should be 2¼ in. from the end. Mark a line across the face of the top rail at this point in from each end. Then lay the bottom rail on the top rail for checking. The ends of the bottom rail should match the marks across the top rail.

Lay one of the legs across the top rail with its outermost edge on the line you have marked across the top rail, and mark along the inner edge of the leg, across the top rail. This will give you two pencil lines, spaced apart by the width of the leg. These are the cutting lines for your first lap joint. As the cut must go only to half the depth of the top rail, mark both edges of the rail along the center line. As nominal 2 in. lumber (like a 2 x 3) is actually 1½ in. thick, this means the line along the edge of the rail will be exactly ¾ in. from the face of the 2 x 3, where you have previously marked for the width of the cut. It is essential to mark a line at the ¾ in. depth on *both* edges of the top rail to serve as stopping points for the saw cuts that come next.

Use the back saw to make the cuts. Start by making a cut on the inside edge of the pencil line across the face of the top rail, first on the outer limit of the lap cut, then on the inner limit—the lines you made to show the width of the lap cut. The saw cuts should go only to the depth shown by the lines on the edges of the 2 x 3, that is, to a depth of ¾ in. After these two cuts are made, make a series of cuts parallel to them, and about ½ in. apart, each to the same ¾ in. depth as the first two.

The series of closely spaced saw cuts across the width of the joint to be made leaves wafer-like sections of wood standing in the joint area. You will find that by wedging your chisel blade between these sections you can click them off easily. Parts of them usually remain at the bottom of the cut. You can shear these off by sliding your chisel (good and sharp) bevel side up across the bottom of the lap cut. A wood rasp is also a good tool for this. It's like a very coarse, sharp-toothed file.

When one cut is completed, as described, in the top rail, try one of the legs in it to check the snugness of the fit. If it's too tight, use your chisel to shave a little from each side of the cut. Push the chisel down the sides of the cut, bevel side out, to take a very thin slice, like a plane shaving. When the leg fits snugly, all's well.

If the initial trial fit shows the leg to be a loose fit, use the sliver trick described earlier, but save the sliver for the final assembly with glue.

Once you have made the lap cuts successfully in the top rail, make the end lap cuts at the tops of the legs in the same manner, and trim them all as necessary for the best fit you can get. The work will go faster on the second top rail and the second pair of legs. If you're making two sawhorses you can lay the second batch of rails together and mark for all the joint cutting at the same time, as in Fig. 8.

The lap dovetails

These are much like simple lap joints except that the inner end of the tenon of the joint is an

ALAN DUNS

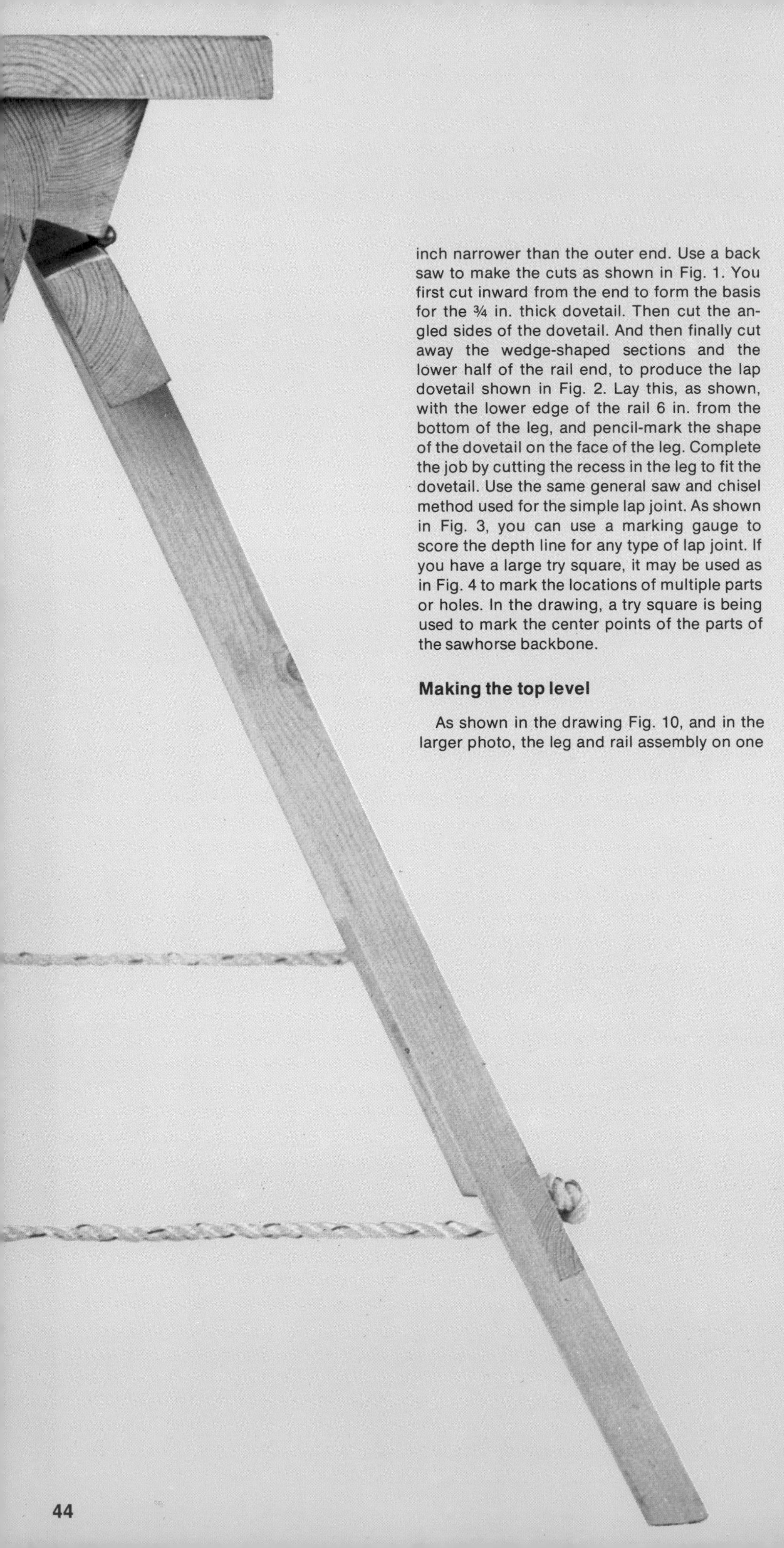

inch narrower than the outer end. Use a back saw to make the cuts as shown in Fig. 1. You first cut inward from the end to form the basis for the ¾ in. thick dovetail. Then cut the angled sides of the dovetail. And then finally cut away the wedge-shaped sections and the lower half of the rail end, to produce the lap dovetail shown in Fig. 2. Lay this, as shown, with the lower edge of the rail 6 in. from the bottom of the leg, and pencil-mark the shape of the dovetail on the face of the leg. Complete the job by cutting the recess in the leg to fit the dovetail. Use the same general saw and chisel method used for the simple lap joint. As shown in Fig. 3, you can use a marking gauge to score the depth line for any type of lap joint. If you have a large try square, it may be used as in Fig. 4 to mark the locations of multiple parts or holes. In the drawing, a try square is being used to mark the center points of the parts of the sawhorse backbone.

Making the top level

As shown in the drawing Fig. 10, and in the larger photo, the leg and rail assembly on one side of the sawhorse is attached rigidly to the backbone with plastic resin glue and 3 in. No. 8 flathead screws. The leg and rail assembly on the other side is hinged to the backbone.

After the leg and rail assemblies and the backbone are completely assembled and glued, fasten the rigid leg assembly to the backbone with glue and screws, as above. Then attach the other leg and rail assembly with hinges as shown in Fig. 10, and stand the unit up in working position. The ¼ in. size rope running between the legs should be knotted to hold the lower tips of the legs about 28 in. apart. As one set of legs is rigidly attached to a side of the backbone, the top surface of the backbone will be tilted. To level it, hold a spirit level across one end of the backbone, and adjust it until the bubble in its vial indicates a true level position. Then draw a pencil line along the top of the level to mark a level line across the end of the backbone. One end of this line should be at the high corner of the backbone, the other end will be at a lower point on the other side. Measure the distance this line is from the top surface, and mark the same distance from the top at the other end. Then connect these points with a line drawn from end to end, as shown by the darkened area in Fig. 9. Make saw cuts across the top of the backbone at the angle indicated, down to the end-to-end line. If these cuts are about 1½ in. apart, it's easy to break out the wood between them with a chisel, as in cutting lap joints. Then use your chisel and plane to smooth off the surface, which will be level.

Use glue and screws, as in Fig. 10, to attach a 30 in. length of 5/4 x 12 lumber to the top of the backbone to serve as a small workbench area when required. The term 5/4 is a lumber term indicating a thickness of 1 5/32 in. If you prefer you can use a 12 in. x 30 in. piece of ¾ in. plywood.

If you make two sawhorses, they can be used to support a larger work platform, built as shown in Fig. 11. For light loads, as when used as a wallpapering table, the framework can be of stock 1 x 2 lumber, covered with ¼ in. or ⅜ in. plywood. Plan the size according to your needs. A 3 ft. x 8 ft. size is good and fits easily over the sawhorses.

The lower ends of the legs may be trimmed to set flat on the floor by marking them as in Fig. 13, along a batten laid against them on the floor. Only one leg need be marked, as a bevel gauge, as shown in Fig. 12, and can be used to

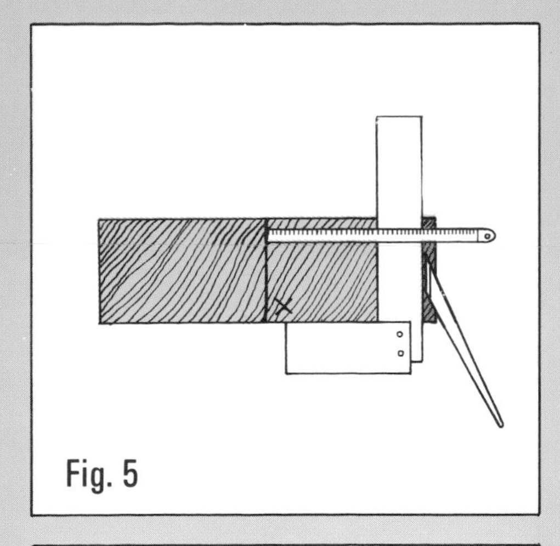

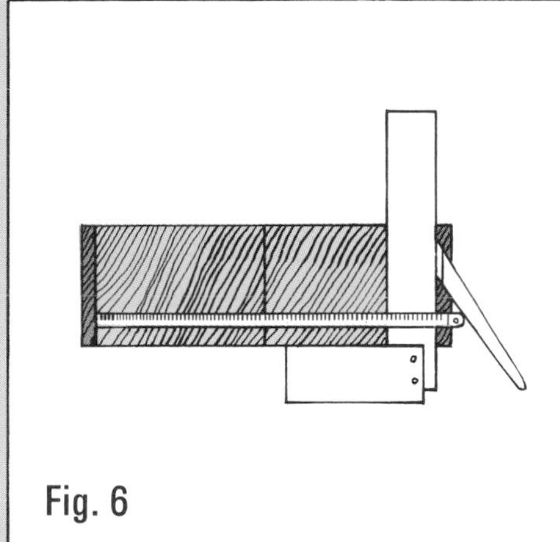

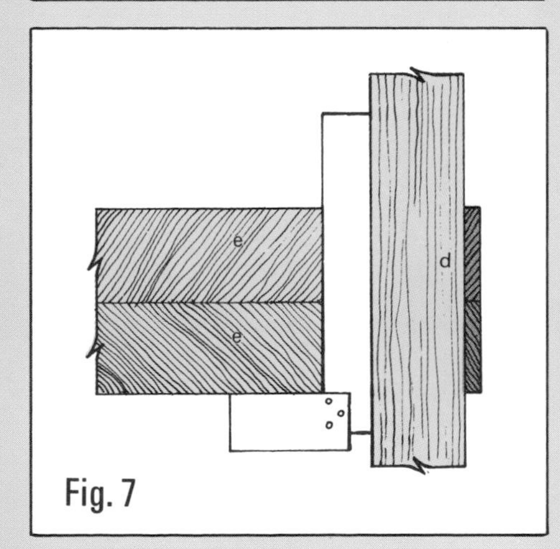

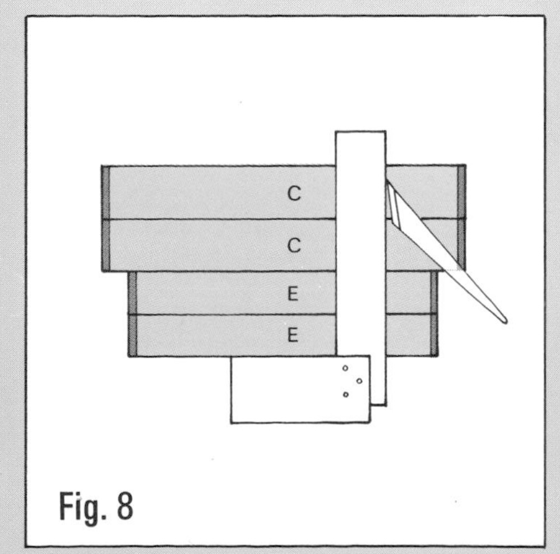

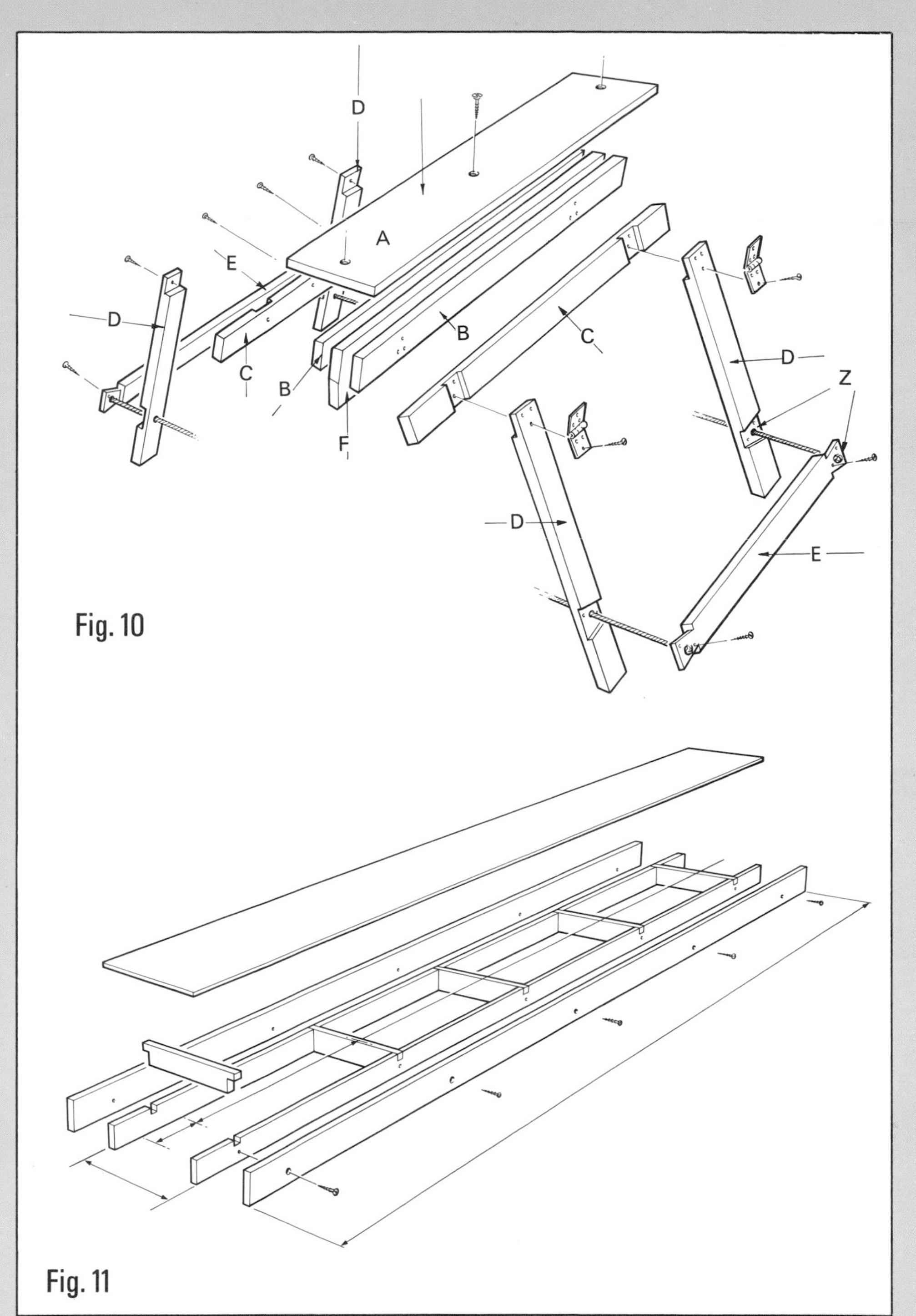

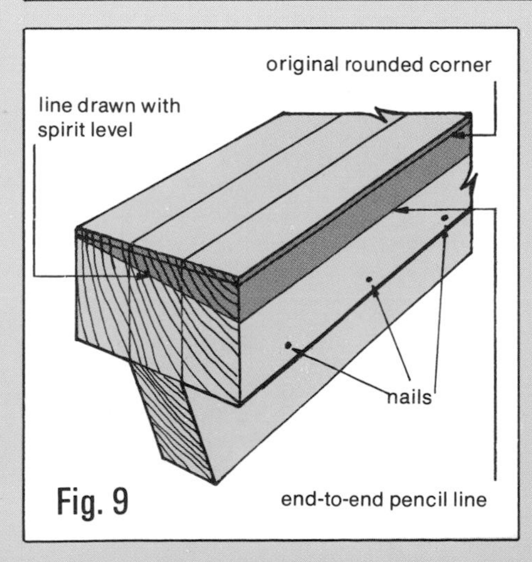

Fig. 5. *How to use try square to mark from center of piece of lumber when end must be cut square.*
Fig. 6. *When one end has been marked for squaring, measure from the mark and set try square to cut required length.*
Fig. 7. *In marking rail and leg assemblies, slide leg to marking position against try square to maintain right angle. Dark areas indicate unsquared ends, as on some lumber.*
Fig. 8. *Marking top and bottom rails with try square.*
Fig. 9. *Markings on sawhorse backbone for leveling with the aid of a spirit level.*
Figs. 10 *and* **11** *are exploded drawings of sawhorse and work table that can be used with paired sawhorses.*

set the same angle on the others. To transfer any angle with this handy tool, simply set the body of the gauge against an edge of the wood, and tighten the blade to correspond to the angled line. Then use it in the same way to mark the same angle on other pieces.

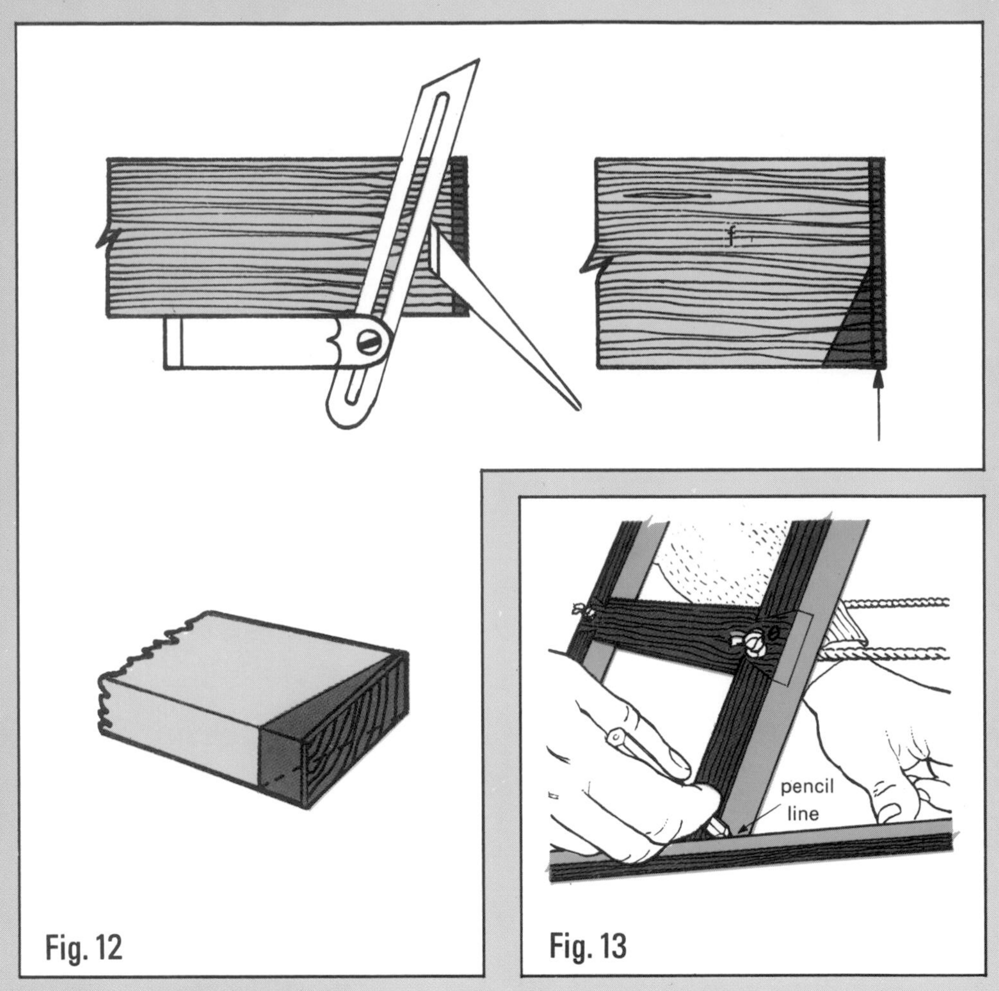

Fig. 12

Fig. 13

Below. *The correct way of planing lumber true to size. Press down firmly at the start of each stroke; slow down as you reach the far end, to avoid "bowing off" there. Order of work on rough wood:* **1.** *Plane one face.* **2. 3.** *Test both ways for straightness.* **4.** *Test for any twisting.* **5.** *Mark "high" areas with pencil.* **6.** *Plane one edge.* **7. 8.** *Test for straightness and squareness with face.* **9.** *Mark "high" edge.* **10.** *Gauge across the thickness, on both sides, and plane the second face to the gauge marks.* **11.** *Gauge across the width, on both sides, and plane the second edge.* **12** *shows how the gauge is tilted to make a fine line and the direction in which it is moved.* **13** *shows one way of holding the gauge when you have to use it across a wide board.*

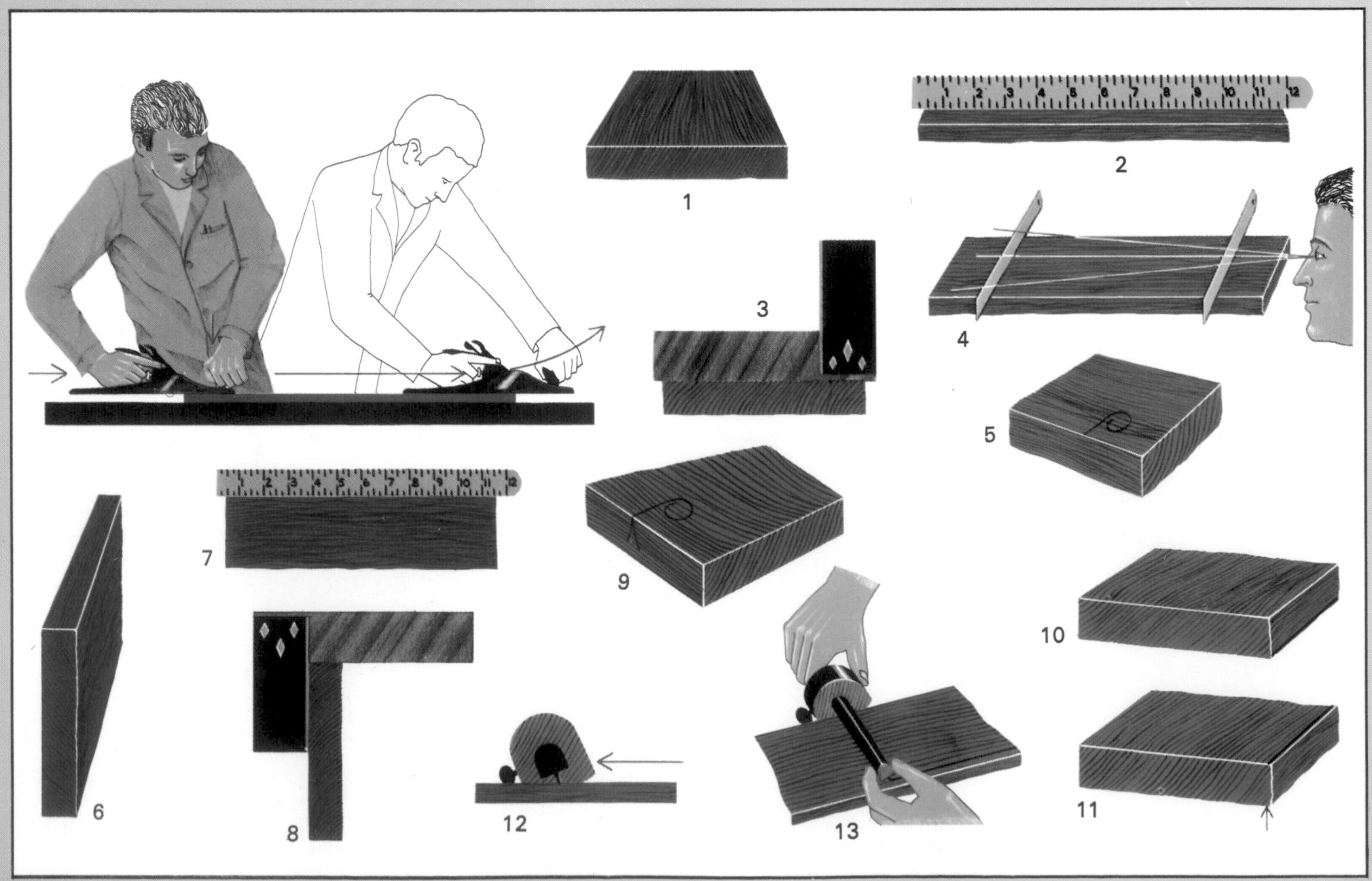

CHAPTER 8

Built-in alcove units

Most old houses, and many new ones, have alcoves in their walls. Too often the space they provide goes to waste in the room layout, when it could be usefully filled with a fitted cupboard.

A fitted cupboard can be built in an alcove of any size from a small niche to a huge recess twenty feet long. If it is built so that its front is flush with the wall on either side of the alcove, it completely hides the alcove and gives the effect of a flat wall. For this reason, it is a popular fitting on either side of an old-fashioned chimney breast, where in a typical older house there is often a pair of alcoves anything from 3 ft. 6 in. to 5 ft. wide x 18 in. deep. In a bedroom, this provides two roomy built-in wardrobes and at the same time simplifies the shape of the room so that it is easier to decorate, furnish and clean.

All built-in cupboards can be made in one of two ways. For reasonably square alcoves with straight walls in good condition, all you need to do is fasten shelves to the walls and face the alcove with a flat frame to carry a pair of doors.

If the walls are irregular, it is easier to build a frame-and-plywood cupboard as described opposite. The frames are made into a box that fits

Below. *A wide, decorative frame with an inset arch makes a feature of this alcove cupboard topped with a display shelf.*

BILL MACLAUGHLIN

the alcove only loosely; plywood covers any gap between frames and adjacent walls.

Scribing

All parts that have to fit against a wall or floor should be scribed to the right shape. The technique is fairly simple—once you understand the principle.

For scribing part of, say, a plinth to a floor, you will need a couple of pairs of *folding wedges*. You can make these yourself. They are just pairs of identical wooden wedges which are laid on top of each other nose to tail. By sliding one wedge over the other, anything resting on the wedges can be raised or lowered.

To scribe a part to the floor, lay it on two pairs of wedges immediately above the position it will occupy when installed. Put a spirit level on top of it and slide the wedges back and forth until it is exactly horizontal. Then take an ordinary school-type pair of compasses with a pencil in, and set them to the size of the largest gap between the wood and the floor.

Now comes the difficult part. Hold the compasses with the point touching the floor and the pencil touching the wood, and the pencil vertically above the point. Then, *without tilting the compasses,* run them along the ground so that the pencil traces the outline of the floor on the wood. If the compasses tilt, so that the pencil is no longer directly above the point, the line will be inaccurate.

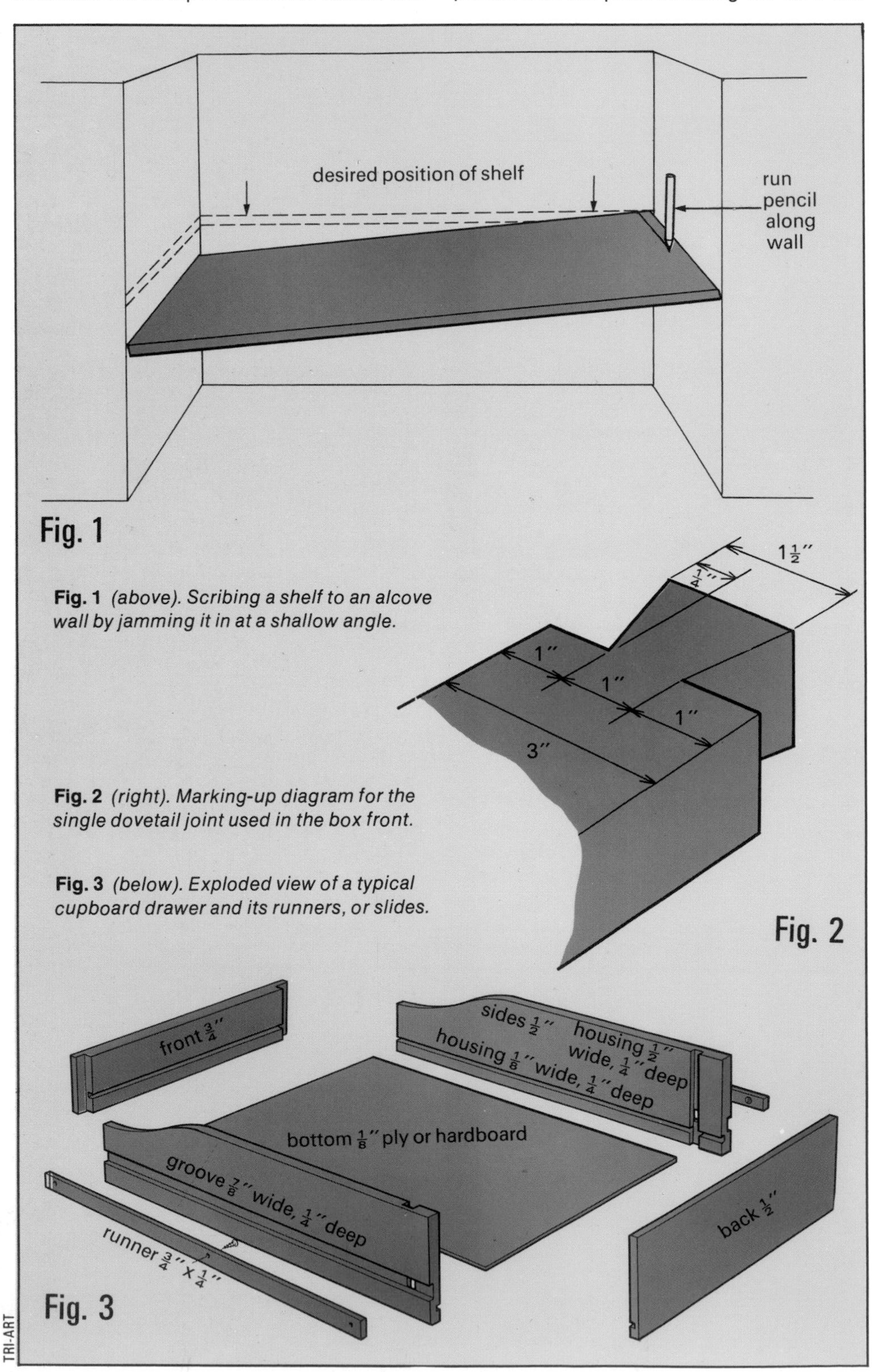

Fig. 1 *(above). Scribing a shelf to an alcove wall by jamming it in at a shallow angle.*

Fig. 2 *(right). Marking-up diagram for the single dovetail joint used in the box front.*

Fig. 3 *(below). Exploded view of a typical cupboard drawer and its runners, or slides.*

You can ensure accuracy by using, instead of a compass, a pencil and a wood block cut to exactly the right height. This arrangement cannot tilt, because the block slides flat along the floor. But you have to cut a new block for every part scribed.

Cut away the part below the scribed line with a coping saw, compass saw, power jigsaw, or, if there is very little wood to remove, a plane or spokeshave. The piece of wood should then fit the floor.

If you are scribing two pieces that fit together, and the floor is more irregular under one piece than the other, you may find that both pieces fit the floor but meet each other at different heights. The remedy is simply to saw or plane a straight strip off the higher piece parallel to the edge, which is easier than sawing the irregular contour of the bottom edge.

To scribe a horizontal board—a shelf, say—into an alcove with total accuracy, you would ideally insert an undersized board into the alcove at the desired height, scribe sheets of cardboard to fit each part of the wall, and pin them to the board so that they formed a template. Then you would trace round this shape on to the board you would use.

This is an ideal method, but slow. Provided the wall is not too irregular, you can use a less accurate but perfectly adequate method that takes much less time (see Fig. 1). Measure the length of the alcove in several places and choose the longest reading. Cut the wood to a bare ⅛ in. longer than this. Do the same with the depth of the alcove, so that you have a board just too long and just too wide.

The side of the board that touches the smoothest and straightest wall should be scribed first, and the most irregular wall last. If all the walls look about the same, do the ends of the board first.

Scribe the board by jamming it into the alcove as nearly horizontal, and as nearly in the right place, as you can. Don't use compasses to rule the line, but a pencil held directly against the wall. Try to remove the absolute minimum amount of wood. If you scribe and cut the ends of the board first, it will then fit into the alcove horizontally, and you can scribe the rear edge normally.

Box cupboard

This design consists of three plywood-lined rectangular frames made of 1 x 3 lumber joined by *blind mortise-and-tenon* joints—that is, joints where the mortises do not go right through the wood. This joint is shown in the center of Fig. 11 of the DATA SHEET on page 39. The only differences are (a) that the mortise does not go right through the wood—make it and the tenon, say, 1 in.—and (b) that the mortise has straight sides, and the tenon is not held in by wedges. Put nails through the joint to hold it instead.

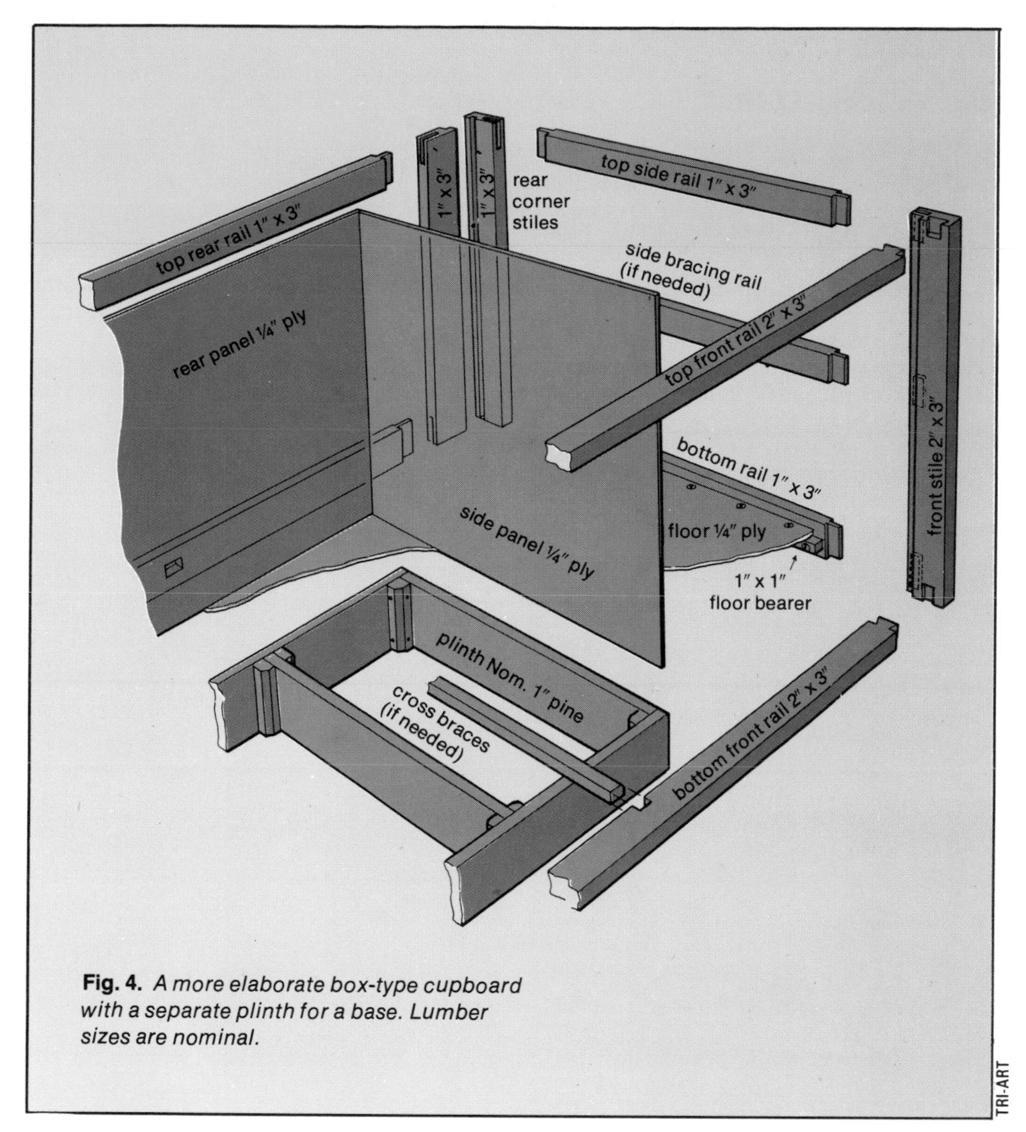

Fig. 4. *A more elaborate box-type cupboard with a separate plinth for a base. Lumber sizes are nominal.*

The three frames—rear and two sides—are joined by tongues and grooves which must be cut with a plow or combination plane.

If the frames are for a full-height cupboard, it is a good idea to put a horizontal brace in each frame.

The front stiles, on which the doors are mounted, are made of 5/4 x 3 lumber. The extra thickness is not strictly necessary, but it does have two functions. One is to make the front stiles the same width as the top and bottom rails, which must be this thick because they are unsupported. The other is to hide the unsightly front edge of the ply lining, which is sunk in a ½ in. deep rabbet on the rear edge of the stile.

The box has a floor made of ¼ in. ply. If you want a flush floor that is easy to sweep clean, it should be mounted on bearers screwed to the bottom rails of the frames to bring the floor level with the bottom front rail, into which it can be rabbeted if you don't want the edge to show.

The cupboard unit stands on a rectangular plinth scribed to the floor to ensure that the unit is level. The plinth is fastened together with glue blocks and nails inside the corners. If it is more than, say, 4 ft. wide, it is a good idea to put a cross-brace in it, and another one in the cupboard to strengthen the floor.

There is no need to screw the cupboard to the plinth, or the plinth to the floor (unless you live in an earthquake zone). The cupboard's weight should be sufficient to hold it steady.

The procedure for making a cupboard of this type is as follows:

1. Measure the alcove carefully all over and draw a plan and elevation of it to scale, showing any irregularities. Determine what is the largest box that will fit the space.

2. Re-draw the sample scale drawings of the complete cupboard and the three frames, so that you can determine what their measurements are.

3. Mark up, cut out and assemble the three frames. Then glue and nail the frames to each other. Nail a temporary batten across the front to hold it to the right width.

4. Cut the top and bottom front rails to 1½ in. longer than the gap they will have to bridge. Square lines across each piece ¾ in. from the ends to mark the shoulders of a single dovetail joint, as shown in Fig. 1. Cut the shapes out and use them to mark the sockets into which they will fit on the side frames. Cut out the sockets and insert the front rails.

5. Line the walls of the box with ⅛ in. hardboard reaching down to the level of the floor of the cupboard. Glue and nail this in place.

6. Glue and screw the floor bearers in place. Install the cross brace for the floor, if there is one. It should be dropped into slots cut in the bottom front rail and the bottom rear rail. The rear slot will be in the shape of a square mortise. Glue and screw or nail the floor to its bearers.

7. Scribe the plinth parts to the floor and make up the plinth. Put it in the alcove.

8. Put the cupboard on its plinth. If it reaches to the ceiling, it will not need a top. If it reaches above head height, you can roof it with ⅛ in. hardboard before installing it. If it is worktop height, scribe a worktop to the alcove walls and screw it to the top of the cupboard.

9. Hang doors on the cupboard, using hinges suited to the decor.

If the cupboard incorporates a vertical division, this should be screwed on from the outside, screwing through the upper and lower rails into the edge of a particleboard or plywood panel. Shelves should be mounted on angle strip, mounted with screws.

Bridging the gap

When the box cupboard is installed there may be some gaps around it between the edge of the frame and the irregular alcove wall. One way to deal with this is to nail narrow quarter-round moldings to the sides of the frame so that they wedge into the gaps and form 2 neat closures all the way. There are other ways to disguise edges but this is the simplest.

Drawers

A typical drawer is shown in Fig. 3. It is a wardrobe "shirt drawer" with a dropped front, but a straight-fronted drawer is made in the same way.

The front is made of ¾ in. thick solid lumber. It is grooved at the bottom to take the ¼ in. plywood or hardboard drawer floor, and rabbeted at the ends to take the ½ in. sides without their end grain showing.

The sides should be made of solid lumber or ½ in. plywood, which has at least three plies. Ordinary ply may make drawers stick on their runners unless wax or drawer lubricant is used. The sides are grooved at the bottom for the floor, vertically at the rear end for the back, and horizontally along the outer sides for the runners.

Since the grooves reach halfway through the wood, the runner grooves may break into the vertical grooves on the other side of the wood. Provided that no glue gets through the hole into the runner grooves, and the drawer back does not project into them, this does not matter.

The drawer back is also ½ in. thick. It is grooved at the bottom only, to take the floor. The drawer runners are slightly smaller than their grooves in the drawer sides. They are screwed to the side wall of the cupboard and to the vertical division, and must be exactly parallel and the right distance apart to work. The screws should be countersunk well below the surface of the wood.

The drawer is assembled by gluing the back to the bottom first, then the sides, then the front. It is impossible to put it together any other way.

CHAPTER 9

Spacious bathroom cabinet

Modern bathrooms are often limited in size—and storage space for the wide range of items that have to be kept readily on hand is, therefore, at a premium. This compact bathroom cabinet affords ample storage space and its simple good looks add an attractive touch to any bathroom.

This bathroom cabinet is of glued and nailed construction made, in the main, from exterior plywood, with the exception of the doors, which are particleboard. Carried out carefully, this job is not beyond the capabilities of even the very inexperienced woodworker.

The joints are simple—either dado joints or, at each of the four outside corners, rabbet joints. The back panel fits into a rabbet cut into the two short sides and is screwed into place. The front edges of the cabinet are covered with a decorative edging; apart from this, the finished cabinet is painted. The cabinet has ample storage space, with two cupboards and two glass shelves which are easily fixed in place using chrome screw eyes. It is fitted with a large mirror. The light above the mirror is optional.

Cutting the sides

All the plywood horizontal and vertical members are 5 in. deep from front to back, with the exception of the short outer sides which are ½ in. deeper to allow for the rabbet into which the back fits, and the center division in the top cupboard, which is 4⅝ in. deep. When the ¼ in. edging is applied to the other members, this central division will accommodate the ⅝ in. thick particleboard doors.

First cut all the horizontal and vertical pieces to the lengths given in the cutting list,

using a back saw. Be sure all cuts are square. If any edges are rough, smooth them with a plane or sandpaper.

Now cut the rabbets into which the back panel will fit, down the edge of the two short sides. The method of cutting rabbets is described fully in CHAPTER 3. Cut the rabbets ½ in. deep to allow the back panel to fit in flush, and ⅜ in. wide, leaving a tongue ⅛ in. wide protruding.

Now mark out the positions of the dadoes. The most accurate method for this is to lay edge to edge the two parts into which a third is to be jointed. Then mark the positions of the dadoes across the faces of both pieces of lumber—this will ensure that the dadoes are perfectly matched. For example, to mark the position where the long vertical piece (piece F in the exploded drawing) is to be jointed into the long sides (pieces A and C), lay the long sides together edge to edge with their ends exactly in line with each other. Mark the position of the dado on both surfaces.

To mark the position where the long horizontal piece (H) is jointed into pieces D and F, butt the edges of these two pieces together. Then slide piece F so that its end is exactly ¼ in. inward from the end of piece D. This is done because the top of piece D in the finished cabinet is ¼ in. above the top of piece F, which is in a dado joint. Measure where the dado is to come from the end of piece D and mark the position on both boards.

When you mark the position where the dividing piece in the top cupboard (G) will joint into pieces A and H, remember that the front edge of this piece is set back ⅜ in. from the fronts of the other pieces. After the ¼ in. edging is added at a later stage, this space will allow the ⅝ in. particleboard doors to fit in flush with the cabinet front.

Now cut all the dado joints and the stopped dado joint for the center partition in the top cupboard to a depth of ¼ in. The method of cutting dado joints is described in CHAPTER 3.

The next step is to cut rabbets on each end of the two short sides to receive the long sides of the cabinet. Cut the rabbet ½ in. deep, so that the long side will fit flush, and ⅜ in. wide. The shape of the rabbet will leave a tongue of wood only ⅛ in. wide projecting. This will allow greater room for you to nail the pieces together by nailing from the longer sides into the shorter. It also reduces the area of end grain to be painted and, therefore, improves the decorative effect of the cabinet.

Applying the edging

The decorative edging can now be applied to all the edges that will be exposed in the finished cabinet, apart from the edge of the central division in the top cupboard. Before doing this, however, stand the pieces on their edges on a flat surface in roughly the arrangement of the finished cabinet. Mark all the exposed edges with a pencil. This will ensure that you do not apply the decorative edging to the wrong edges of the pieces by mistake, and also check that you have cut all the components accurately.

The finished size of the edging will be ¼ in. x ½ in., but it should be bought and applied slightly wider than ½ in. to allow for final cleaning up. The type of wood used is a matter of personal taste—the cabinet in the photograph is edged with rosewood. At most lumberyards you will have to cut the edging from ¼ in. thick pine "lattice strip." This will make two edging strips when sawed down the center and planed to width.

The edging is glued with a white glue to the front edges of the cabinet components. It then has to be clamped in place while the adhesive dries.

There are three ways of doing this. The first is to use edging clamps. These are shaped like C clamps but they have an additional threaded bar through the long side of the C which will apply pressure to the narrow edge of a piece of wood. Clamp the edging as shown in Fig. 3. The second method uses bar clamps. Take matching pieces, for example, the two short sides of the cabinet. Butt the two edges where the decorative edging has been applied. Clamp the components together using bar clamps (see Fig. 4). Either way, be careful to wipe off with a damp cloth any smears of adhesive.

The third way of applying pressure is to nail through small plywood blocks and the edging, into the component. Use 1½ in. panel pins or 1 in. brads and small blocks of ¼ in. ply, about 1 in. square. Place the blocks on the edging and nail through them. Angle the nails as in dovetailed nailing, which is described and shown in CHAPTER 2 (see Fig. 9). The disadvantage of this method is that it leaves nail holes in the edging. These are, however, only the size of the nail and can be filled with a matching filler.

When the adhesive has dried, use sandpaper to smooth off the edging so it is level with the plywood components.

Next miter the edging on the ends of the long and short sides. Arrange the four sides on the floor in the shape in which they will finally fit together. Set a bevel gauge to an angle of 45°. On the ends of each piece mark a diagonal line from the outer edge along the blade of the bevel gauge. Carefully saw through the edging down these lines until you reach the plywood. Remove the waste edging.

Now, at the edged ends of each internal dividing component, except piece G, cut a small step. The application of edging strip to the front of the components has, in effect, created stopped dado joints at the points where through dado joints were cut. Thus a small step, the depth of the dado, has to be cut through the edging on the pieces which fit into these joints. On the edged side of each dividing member, except piece G, measure inward ¼ in.—the depth of the dado. Square a line through this point, around the edging. Carefully cut down this line through the edging until you reach the edge of the plywood board. Do this at both ends of each internal component, except piece G. Remove the waste edging.

Assembling the cabinet

Now assemble the cabinet. The long and short sides are glued and skew nailed together (described in CHAPTER 2) and the internal divisions are glued into the dadoes. Wipe any excess adhesive from around the area of the joints with a soft cloth.

Check that the cabinet is square by measuring each set of diagonals. The measurement of each diagonal should exactly match that of its intersecting diagonal.

The cupboard doors

The next step is to cut the cupboard doors and fit them. Mark out and cut the doors to the size indicated in the cutting list. One door is ¼ in. wider than the other, to allow for a matching rabbet down the edges of the two doors in the top cupboard where they meet at the center division. To cut the rabbet, set the rabbet plane to a width of 5⁄16 in.—half the thickness of the doors (see Fig. 2).

Before fitting the doors, remember that there should be a thin coin's width between the edges of the doors and the cabinet sides. Plane the doors to a size that will suit but do not, of course, plane the rabbeted edges of the top cupboard doors.

Mount each door with a pair of 1½ in. brass butt hinges and brass screws. Remember, though, that when using brass screws the thread hole in the wood should be cut first with steel screws. This will avoid the danger of the brass screws breaking in the hole.

The cabinet back

The back of the cabinet can now be fitted. The back is made from ½ in. plywood and is screwed into place. This enables it to be easily removed when you want to repaint the interior of the cupboard. Use No. 6 flathead brass screws to attach the back. Space them about 7 in. apart and screw into the rabbeted (back) edges of the two short sides and into the edges of the long (horizontal) sides. Be careful to drill straight; you have little margin for error.

The mirror frame

The next step is to make the frame for the mirror. This is a simple box construction, made from nominal 1 x 2 lumber, rabbeted at the corners. The four pieces are glued and skew nailed together. So that the mirror screws can be set in from the corners of the mirror, two pieces of the same lumber are fixed to the inside of the box. Position them vertically, with their wider surfaces facing outward (Fig. 6) to provide the greatest possible area for the screws.

When the mirror frame is made, it has to be glued and screwed in position. It is attached centrally in the bottom left-hand panel of the unit—that is, 2½ in. from the side and from the bottom.

The most difficult part of this job is that, although the screws have to be driven through the back of the back panel, the only way to site the frame accurately is to work from the front. To overcome this, place the frame in position, and draw a light penciled line around it. Remove the frame, spread white glue over its

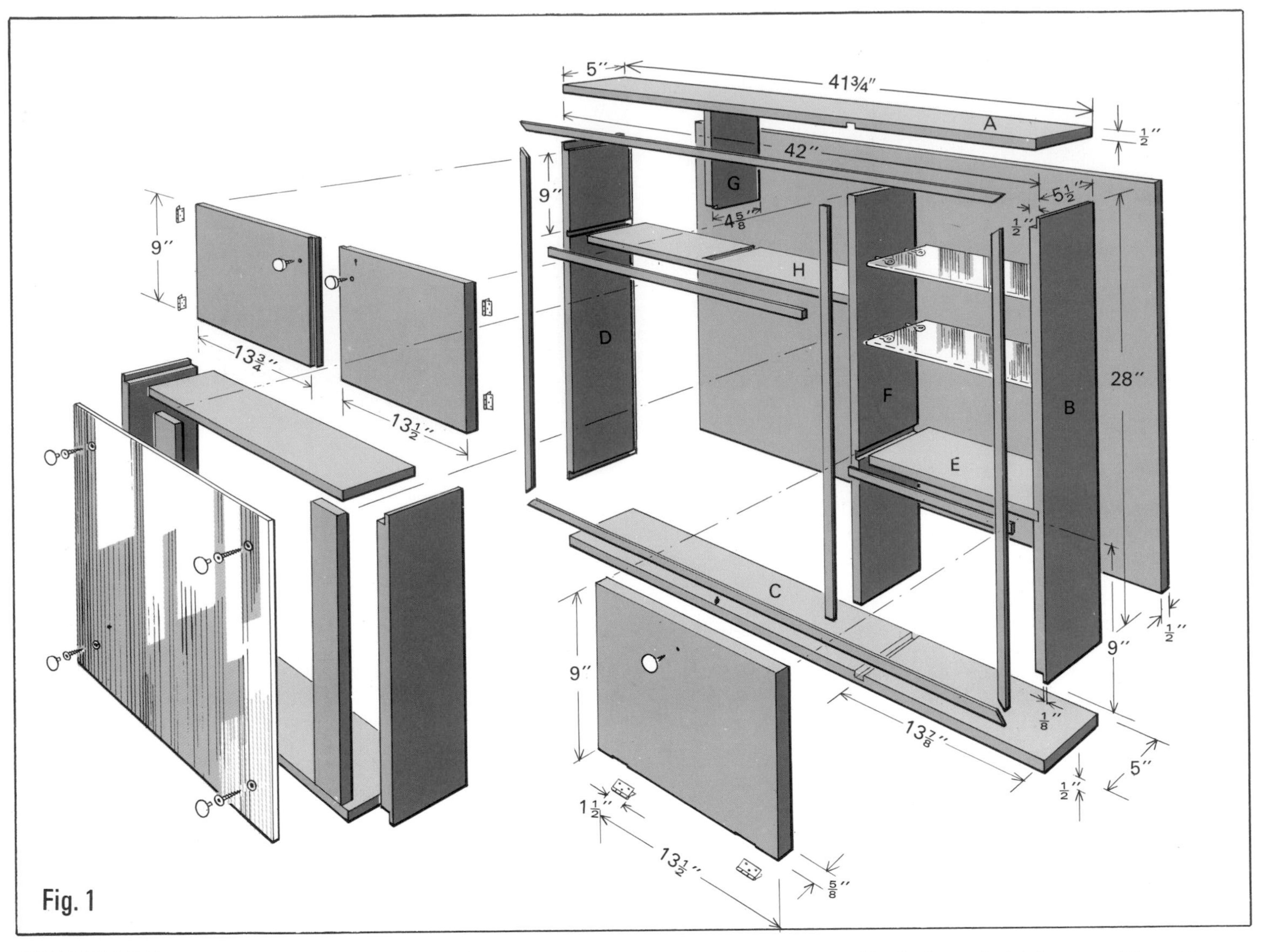

CUTTING LIST

Part	No.	Size
Short sides	2	28 in. x 5½ in. x ½ in.
Top side	1	41¾ in. x 5 in. x ½ in.
Bottom side	1	41¾ in. x 5 in. x ½ in.
Back	1	41¾ in. x 28 in. x ½ in.
Vertical division	1	27½ in. x 5 in. x ½ in.
Horizontal division	1	14 in. x 5 in. x ½ in.
Horizontal division	1	27½ in. x 5 in. x ½ in.
Top cupboard division	1	9½ in. x 4⅝ in. x ½ in.
Mirror "bearers"	2	11½ in. x 2 in. x ¾ in.
Short sides of mirror frame	2	13 in. x 2 in. x ¾ in.
Long sides of mirror frame	2	22 in. x 2 in. x ¾ in.
Edging strip (for sides)	2	42 in. x ½ in. x ¼ in.
Edging strip (for sides)	2	28 in. x ½ in. x ¼ in.
Edging strip for internal divisions	2	27½ in. x ½ in. x ¼ in.
	1	14 in. x ½ in. x ¼ in.
Door	2	13½ in. x 9 in. x ⅝ in.
Door	1	13¾ in. x 9 in. x ⅝ in.
Other requirements:		
Shelves, plate glass	2	13½ in. x 5 in. x ½ in.
Mirror	1	22 in. x 13 in. x ¼ in.
Shelf supports	8	Chrome screw eyes
Brass butt hinges	6	1½ in. x 1¼ in.

You will also need 3 knobs for the cupboard doors, and a door stay and magnetic catch for the lower cupboard door.

The sides, internal divisions and back of the cabinet are plywood, the mirror frame and bearers are softwood, the doors are made from particleboard and the edging strip is white pine. The measurements given do not include any allowance for waste.

Fig. 1. *An exploded drawing of the cabinet.*
Fig. 2. *A cross section through the top cupboard showing the matching rabbet in the cupboard doors.*

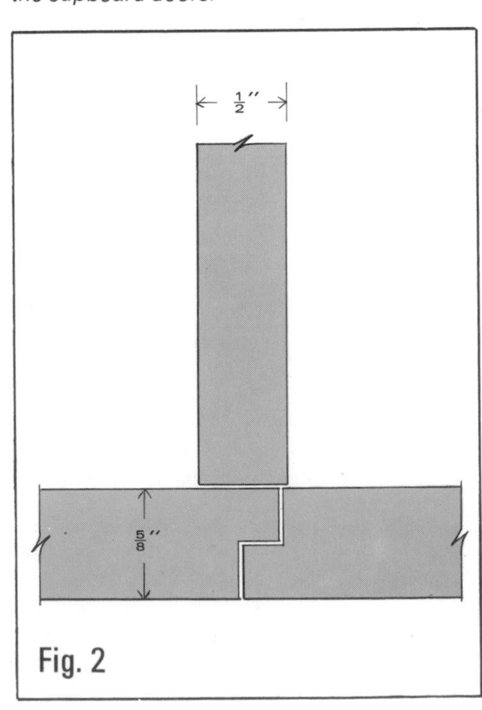

Fig. 3. *One method of clamping the white pine edging to the cabinet components once it has been glued in place is to use edging clamps. These force the edging up against the edges of the components while the glue dries.*
Fig. 4. *Another method is to take matching parts of the cabinet, for example, the two short sides, and butt the edges to which the edging has been applied. The two parts are then sash clamped together.*
Fig. 5. *The third method of clamping the edging is to nail through scrap plywood blocks into the edging and the cabinet component. The panel pins are angled as in dovetail nailing to provide greater clamping pressure.*
Fig. 6. *A simple box construction, the mirror frame is rabbeted at the corners and the pieces glued and dovetail nailed together.*

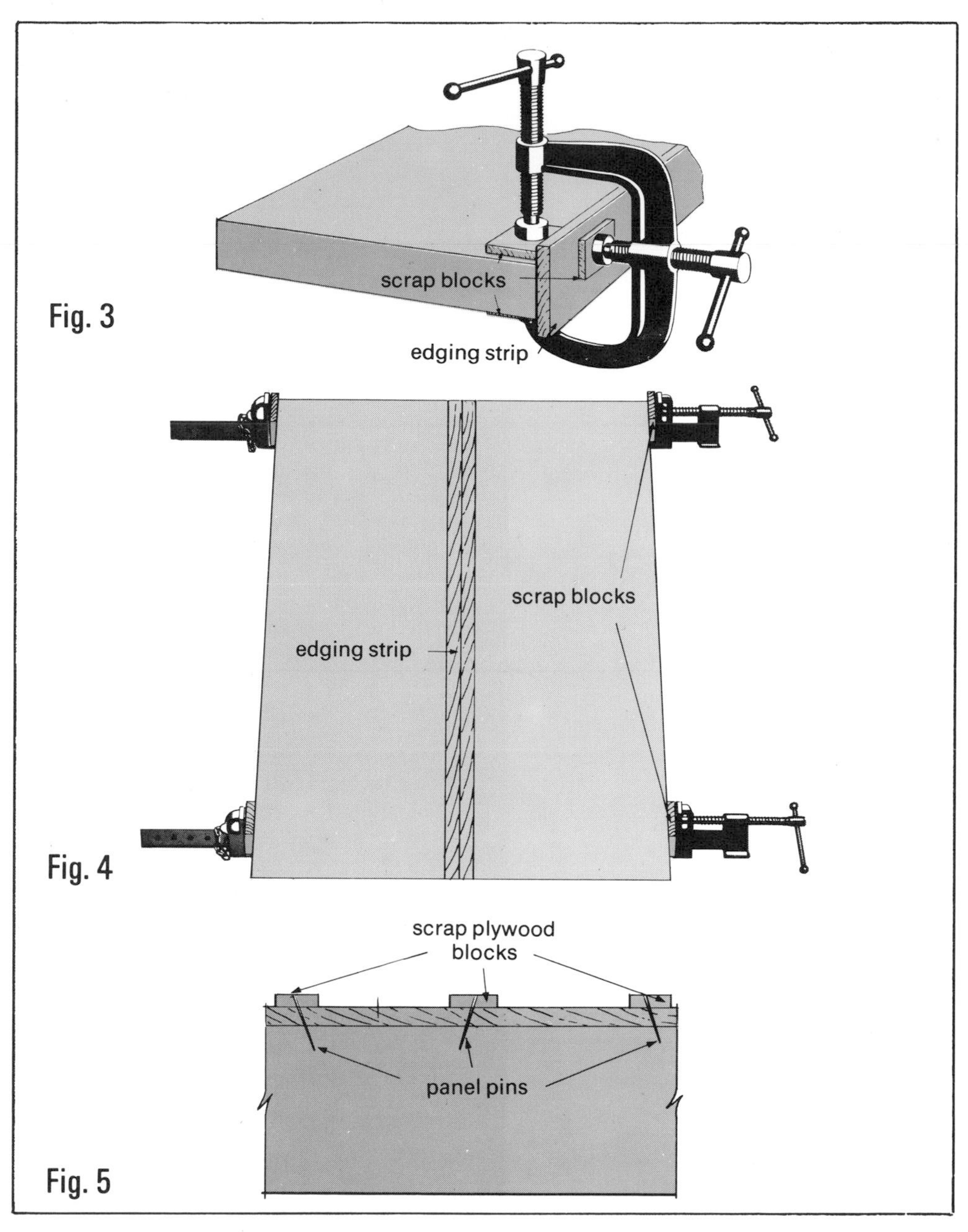

back edge, replace it, and weight it down with a few books until the adhesive dries.

Next remove the back panel from the unit, and turn it upside down. Measure carefully from the outside edges to give accurate screw positions. Drill for, and drive, No. 6 flathead screws at about 7 in. intervals. Once again, use No. 6 steel screws first to cut the thread. Then remove the steel screws individually and replace them with brass screws of the same size.

The glass shelves

Next attach the supports for the two glass shelves. The shelves are plate glass with ground edges—your glass dealer will probably prepare these for you. The shelf supports are chrome screw eyes. Mark the position of the supports on the short side B and on the long vertical dividing piece F. Bore small holes at these points with an awl and screw in the supports.

Finishing the cabinet

The cabinet can now be cleaned up and painted. Remove the shelf supports and the back panel with mirror frame. Sand the face of the decorative edging to remove any irregularities at the corners, and place masking tape over it. Paint all the component parts. For the very best results use a primer, followed by an undercoat and finished with the topcoat.

When the paint has dried, reassemble the cabinet. Remove the masking tape and to the decorative edging apply two coats of clear polyurethane. Fix knobs to each of the cabinet doors and a brass door stay and magnetic catch on the lower cupboard door.

Now set the mirror in place. Your glass dealer will supply a mirror cut to size. The mirror is mounted with four mirror screws each set in 1½ in. from the corners as shown in the exploded diagram. It is fairly easy to drill holes in glass with a slow speed drill but your glass dealer will probably be willing to do the job for you for a reasonable charge.

Screw the finished cabinet to the bathroom wall. Position the screws so that they are as inconspicuous as possible.

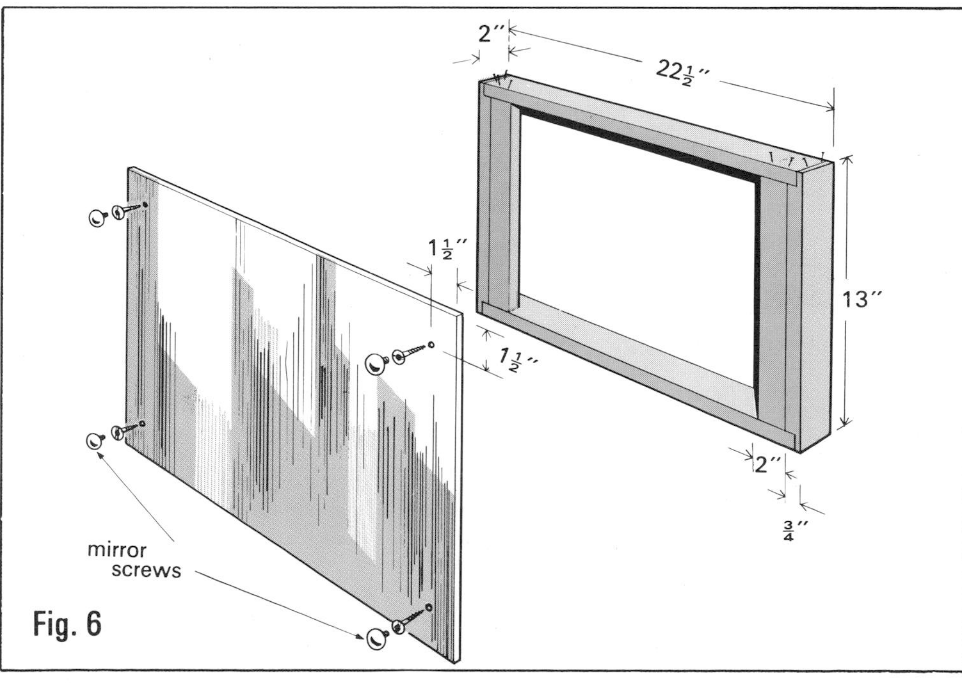

Nails

The number of different kinds of nails and other hammered-in fasteners is staggering—and each one has a different function. Here is a list of all the main types, and what they are generally used for.

Commonly used nails:

1. Finishing nail. Head can be punched below surface for a neater finish in fine work.
2. Flooring brad. For work with hardwood strip and parquet flooring.
3. Oval wire nail. Oval cross-section makes nail less likely to split wood. Not readily available.
4. Shingle nail. Large-headed, for nailing roofing felt, etc., to wood.
5. Glazing point. Headless, holds glass in window frames; covered by putty.
6. Tileboard nail. Small nail for securing enameled hardboard to walls. Inconspicuous head.
7. Common nail. For rough carpentry work; large, ugly head ensures a firm grip.
8. Masonry nail. Hardened steel nail for fixing wood direct to masonry. Several forms.
9. Clincher nail. Point can be "clinched" (turned over) for extra grip. Not readily available.
10. Hardboard nail. Unusual head shape countersinks itself in hardboard, can be filled over.
11. Upholstery nail. Decorative head for tacking leather, etc., in upholstery work.
12. Tack. Small nail with broad head, for tacking down carpets and fabrics.
13. Staple. For securing wire, upholstery springs, etc., to woodwork.

Special-purpose nails:

14. Corrugated fastener. For butt-joining lumber quickly and easily; not very strong.
15. Screw nail. For fastening sheet materials to wood. Great holding power.
16. Cut flooring nail. Holds down floorboards. Great holding power, unlikely to split wood.
17. Joiner's brad. Small carpentry nail not readily available.
18. Cut clasp nail. Obsolete general-purpose nail superseded by modern forms.
19. Needle point. Steel pin for fixing small moldings. Not readily available.
20. Annular nail. Used like the screw nail (15); used in boatbuilding.
21. Duplex head nail. For concrete formwork; double head permits easy removal.
22. Dowel nail. For end-to-end hidden joints in high-quality work. Special item.
23. Wood connector. For joining corners of frames where strength and appearance are unimportant.
24. Insulated staple. For securing telephone and other low-voltage wiring.
25. Saddle tack. For wiring; tacked, folded, and clinched. Not readily available.
26. Corrugated roofing nail. For securing corrugated metal roofing to wooden roof.
27. Chisel point concrete nail. For fastening direct to masonry.

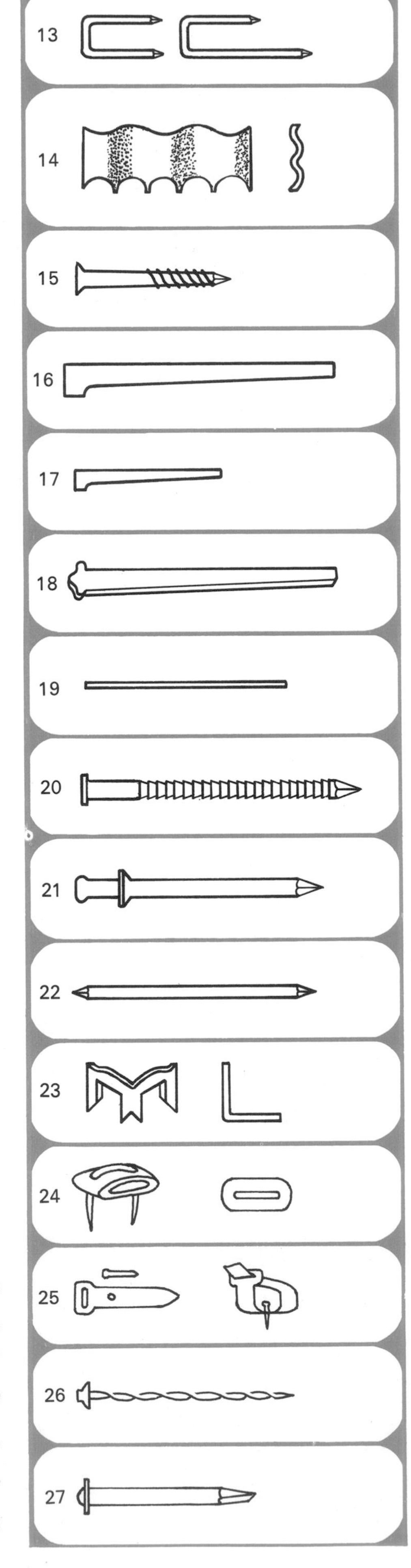

CHAPTER 10

Easy-to-build workbench

Every carpenter needs a workbench. Without it, he is severely restricted in the range of jobs he can tackle; with it, he can build advanced projects quickly and easily. Here is a design for a fullsize professional-type bench—and the best thing about it is that you don't need a bench to make it.

This workbench has three important advantages over other benches. First, it is extremely simple to make. No advanced carpentry techniques are required, and yet the result is just as strong and long-lasting as a more complicated structure.

Second, it has been designed not to require expensive equipment to make it. The main parts of the frame are held together by bolts, so no bar clamps are used in assembling them. Everything can be done easily and quickly with ordinary hand tools.

Third—and unlike most other workbenches—it has a replaceable top. The design calls for high-density flooring grade particleboard to be used, and this should last for years; but when it does wear out, all you have to do is undo a few screws and put in a new piece. It is not expensive. If you can't obtain this type of material, use tempered hardboard over plywood to make up the same thickness.

This project introduces the simple but useful technique of rubbing "glue blocks" into the corners of a frame—a technique much used in furniture-making to give a square frame extra rigidity and strength.

For the lumber and accessories you will need, see the materials list, Fig. 15. The vise shown in the pictures is a quick-release type extremely suitable for the amateur woodworker, but any woodworking vise can be fitted to the bench.

A *bench stop* is also a recommended fitting. This is a small projection that can be raised from the flat surface of the bench top, and used to rest wood against when sawing or planing it. Various kinds of stops are made, but one of the hinged type shown in Fig. 14 is easiest to fit. It is just let into a shallow, chiseled-out recess in the particleboard and secured with screws.

To save time and trouble in construction, if you're working with hand tools, ask your lumber dealer to cut as much of the lumber as possible to size or length. You can list the sizes from the materials list. He can do this work a lot faster and more accurately than you can. Ask in advance what his charges will be for this service. It might not be a bad idea to have the plywood and other panel materials cut to exact size as well—though you will then have to work accurately to be sure they fit.

The end frames

The "legs" of the bench are two simple upright frames, one at each end, made of two heavy pieces of lumber screwed to a sheet of plywood. Mark out the plywood by drawing a pencil line down each of the long sides 1¼ in. in from the edge. Then make crosses on the line 2 in. from each end, and another two crosses 8⅛ in. further in from the first two. These crosses mark screw holes. As a check on their accuracy, the second pair of crosses should be 8⅛ in. from each other, as shown in the exploded drawing. Bore and countersink through the crosses for 1½ in. No. 10 screws.

Mark out all four legs together to a length of 30½ in. Square the lines right around each end and saw them to length.

Put each frame together "dry," lining up the legs carefully so that their outer edges are flush with the edge of the plywood panel, and their tops flush with its top. Mark the position of the screw holes through the holes already drilled in the ply. Take the frame apart and drill the "pilot" holes in the legs, using a depth stop (such as a piece of adhesive tape) on the drill bit to ensure that the holes are not more than 1 in. deep.

Glue and screw both frames together, doing the screws up very tightly so that they hold the pieces together firmly while the glue dries. Wipe off any surplus glue with a damp cloth and set the frames aside on a flat surface to dry. Use white glue unless the bench will be in a damp place. For dampness or high heat, use plastic resin glue.

The top and bottom frames

The top and bottom frames are almost identical, and their parts can be marked up together. There are two differences between them: the top is 5½ in. deep, while the bottom is only 3¾ in. deep; and the side members of the bottom frame are 8½ in. shorter than those of the top frame. The first difference does not affect cutting out the pieces, because they are already ripped to their finished width. As for the second, it is easiest to mark up all four side members 4 ft. 6 in. long, and cut 4¼ in. off each end of the bottom frame after you have assembled it.

Put the wood for the six cross-members together in the vise, and mark them all up at the same time to a length of 23¾ in. Do the same with the four side pieces, marking them up 4 ft. 6 in. long. Cut all the pieces to length.

Mark the positions of the dado joints on the side pieces, where the cross-members fit into them. The outer edge of the two outside cross-members is 7 in. from the end of the side piece—provided you have followed instructions and cut them all to the same length—and the third cross-member is mounted centrally. Mark the width of the dadoes directly from one of the cross-members by the simple method of holding it against the side piece, squaring it with a try square, and drawing a marking knife down both sides. Using a ruler leads to inaccuracy in cases like this.

Set a marking gauge to ⅜ in. and mark the depth of the dadoes on the side of the wood. With a back saw, saw down to the depth marked, making the cuts very slightly inside the marked lines to make the joints a tight fit. Clean out the dadoes with a ¾ in. paring chisel, working from the ends to the middle to avoid splitting the wood. A router, if you have one, speeds up the job considerably, but it is not necessary.

It is a useful check on accuracy to make one complete joint before starting any of the others, and fit it together dry. Then you can adjust your sawing of the other dadoes.

Glue blocks

Before you finally assemble the dado joints, make sixteen glue blocks to brace them. These can most easily be made from a single strip of ¾ in. x ¾ in. lumber 4 ft. 8 in. long, and then cut it into eight 4½ in. and eight 2½ in. long blocks.

Before cutting, plane a chamfer off one corner of the strip along its whole length, so that it loses ¼ in. off each side (see Fig. 1). Plane a much smaller chamfer of about 1/16 in. off the corner diagonally opposite. This corner fits into the inside corner of the dado joints, and the tiny chamfer helps it to fit well.

Cut this long strip into separate glue blocks. Then assemble the frames by gluing and pressing together the dado joints, and fastening them firmly with nails driven in diagonally as shown in Fig. 3, and with their heads punched below the surface. Three nails a side is enough. Check the frames for squareness, and if satisfied, spread more glue in the corners of the joints—the inside corners of the two outer cross-members, and both sides of the center cross-member. The glued area should extend about 1 in. from the corner on either side, along the whole length of each dado.

Spread glue on the two surfaces of the glue blocks adjacent to the smaller chamfered corner. Then press a glue block into the inside corner of a dado joint—the longer glue blocks fit into the wider top frame, the shorter ones into the bottom frame. Rub the block backward and forward to get tight contact in the joint, and to make the block sit firmly. Then use finishing nails to nail the block to the side-piece and to the cross-member (see Fig.4). Use the minimum number of nails—one each way if this makes a tight fit—and, if the first block shows signs of splitting, drill nail holes through the others. Repeat this with all the other glue blocks, then recheck both frames for squareness and set them aside to dry.

Assembly

While the glue is drying on the top and bottom frames, drill holes for the carriage bolts in the end frames. There are two rows of three holes at the top, and one row at the bottom. Drill the outside holes of each row 2 in. in from the inside edge of the legs, and the center holes in the exact center of the frame. The top row of the holes should be 2 in. from the top edge of the frame, and the second row another 2 in. below the first row. The bottom row should be 2 in. up from the lower edge of the plywood panel. The size of the drill should, of course, be the same as the size of the shank of the bolt.

When the glue on the other frames is dry, assemble the whole framework without glue and mark the position of the bolt holes on the top and bottom frames by poking a pencil through the holes in the end frames. Also mark how much wood you will have to cut off the projecting ends of the bottom frame to make them flush with the legs of the end frames. This should be about 4¼ in.

Take the framework apart, drill out the bolt holes you have marked, and trim the end frame to length. Then spread glue on all the contact surfaces of all the frames, and bolt them together with the carriage bolts. As each bolt is put into its hole, its head should be hammered to make the square collar under it sink into the wood. Put a "Fender" washer on the other end of the bolt and then do up the nut really tight to hold the glued joint together. When you have tightened all 18 bolts, check the complete framework for squareness and leave it to dry.

Glue and screw the 1½ in. x ⅞ in. edging to the larger of the two pieces of particleboard. It should be positioned so that its *lower* edge is flush with the under-surface of the particleboard and its top edge projects ¾ in. above it. Put the edging on the ends of the board first, plane the ends of the edging flush with the sides of the board, then apply the edging to the sides and plane its ends flush. It is not worth the trouble to miter the corners.

Screw the particleboard sheet on to the top of the frame. The sheet and edging together are ⅞ in. wider than the frame (though exactly the same length) and should be allowed to project this distance off the *back* of the table, so that the front of the edging is flush with the front of the frame.

The vise

At this stage the vise should be installed. The type of vise you choose is, of course, up to you, but the instructions here are for a typical model, and you can adapt them to the vise you choose.

Woodworking vises may come without the wooden lining that goes inside their jaws, so you will have to make it yourself. The dimensions depend on the vise.

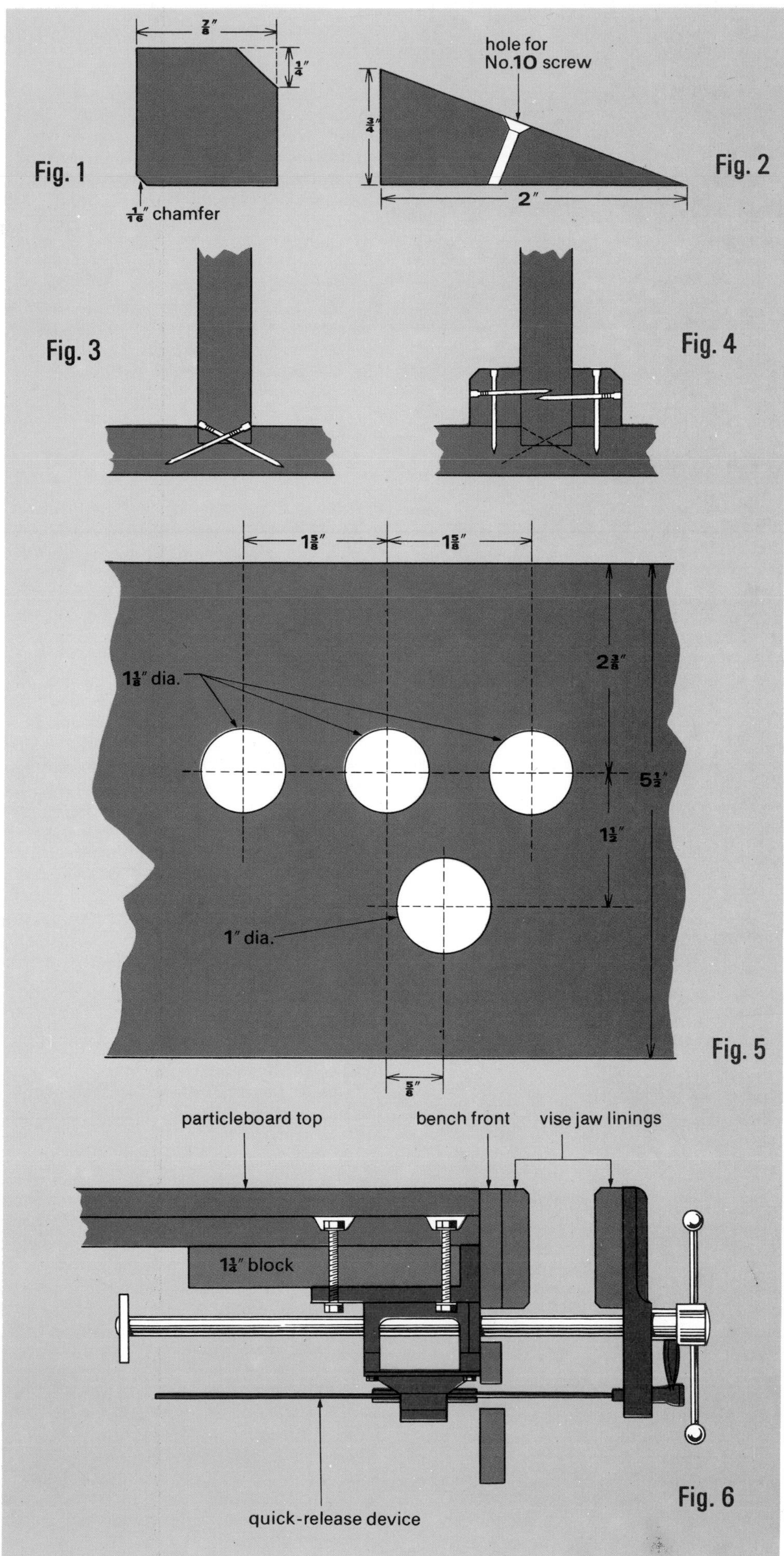

A cross-section of the vise mountings is given in Fig.6, and a marking-out diagram for the four holes that have to be drilled is in Fig.5. Note that the holes are very large—1 in. and 1¼ in. in diameter—and will have to be drilled out with a very large flat bit or a hole saw. (See CHAPTER 33 if in doubt. A similar diagram may accompany the vise you buy. If so, follow it carefully, as hole sizes and positions vary with the vise.)

The recommended procedure for mounting this particular vise is as follows: Mark out the front edge of the frame, anywhere along its length, as shown in the diagram and drill the holes. Make sure the rods and screws on the outer jaw of the vise fit into the holes.

Place the inner jaw lining piece against the front of the bench at the place where you want to install the vise, and with its top edge level with the top of the bench. Make sure it is the right way up. Screw it firmly to the front of the bench with four screws, one at each corner, out of the way of the vise's outer jaw. The screws should be countersunk so that their heads are slightly *below* the surface of the wood, to avoid damaging objects clamped in the vise.

Prop the bench up on one end. Poke the outer part of the vise through the holes from outside and attach the inner part temporarily to it. With most vises, you will need to put wood blocks between the inner part of the vise and the underside of the bench top to mount it firmly. For this vise, the blocks were about 1¼ in. thick.

Mark the position of the inner part of the vise clearly; its front edge should lie against the inside of the bench frame. Remove the outer part of the vise. Cut out some blocks of a suitable size, put the inner part of the vise on them in the correct position, and drill a hole the size of the vise mounting bolts through the mounting holes on the inner part, the blocks and the particleboard bench top.

Turn the table right side up and chisel a small recess around each hole to allow the bolt heads, with a good, large washer under them, to be let in flush with the surface of the particleboard. Do not over-recess the heads or you will weaken the mounting. Bolt the inner part of the vise and the blocks firmly to the table, passing the bolts through from the top.

Screw the outer jaw lining to the outer part of the vise at such a height that, when the vise is assembled, the top of the wood will be level with the inner jaw lining and the edge of the bench. Then assemble the vise.

The vise shown has a quick-release device that enables you to move the jaw without turning the screw. There is also a cheaper version without this device. It only needs the three 1 in. holes drilled to mount it, and not the single 1¼ in. hole. This applies to many other makers' vises as well.

Final assembly

Once the vise is installed, there is not much else to do. The smaller piece of particleboard should be dropped into the top of the table at the front, so as to leave a 5⅞ in. wide well at the back, which can be used to stop tools from

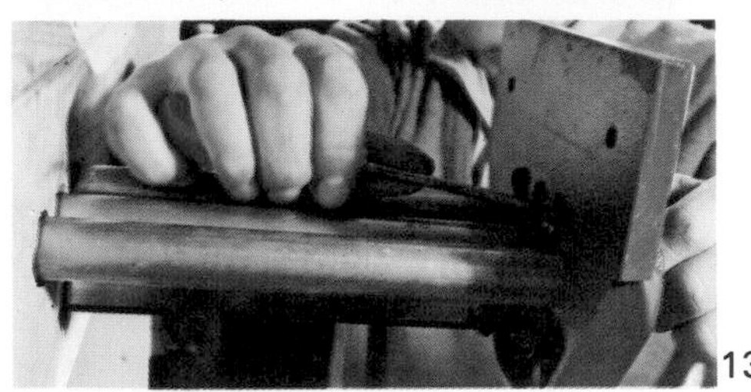

Fig. 7 *(left). The top and bottom frames are similar in width and length, and should be marked out and cut to length together.*
Fig. 8. *Fastening the end frame to the top frame with glue and carriage bolts.*
Fig. 9. *The complete frame assembly, finished except for installing the top.*
Fig. 10. *The replaceable front part of the top fits over three-quarters of the width of the lower part, and is wedged in place with triangular blocks to form a well behind it.*
Fig. 11. *Nailing on the tool rack at the rear of the bench. Tools and unfinished parts can also be stored in the well on the top.*
Fig. 12. *The top of the bench raised to show how the rear part of the vise is mounted. Only the first three holes have been drilled.*
Fig. 13. *Passing the other half of the vise through the holes in the front of the bench.*
Fig. 14. *The metal bench stop in use.*
Fig. 15. *Exploded view of the bench.*

MATERIALS LIST

Parts		Dimensions	Nearest Stock Size
4 legs		30½ x 1⅜ x 2½	2 x 3
top frame:	2 rails	54 x ⅞ x 5½	1 x 6
	3 cross rails	23¾ x ⅞ x 5½	1 x 6
	8 glue blocks	4½ x ⅞ x ⅞	¾ x ¾
bottom frame:	2 rails	54 x ⅞ x 3¾	1 x 4
	3 cross rails	23¾ x ⅞ x 3¾	1 x 4
	8 glue blocks	2½ x ⅞ x ⅞	¾ x ¾
top edging:	2 sides	54 x ⅞ x 1½	1 x 2
	2 ends	23⅞ x ⅞ x 1½	1 x 2
tool rack:	3 blocks	3 x ½ x 1½	½ x 2
	1 rail	54 x ½ x 1½	½ x 2
2 well blocks (approx. size)		7 x ¾ x 2	cut from 1 x 3
½ in. plywood			To eliminate special millwork you can use the standard nominal sizes listed above. Lengths of parts, and basic construction remain unchanged. Adjust dado joint size to ¾ in. thickness if standard nominal lumber sizes are used.
2 end panels		28½ x 23	
1 bottom panel		40 x 24¾	
¾ in. high-density particleboard			
top (permanent part)		52¼ x 23⅞	
top (replaceable part)		52¼ x 18	
vise jaws, dimensions depend on the vise.			

Also needed
18 carriage bolts 1¾ in. x 5/16 in., 18 large fender (large diameter washer with small hole) washers, 3 dozen 1½ in. No. 10 flathead screws, and a bench stop, vise and mounting bolts to choice.

rolling about. The two well-blocks with a triangular cross-section (see Fig.2) fit at each end of this well. They have two functions: their sloping shape allows the well to be swept out when it gets full of sawdust, and they stop the replaceable particleboard top from slipping backward. They should be a very tight fit, so cut them out over-length and trim them to length on the spot. Then screw them to the lower piece of particleboard with two 1 in. flathead long screws. Screw the lower piece of particleboard to the top replaceable piece from underneath with a few 1½ in. screws. When the top needs replacing, all you will have to do is remove these screws. Use galvanized screws so they don't rust in place.

The tool rack on the rear edge of the bench should be fitted by gluing and nailing the three blocks in place, then gluing on the rail and screwing it to the bench through the blocks with 1½ in. screws. It provides a useful storage space for chisels, screwdrivers and saws.

Glue and nail the remaining sheet of ½ in. plywood to the top size of the lower frame. This will provide a storage shelf for larger tools and half-completed projects.

To finish the bench, give it two coats of polyurethane varnish. Give the top a third coat, just to make it last longer.

The bench is now complete, unless you want to install a bench stop (which is a good idea). There are many types of stops, but all are simple to fix. The type shown here (see Fig. 14) can be fastened on in a few minutes by laying it on the table, drawing round it with a pencil, and chiseling out the outlined area to the required depth. A hole is also drilled to give clearance to a large screw that projects below the rest of the stop. Then the stop is dropped into the recess; it should fit flush with the bench top. It is held in with two ¾ in. No. 6 countersunk screws, and can be raised or lowered by turning the large screw.

If you do not want to use a proprietary bench stop, buy a piece of hardwood about 1½ in. square and about 6 in. long, and install it in a square hole chiseled right through both pieces of particleboard.

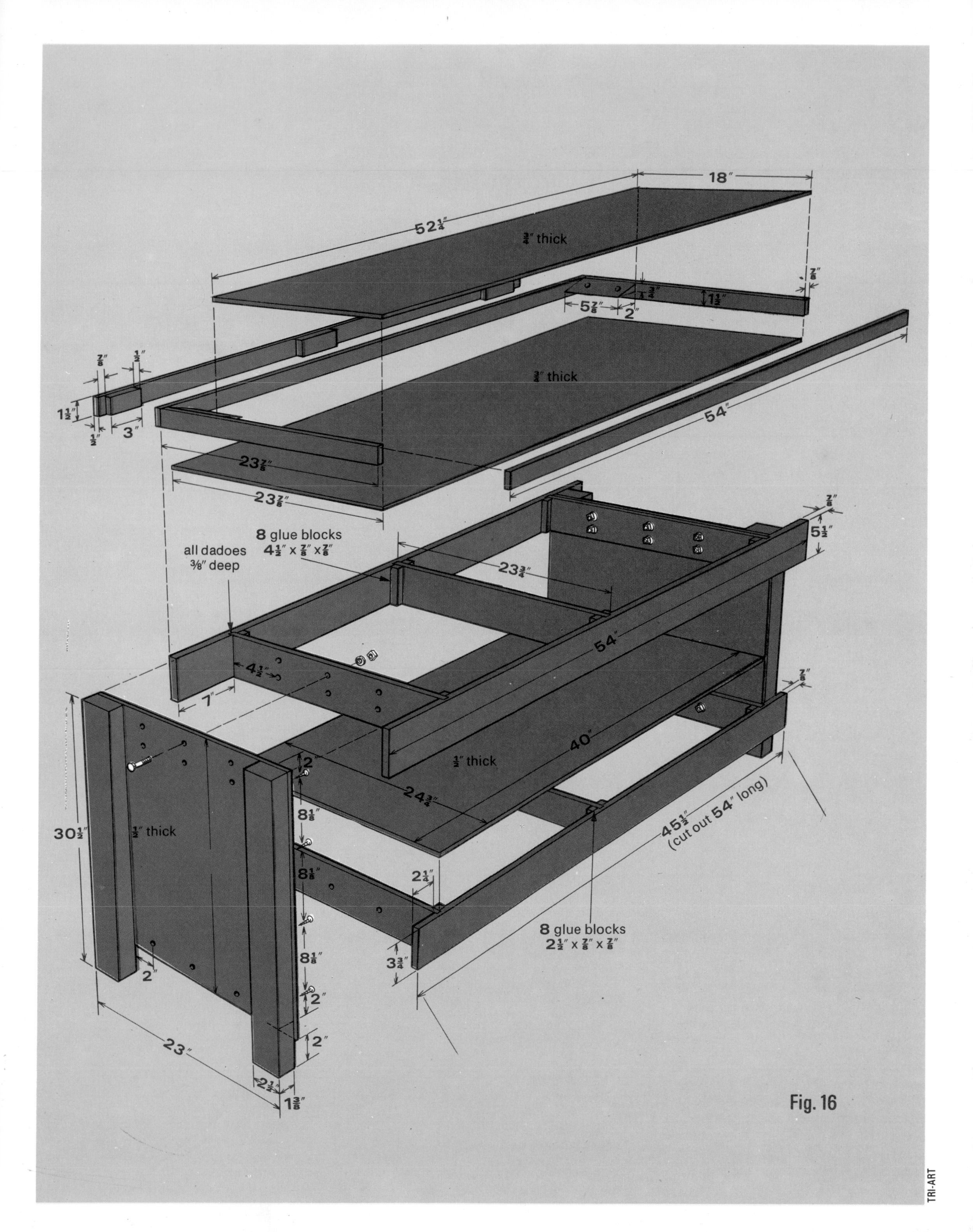

Fig. 16

TRI-ART

CHAPTER 11

Build a toolbox and step stool

There are a lot of jobs around the house that are done just above head height—putting up valances, repairing door and window frames, and so on. A stepladder is too tall for this kind of work, and a chair unsafe. What you need is a stable, horizontal support—and here's how to make one.

This versatile unit combines a roomy storage box for your carpenter's tools with a step stool—a stand that you can use for supporting the end of a plank, or just for standing on. If you have already built the sawhorse which is described in CHAPTER 7, add this unit, and you'll be equipped for most household fix-it jobs.

The project introduces the *through dovetail* joint, one of the strongest joints in carpentry—and the most impressive looking. In fact, it is much easier to make than it looks.

Materials needed

The sides of the box and the stool frame are made of ¾ in. pine. You will need: 2 pieces 21½ in. x 7 in., 2 pieces 20 in. x 6¼ in., 2 pieces 18¼ in. x 5 in., 1 piece 25 in. x 6 in. and 2 pieces 25 in. x 1¾ in. (You can cut these parts from the next greater standard width.)

The lid is made of ¾ in. birch plywood:

2 pieces 18¼ in. x 8¼ in., 1 piece 18½ in. x 5 in.

The bottom is made of a piece of ⅜ in. birch ply 21½ in. x 20 in.

You will also need four 9 in. and two 21 in. pieces of ¾ in. x ⅜ in. strip, and two 20 in. pieces of ¾ in. x ½ in. strip.

The hardware for the unit includes:

4 "strut" hinges—which are easier to get from a leathergoods shop than a hardware store—and some ¾ in. No. 6 screws for mounting them, 8 carriage bolts 2 in. x 3⁄16 in. with 8

nuts and 8 washers, 4 rubber "tack bumpers" (the small rubber "feet" used on furniture), two dozen 1½ in. No. 10 screws, 1¼ in. finishing nails (for fastening the bottom of the box), 1 in. brads (for fastening the strip) and adhesive. If you can't get strut hinges, you can use a snap-open spring type instead.

If the box is to be used out of doors—as it probably will—all metal fittings should be brass, or lacquered, plated or galvanized. The adhesive should be a waterproof plastic resin type.

All sizes given for lumber are exact finished sizes. They can be cut from the next wider standard size.

Marking out for a dovetail joint

The corners of the toolbox part of the unit are dovetailed together. Since this type of joint must fit perfectly to work, it is essential that the ends of the boards are cut exactly at right angles to give you an accurate start.

Mark out the pieces for the opposite sides of the box together in pairs, to ensure that they are exactly the same length as each other. Then sand or plane the ends dead square with a shooting board, leaving ⅛ in. extra length overall—1⁄16 in. at each end of the board. Check the angle of the ends to the sides and edges of the wood with a try square.

Choose the best side and edge of each piece and arrange them as they will be in the finished box. Then mark and number the adjacent corners as shown in Fig.1, using a pencil. This is important, for it is all too easy to make a dovetail joint backwards if you don't know which way around the wood is to go.

The shape that the dovetail joint will eventually have is shown in Fig.7. Each side of the joint is different, and also assymetrical in its own right. There are two shapes of interlocking teeth: *pins* and *tails.* The tails are made first and the pins marked directly from them to ensure accuracy.

First mark a line to indicate the depth to which the dovetails will be cut. Set a marking gauge to the thickness of the lumber plus 1⁄16 in. for the waste—this comes to 13⁄16 in. Then, working from the squared end of the box sides and the ends, *lightly* mark the line all the way around the end of each piece.

The sides of the pins and tails are sloped to lend greater strength to the joint, and the slope in this particular case is 1 in 7. This should be marked with an adjustable bevel gauge, which must be very accurately set.

To set the gauge, take one of the 7 in. wide pieces of wood and lightly square a line across it anywhere along its length in pencil (see Fig.8). Then mark a point on the edge of the piece 1 in. along from the squared line, and draw a sloping line from the mark to the far end of the first line. This sloping line and the edge of the wood form the angle to set your bevel gauge to. Erase the lines afterward.

Take the two narrower pieces of wood—the box ends—which will have the tails of the joints on them, and mark them up carefully with the bevel gauge as shown in Fig.6. Mark on both sides, using the bevel gauge for each slope and squaring the lines across the end of the wood with a try square. Do not mark the pins yet.

Cutting for a dovetail joint

As in all carpentry, the saw cuts for the joints must be made on the waste side of the marked lines. It is easy to forget which *is* the waste side when cutting out a complex shape, so draw diagonal pencil lines through the waste wood to identify it.

The most accurate way to cut the tails is put the timber upright in a vise and make all the cuts sloping one way first (see Fig.2), then turn the wood around and make the other halves of the cuts. In this way you will always be tilting the saw blade the same way, and so you will be able to gauge the angle you are cutting at with greater ease. A right-handed person should cut on a slope from top right to bottom left for the greatest accuracy; a left-handed person the other way. This will enable you to see the line you are sawing to.

Cut out the majority of the waste with a coping saw (see Fig.3), and then trim back to the marked line with a ½ in. *bevel-edged* chisel. A square-edged chisel blade cannot cut right into the acute-angled corners, so it is no use here.

Once all the tails are cut to your satisfaction, transfer their shape to the pins in the following way. Put the piece in which the pins are to be cut upright in the vise. Lay on top of it the adjacent piece with tails to which it will fit, check the numbers you have marked on both pieces to ensure that they are the right way round.

The top piece must be square on to the bottom piece so that the markings can be accurately transferred. The best way to do this is to prop it up on a big wood block—for example, a wooden plane on its side, and put a weight on top to hold it steady (see Fig.9). Then you can move the other piece up or down in the vise until you have the two pieces correctly arranged.

The piece on which the pins are to be marked is slightly wider than the piece with tails in this unit, so make sure that the extra width is at the side that will form the top edge of the box. When everything is lined up right, trace around the outline of the tails with a marking knife (see Fig.4). Then remove the wood from the vise and square the marks

Fig. 1 *(top). Mark the adjacent corners of all the pieces that are to be dovetailed together. They should be numbered as well.*

Fig. 2 *(second from top). It will improve the accuracy of your sawing if you cut all the right-hand slopes, then turn the wood around and cut all the other slopes.*

Fig. 3 *(third from top). When all the slopes are cut, remove most of the surplus wood with a coping saw, and chisel out the rest.*

Fig. 4 *(bottom). Transfer the outline of the tails to the other piece with a marking knife pressed well into the corners.*

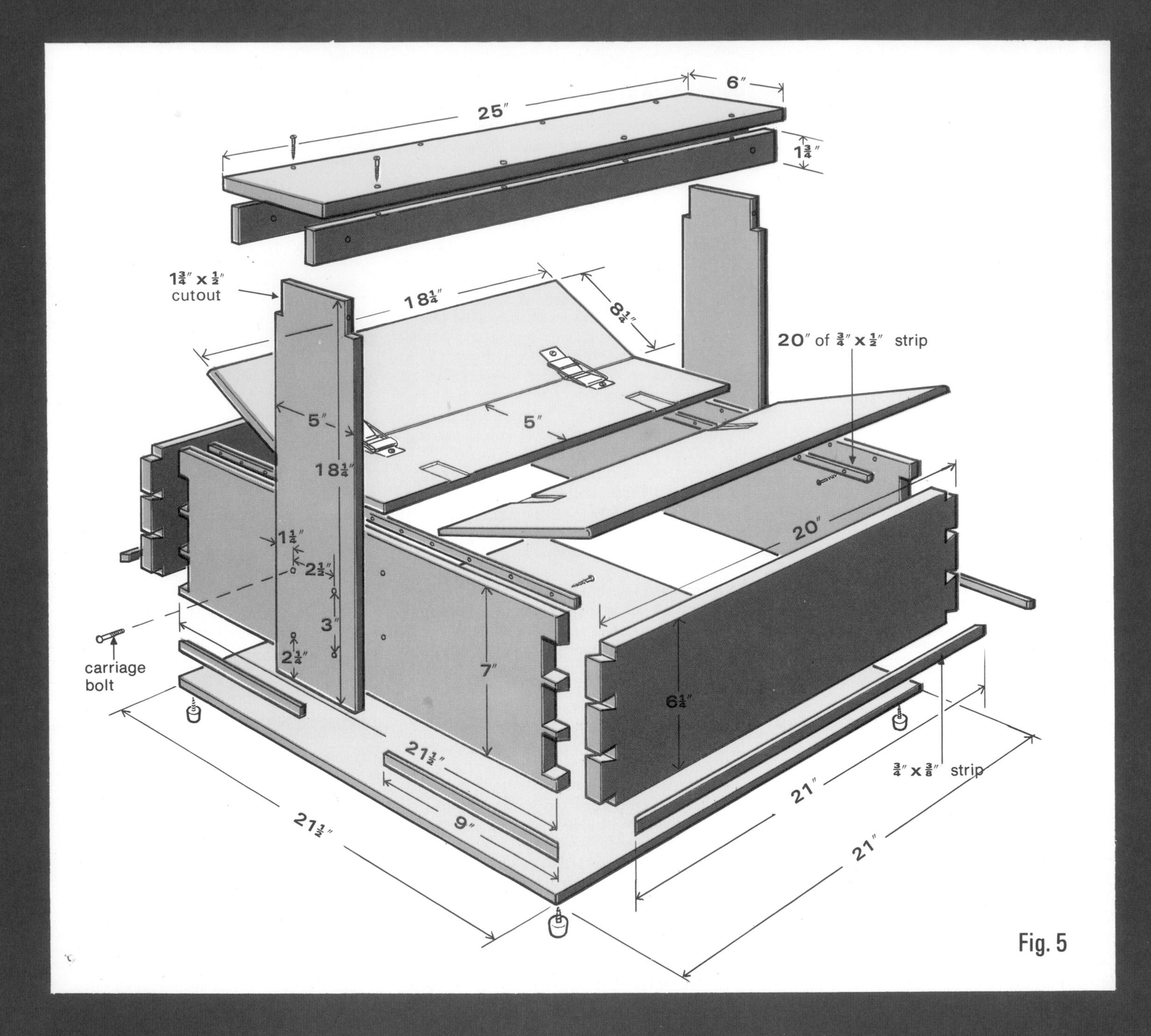

Fig. 5

down each side as far as the gauge line.

Saw out the sides of the pins as before, the only difference being that the slope is the other way—tilted in relation to the edge of the wood instead of to the end. Cut out the waste with a coping saw as before, and finish with a chisel—a 1 in. bevel edge is better than a ½ in. one, since there is more wood to remove this time. Be very careful not to cut through the pins.

Test the first joint you finish by tapping the pieces together with a block of wood. If it is a very tight fit you may be able to improve it by chiseling away a little wood from the tails. If it is too loose, you can fit slivers of wood to fill the gaps.

Assembly of the box

Once the dovetail joints are cut, everything else is fairly straightforward. Glue the joints together—if they are well made they should not need clamping. If they are a bit loose, chisel out pieces of scrapwood to press only on the tails (not the pins), and put them under the jaws of the clamp to press the tails down on to the pins (see Fig.10). You will have to use bar clamps. Check the diagonals to make sure the box is square and leave it till the glue sets.

Plane the bottom edge of the box frame lightly to level it off, then glue and nail the ply base in position. Plane off the excess ply, and then plane down the 1⁄16 in. waste sticking out of each side of the dovetail joints.

Glue and screw the two thicker pieces of strip to the inside of the longer sides of the box, with their top edges ¾ in. below the top edge of the box. These pieces act as bearers to hold up the lids. Glue and nail the center fixed part of the box top in place, resting on the bearers and secured by finishing nails driven into its ends through the outside wall of the box.

Plane the moveable lids for the box to shape. They should have at least 1⁄16 in. gap down each side to stop them from jamming shut when varnished. If you can get a thin coin between each side of the lid and the box wall, that is about the right clearance. Plane the outer edge of the lid off at an angle (see Fig.11) so that you can hook your fingers under it to lift it.

The stool

The stool is a simple square frame bolted to the sides of the box. There are two reinforcing strips under the horizontal top to stop it from sagging when you stand on it. Make a ½ in. x 1¾ in. cutout in each side of the top of the two

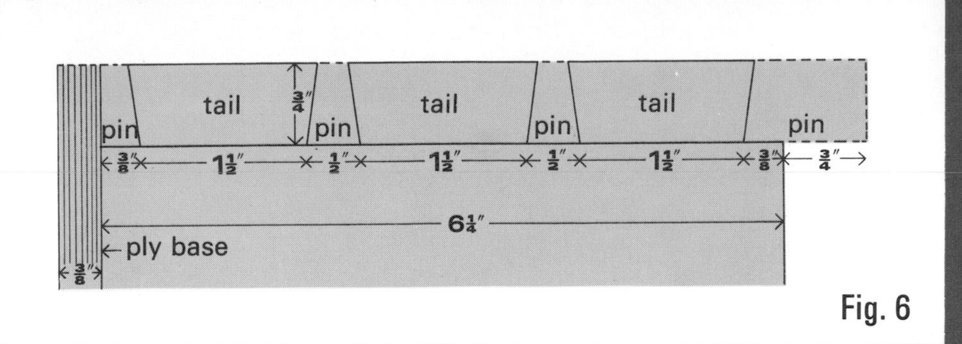

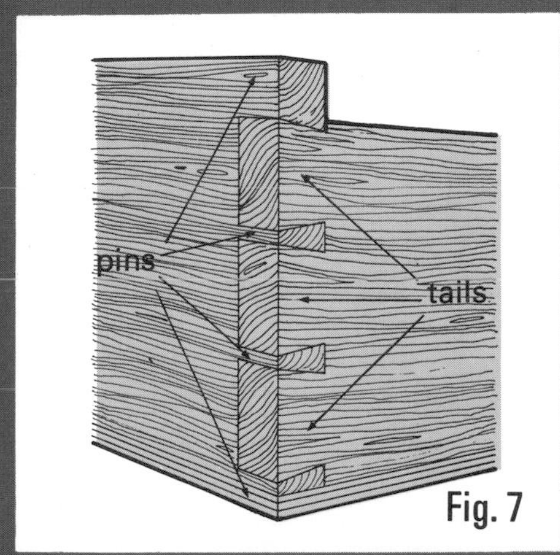

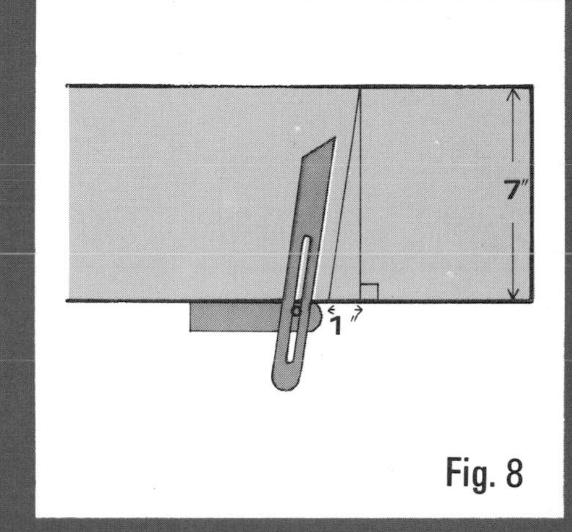

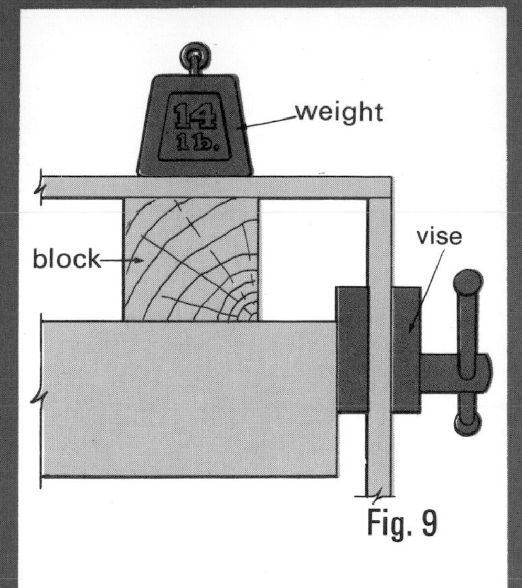

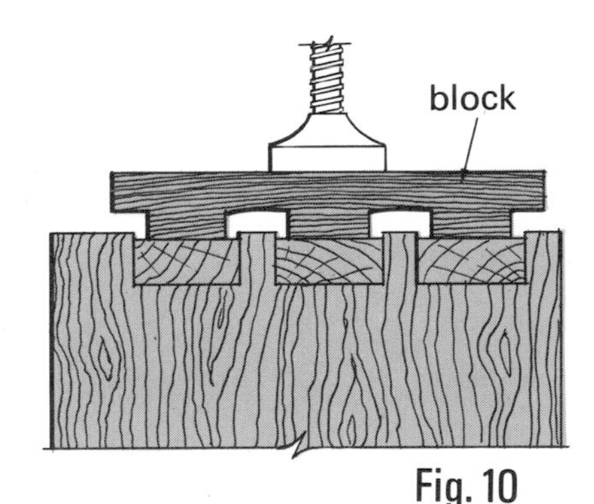

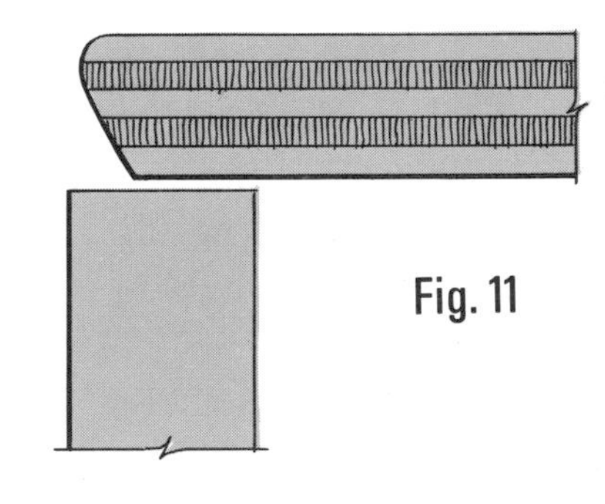

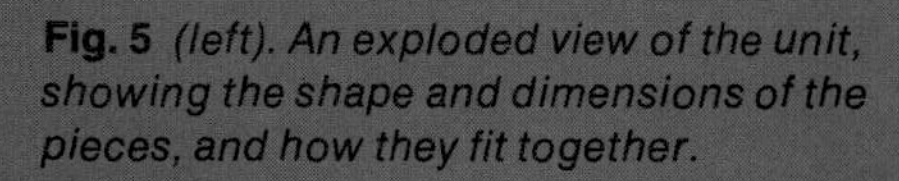

Fig. 5 *(left). An exploded view of the unit, showing the shape and dimensions of the pieces, and how they fit together.*

Fig. 6 *(top). The measurements of the dovetail joints, seen from the tail side of the joint. Use this diagram as a guide for marking out the tails (it is not to scale).*

Fig. 7 *(above, left). The same joint seen in perspective, to show how it fits together.*

Fig. 8 *(above, right). How to set the bevel gauge to a 1 in 7 slope. It is vital that the line running straight across should be at right angles to the edge of the wood.*

Fig. 9 *(top, far right). Prop the two halves of each joint up in this way when transferring markings from one piece to another.*

Fig. 10 *(second from top). If the dovetail joints need clamping while the glue dries, make four of these special blocks to press down only on the tails of the joints.*

Fig. 11 *(third from top). The outside edge of each lid should be planed off at an angle to allow you to lift it easily.*

Fig. 12. *(bottom). A cross-section of the base of the unit, showing the position of the rubber foot and the strip.*

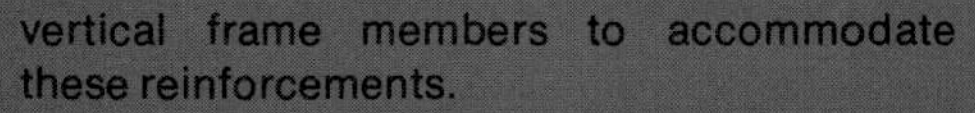

vertical frame members to accommodate these reinforcements.

Assemble the three pieces of the top first, gluing and screwing them together with four screws a side for extra strength.

While the glue is drying, bolt the vertical pieces to the sides of the box. This is done by laying the pieces against the box in the correct position and drilling straight through vertical piece and box at once with a drill the same size as the shank of the bolt. There are four bolt holes a side, as shown on the exploded view, Fig.5.

Press each bolt through from the outside and hammer the head until the square flange under it has sunk completely into the wood. Then apply a washer and nut to the shaft on the inside. Before you tighten the nuts, check that both vertical frame members are upright and parallel.

Slide the horizontal top down on to the vertical frame members. Center it and screw it on with screws passed through the reinforcing strips into the edge of the vertical members.

Glue and nail the strip to the lower edge of the box (see Fig. 12.) This can be mitered at the corners if you prefer, but an easier method is to butt-joint it.

Complete the assembly by screwing the lids on to the box with strut hinges 3 in. from each outside edge. This type of hinge holds the lid open and also prevents it from falling over backward and getting in the way of the other lid. Make sure that the planed finger-grip on each lid is the right way up. Screw or nail a rubber foot to each corner of the box base.

Finally, plane down any rough spots, lightly sandpaper the unit all over and finish it with three coats of polyurethane varnish, rubbing down lightly between coats.

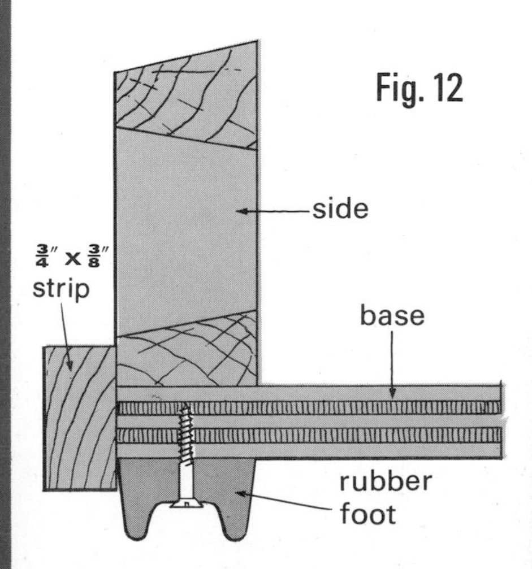

CHAPTER 12

Making picture frames

Have you ever wanted a special picture frame for one of your favorite paintings? Shops have only a limited range of frames, but you can quickly and easily make a frame to suit any picture and any room decor—at very little cost, using nothing but inexpensive moldings.

This project introduces *mitering,* a technique used for making a neat-looking L-shaped joint. The ends of two pieces of wood are cut at 45° and fastened together so that their two angles make up a 90° right angle. The mitered joint is symmetrical, and so is particularly convenient where a rabbet or molding has to be carried around a corner without interruption (see Fig. 7).

Great accuracy is essential when making a miter, particularly in angling the 45° cut. If one or both of the angles is not 45°, the resulting corner joint will not be a right angle, or there will be an unsightly gap in it. For this reason the angled cut is made with the wood held in a simple sawing guide called a *miter box* (Fig. 1). You can make your own out of scrap lumber, but since professionally-made ones cost very little, you'll find it is easier to buy one at a hardware store. More accurate, but expensive, metal types are also made.

Mitered frames are normally glued and nailed together, and must be held in clamps while the glue dries. Ordinary clamps are not suitable. You can buy metal corner clamps like those in Fig. 2, which often have a saw slot so that they can be used for cutting miters as well as holding them. But they tend to be expensive, particularly if you have to buy four. A far cheaper alternative is the simple plastic device illustrated in Fig. 5, which consists of four corner pieces, a length of nylon cord and a clip for holding it tight. Similar metal types are also available.

Buying moldings

The choice of moldings used to make your frame depends on personal taste and the type of picture you are framing. A plain, narrow frame can be made from a single molding, or you can make up a complex one from a lot of moldings stuck together. Some ideas are shown in Fig. 8.

The only feature all frames must have is a rabbet around the inside rear edge to accommodate the thickness of the picture, a hardboard backing and the glass (if any), plus about ⅛ in. extra depth. This rabbet can be cut with a rabbet plane, though you may find it less trouble to arrange the moldings so that a rabbet shape is built up at that point without the need for cutting. It is difficult and tiresome to cut a long rabbet in a narrow molding.

The amount of molding you will need for each picture can be found by measuring the perimeter of the picture and adding eight times the width of the molding itself, plus a generous allowance for waste.

For a frame made up of several moldings, use this formula for determining the amount of each molding.

You will also need a piece of hardboard (or cardboard) the same size as the picture, and some thin glass if the picture is to be covered. The only other materials you will need are some small brads, finishing nails (small ones, except for a huge frame), some white glue, and a roll of brown-paper tape.

Preparation and measurement

An oil painting, or a picture on a stiff backing, will need no preparation before you start work. A picture or print on paper should be stuck carefully on to a stiff card backing. Use rubber cement from a stationery store, which will not damage the paper and which allows the picture to be peeled off if you lay it down crooked. Try a very small drop on the back of the picture first, to be sure it has no ill effects. Otherwise, use a white paper glue.

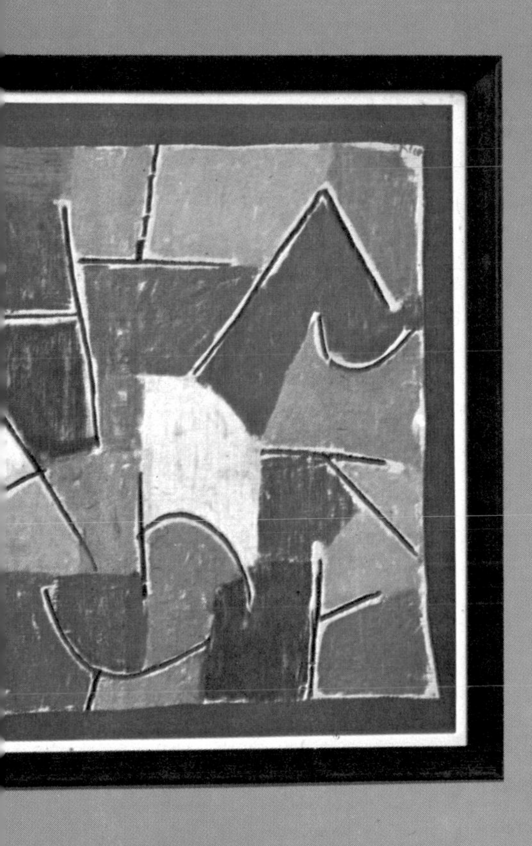

Now decide whether you want any of the mount or backing around your picture to show when it is framed. If not, trim it down until it is ¼ in. larger all around than the size of your picture, using a handyman's knife (this does not apply to oil paintings). Very small pictures can be trimmed to within ⅛ in. of their final size. This margin will also be the width of the rabbet on the back of the frame. It is impossible to frame an oil painting without covering up a little of the edge.

As a general rule, oil paintings should not be covered by glass—their varnish protects them. Watercolors are generally covered with glass and pastels must be, because they smudge easily. Covering prints is optional, but it is not recommended where the glass might catch the light from a window and hide the picture with reflections.

Before starting work, check that the corners of your picture and mount are perfectly square, and that the opposite sides are the same length. Mark the width of the rabbets clearly at each corner.

Assembling the moldings

If your picture frame is to be made out of several moldings stuck together, you should assemble them before cutting the miters. The best way to do this is to fasten, with glue and temporary brads, a whole length of molding in the size in which you bought it. In brief: assemble first, cut later.

The brads cannot be left in because they may get in the way of the saw later. In any case, picture frames do not have to be particularly strong, so glue alone will do for the moldings. The best way to hold the glued joints together is with finishing nails driven through from the back, so that the holes do not show later. If the frame is to be painted, you can put nails in the front too, and fill the holes with filler after you have removed them.

Put a small piece of scrap plywood or hardboard under the head of each nail. This will protect the wood from hammer marks, and also make the nails much easier to remove.

Leave the glue to dry to its full strength (overnight at least) before doing anything else, then remove the nails.

Cutting the rabbets

The rabbets are cut with a plow plane, which cuts a deep narrow groove to the depth required along one edge of the wood. Then another groove is cut at right angles to the first. The two grooves, when they meet, remove a square or oblong section of wood to leave a step-shaped recess, or rabbet. (You can also do the job in a single step with a rabbet plane, if you have one.)

To ensure uniformity of your rabbets, set your marking gauge first to the width required and mark all your lengths of molding. Then re-set the gauge to the depth required, and again mark all the lengths. This avoids any error through repeated re-setting.

In cutting, the most important thing is to set the fence, or guide, on the plane to the correct width. The next most important thing is to start cutting toward the *far* end of the line, pushing the plane away from you in a series of short, overlapping strokes. (Further details on rabbeting are in CHAPTER 3.)

Cutting miters

From your standard molding or collection of glued-together moldings, cut out the four sides of the frame, leaving a generous waste allowance in addition to the allowance for the width of the miter at each end (see Fig. 9).

Then transfer the marks you have made at each corner of the picture to the inside edge of the molding, using a pencil and aligning the pair of marks carefully on each piece (see Fig. 10). These marks indicate the length of the shorter inside edge of the frame.

Put a piece of molding, front side up, in the miter box so that the mark at one end of the molding lines up with one of the 45° slots on the top of the miter box, and the marked inside edge lies along one side of the box (see Fig. 11). Fasten it firmly in place with a C-clamp, its top jaw padded with scrap wood so as not to damage the molding.

Place a back saw blade in the appropriate 45° slot and cut through the molding. Repeat with the other seven ends, making sure that the miters are being cut the right way around. The back saw you use must have a blade wide enough to reach to the bottom of the box, and it must be *very sharp.* (Back saws remain sharp for a long time in normal use.)

If you are not confident of your ability to do this absolutely right first time, you have two choices. One, is to do a trial run on a short offcut of the molding. The other is to leave 1/16 in.–1/32 in. waste on the wood when sawing,

Fig. 1. *The miter box: a channel-shaped guide to take the molding, with pre-cut slots to hold the saw at the correct angle.*

Fig. 2. *A more expensive and accurate metal miter guide with built-in screw clamps to hold the wood steady.*

Fig. 3. *The miter shooting board ensures 100 percent accuracy in cutting a miter at exactly the right angle.*

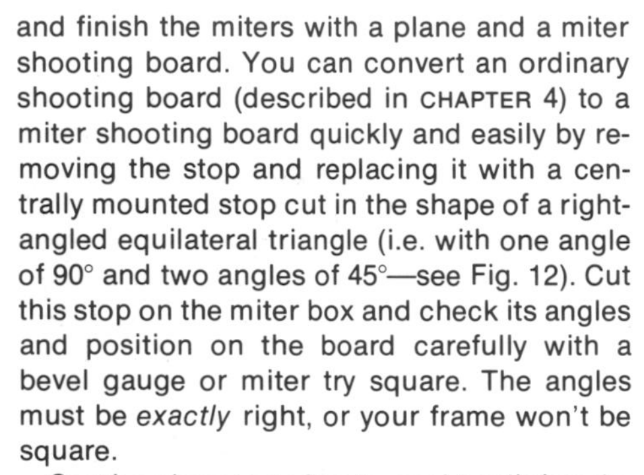

and finish the miters with a plane and a miter shooting board. You can convert an ordinary shooting board (described in CHAPTER 4) to a miter shooting board quickly and easily by removing the stop and replacing it with a centrally mounted stop cut in the shape of a right-angled equilateral triangle (i.e. with one angle of 90° and two angles of 45°—see Fig. 12). Cut this stop on the miter box and check its angles and position on the board carefully with a bevel gauge or miter try square. The angles must be *exactly* right, or your frame won't be square.

Set the plane very fine to avoid splitting the moldings, and use the shooting board like an ordinary one—though you will have to move the plane the other way for half the miters, because you always plane *toward* the stop. Hold the work extended just enough to avoid planing the shooting board.

Assembling miters

Glue and nail the miters together. The correct way to drive in the nails is to put one side of each joint upright in a vise and to hold the other horizontally above it, as in Fig. 13. Note that the top piece is a little too far "uphill"—about 1/16 in. This is because the pin will inevitably slide a little to the side as it is driven in.

The bottom corners of the frame should be made with the nails driven in horizontally on both sides, to prevent the weight of the picture from pulling the bottom piece of the frame away from the sides.

With very narrow moldings, there is a risk of the nails splitting the wood. To prevent this, drill a narrow pilot hole in the first piece of wood the nail passes through. If you do not have a drill bit of the correct size, cut the head off one of your nails and file one end to a wedge shape. Do not drill the second piece, because the pin will be going into end grain, and there is little danger of splitting the wood.

Clamp up the frame, check the diagonals to see that it is square, and leave it to dry.

Finishing the frame

The frame can be painted or varnished. Wide flat areas on a frame can be covered in cloth toning with a color in the picture, or matching your room decor. It is a good idea to make a painted frame first of all, so that any gaps or mistakes made during construction can be filled.

Gloss paint generally does not look good on frames. It is a better idea to use three or four coats of flat paint, allow it to dry thoroughly and smooth with fine steel wool.

Mount the picture on the frame in the following way: cut the glass (if any) and hardboard backing to the right size to fit neatly into the rabbeted back of the frame. Put the glass in first, then the picture, then the backing. Secure this "sandwich" in place with brads driven into the edge of the frame almost horizontally, as shown in Fig. 14. There should be at least one brad every three inches. Finally, cover the joint between frame and backing with strips of brown-paper tape for a professional finish and to keep out dust.

The completed framed picture can be hung up invisibly by putting screw-eyes in the back of the frame sides, three-quarters of the way up, and stretching picture wire between them. The wire can then be hooked over a nail or a special picture hook.

Fig. 4. *An alternative method of holding mitered corners together—corrugated fasteners driven into the back.*

Fig. 5. *This beautifully simple string-and-plastic device holds a frame together as stongly as more expensive corner clamps.*

Fig. 6. *To color the frame but leave the natural grain of the wood clearly visible, use waterproof drawing ink.*

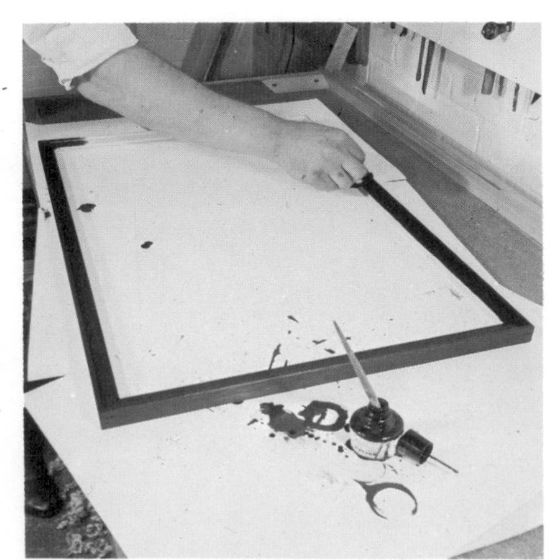

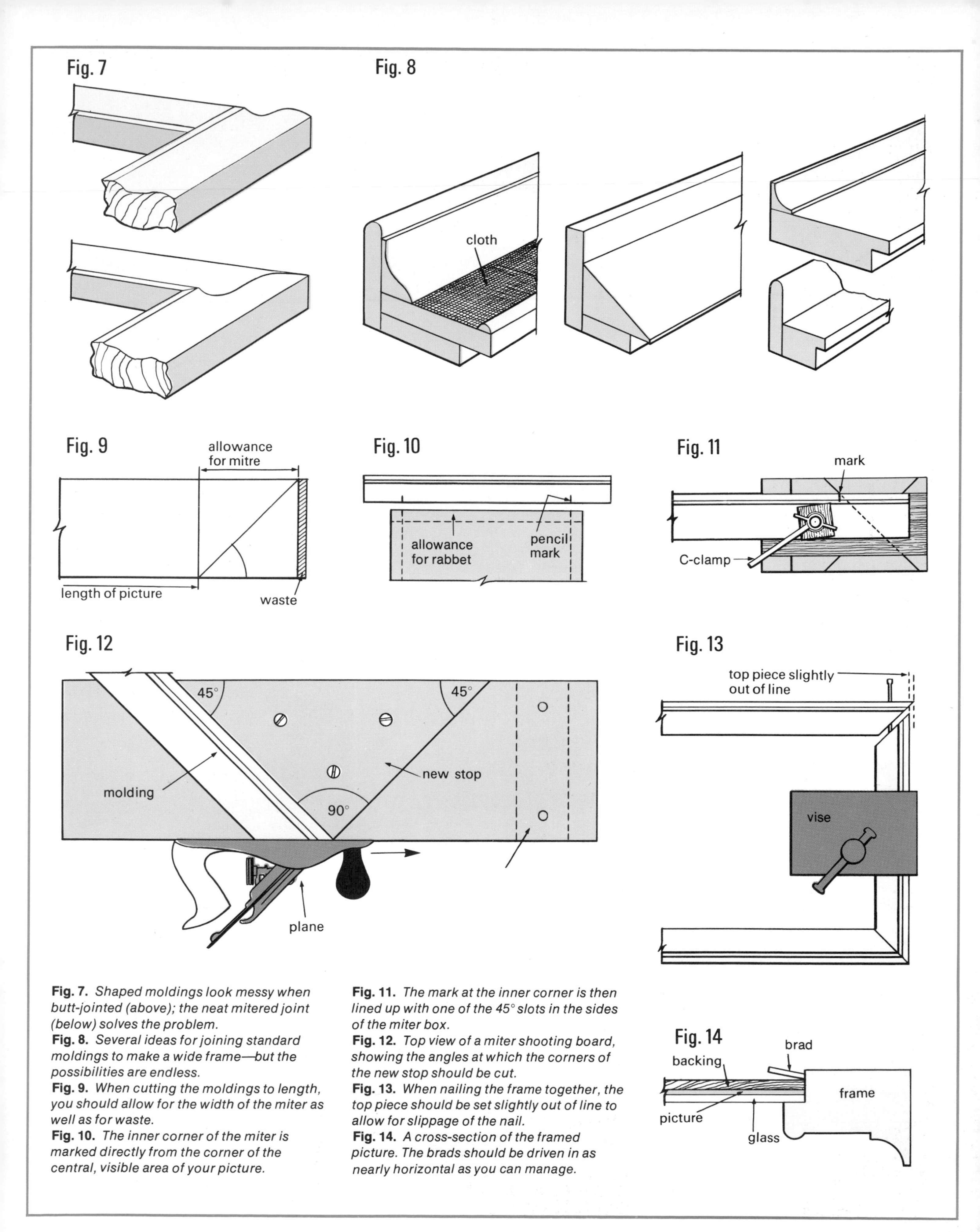

Fig. 7. *Shaped moldings look messy when butt-jointed (above); the neat mitered joint (below) solves the problem.*
Fig. 8. *Several ideas for joining standard moldings to make a wide frame—but the possibilities are endless.*
Fig. 9. *When cutting the moldings to length, you should allow for the width of the miter as well as for waste.*
Fig. 10. *The inner corner of the miter is marked directly from the corner of the central, visible area of your picture.*
Fig. 11. *The mark at the inner corner is then lined up with one of the 45° slots in the sides of the miter box.*
Fig. 12. *Top view of a miter shooting board, showing the angles at which the corners of the new stop should be cut.*
Fig. 13. *When nailing the frame together, the top piece should be set slightly out of line to allow for slippage of the nail.*
Fig. 14. *A cross-section of the framed picture. The brads should be driven in as nearly horizontal as you can manage.*

CHAPTER 13

A classic Welsh dresser

The clean lines of this Welsh dresser add a modern touch to the traditional design and enable the unit to blend well with any style of decor. An added attraction is the ample storage space the dresser provides with its spacious cupboard, shelves and drawers and the sturdy work surface.

The Welsh dresser in the photograph is made from pine, but any softwood that can be sanded to a smooth finish will do. The dresser has two spacious drawers and a large cupboard with a shelf. The doors of the cupboard and the short sides of the dresser are made by the same techniques. The unit has a large, solid work surface and an upper part, with two shelves for the storage of plates or jars.

The construction is simple, the joints being either through or stopped dadoes or blind doweled joints. Some grooves have to be made, but if you have the tools, this is easy work.

Another advantage of the type of construction used in the Welsh dresser is that nearly all the components can be cut out and partly assembled before any major assembly work has to be undertaken. For example, while the adhesive that joins the boards that form the work surface is drying, you can be making another component. In this way the dresser can be made quickly—though you should be careful not to spoil the job by working *too* quickly.

Because most of the components can be made independently of each other, the order in which they are made is not really important.

The cupboard frames

These are the easiest components to construct. They form the skeleton of the front and back of the cupboard part of the dresser. The drawers fit into the front frame, and the cupboard doors are hinged to it. The plywood panel that forms the back of the cupboard is nailed to the near frame.

The arrangement and dimensions of the components used in the frames are shown in Fig. 2. The components of the frame are butted together and strengthened with one dowel nail in each joint. CHAPTER 15 describes the blind doweling method that is used.

When making these frames it is essential that the final construction is square. You can test this with a large try square or by measuring the diagonals. The components are glued together and bar clamped while the glue dries, so if the frame is slightly out of square you can often correct it by doing up one bar clamp more tightly than the others.

Left. *A functional and attractive unit, the Welsh dresser has ample storage space for food, dishes and table linen.*

The sides and cupboard doors

The construction of the two side panels of the cupboard and of the cupboard doors is the same; only the dimensions are different. Each door is 23 3/16 in. x 17 in. and the finished size of the side is 31 1/8 in. x 21 in. The rails and stiles of doors and sides alike are made of 2 3/4 in. x 1 5/32 in. lumber. This lumber thickness is designated as "5/4" in lumber terms. Ask for "five quarter" thickness when you buy it.

The rails and stiles are fastened together at the corners by lap joints. A 1/2 in. thick infill of T & G paneling is fitted into grooves cut in the inside edges of these pieces.

First cut the rails and stiles a little oversize. Mark a point in the center of each piece along its length and take all measurements for the tenoned joints from this point. Mark out the area of the joints with the four rails of the two doors laid together with their long edges butting. Do the same for the four rails of the two sides. Mark out the stiles for the doors and the stiles for the sides in the same way. Then cut the joints in the manner described fully in CHAPTER 5.

When the joints are cut, lay the pieces together in roughly their finished position. Mark the top faces. This ensures that when you cut the grooves into which the paneling fits, you cut from the correct side of all pieces.

Now cut the groove on the inside edge of the stiles and rails. You can do this with a rabbet or plow plane, or more easily with a router, as shown in Fig. 4 of CHAPTER 24. A small router like the Stanley "Routabout" will do. You will also have to make stopped grooves on some of the pieces to avoid encroaching on the lap joints. You can make most of each cut with the plane and finish the ends with a chisel. Cut the groove to a depth of 3/4 in. and a width of 1/2 in. Make the cut from what will be the front of the doors or sides.

The next step is to fit the paneling. This is 3 in. x 1/2 in. tongue-and-groove boarding; the length of the boards is the distance between the inside edges of the stiles plus 1/2 in. at each end. It is fitted so that it is parallel to the rails of the doors and side panels. (Other board widths may be used.)

Before you fit any paneling, work out the widths of the first and last pieces. These should be of equal width but they should not be too narrow. You can find the width easily by dividing the width of the paneling into the distance between the inside edges of the rails an exact number of times and halving the remainder.

Cut the bottom piece of paneling to size. If you need to reduce its width, cut from the tongued side. In any case remove the tongue from the first piece. Before fitting it, glue the bottom rail of each door or side piece and the two stiles together. Clamp the assembly with a bar clamp running along the rail and parallel to it. Then fit the first piece of paneling before the adhesive dries. The ends of the pieces and the faces of the rabbet into which they fit should be glue-coated.

Push the second piece of paneling into the recess and lightly tap it into place with a mallet and scrap wood block. Fit this piece tongued side down. Fit each subsequent piece in the same way, but exert a little more force with the mallet to push the tongue of the new piece into the groove of the previous piece. Push the last piece of paneling into place. Apply glue to the joint surfaces at the top of the stiles and top rail. Jam the top rail into place on top of the last piece of paneling. Release the clamp at the bottom rail and check that the assembly is square. If it is, clamp it together again firmly.

When the adhesive has dried, plane off the small pieces of the rails and stiles that protrude from the lap joints.

Cutting the finger holds

The doors (but not the sides of the cupboard) have a finger hold cut into the outside edge of one stile of each door. The stile is, of course, the opposite one to the one that has a hinge fitted to it later.

The finger hold is 5¼ in. long and starts 2¾ in. down from the top edge of the stile. It enters the edge of the stile to a depth of ⅝ in., and the wide inner face of the stile to the same depth. First mark from the top of the stile a distance of 2¾ in. and from this point mark off 5¼ in. down the stile. Then cut out the finger hold with a sharp chisel.

The worktop

The dimensions of the cupboard top are 42³⁄₁₆ in. x 21 in. x ¾ in. You will not be able to buy a single board wide enough for this job so you will have to join several boards along their edges and glue them together as in the case of the dresser in the photograph.

First lay the boards on the floor and choose the most attractive combination of grain patterns. Mark the surfaces of the boards with a pencil so that you do not accidentally reverse one of them later.

To join boards along their long edge, first plane the edges that are to butt exactly square. Then spread adhesive on these edges and put the boards together. Clamp them with bar clamps or wedges and wipe excess adhesive off of the surface with a damp cloth.

All of this must be done on top of several layers of newspaper so the work won't stick to the floor.

The cupboard drawers

The drawers are made from ¾ in. thick lumber. Fig. 4. shows the dimensions of the finished drawers. The sides, back and front pieces are fitted together with dado and rabbet joints. The front piece has stopped grooves cut into it to accommodate the sides. This is so that the grooves are not visible on the top front edge of the drawer. The bottom of each drawer is a ¼ in. plywood panel which is glued in grooves cut near the bottom edge of the side, front and back pieces. The bottom of the inside face of each front piece has a finger hold cut into it.

Cut the side, front and back pieces to the dimensions given in Fig. 4. Note that the sides are 17¾ in. long when tongues are included. Square the ends with a plane, but take care not to damage the corners of the pieces. Clamp a scrap of wood to the edge to which you are planing to prevent this.

The next step is to mark out and cut the tongues and grooves of the joints. Cut rabbets along the ends of the side pieces to leave a tongue ⅜ in. long and ⅜ in. wide protruding. On the back piece, cut the groove to accommodate the tongue. The face of the groove nearest the end of the pieces should be ⅜ in. away from the ends. The front of the drawer is longer than the back piece. Here the stopped grooves are cut 1⅞ in. away from the ends. Stop them ½ in. away from the top edge of the drawer front.

Next cut the groove near the bottom edge of the four pieces that accommodates the bottom ¼ in. plywood panel. The groove is ¼ in. wide and ⅜ in. deep. It begins ¼ in. from the bottom edge of all pieces except the front. On the front it starts ¾ in. from the bottom edge. The groove can be cut the whole length of the side and back pieces. On the front piece, however, stop the cut 1⅞ in. from the ends, so that it does not show.

The next step is to cut the finger hold at the bottom of the inside faces of the drawer fronts. The method of marking out and cutting is the same as that for the door finger holds, but the dimensions are different. The finger hold is in the center of the bottom inside edge of the front piece. It is 7¼ in. long, enters the bottom edge of the front piece to a depth of ⅝ in. and the inner face to a depth of ¼ in.

Cut the bottom plywood panel to 14¼ in. x 17¾ in. Glue the front and sides of the drawer together. Slide the bottom into the grooves and glue the back of the drawer in place. Check that the drawer is square and clamp the assembly together.

The drawer runners

The drawers slide on wood runners—pieces of nominal 1 x 2 with a ⅜ in. x ⅜ in. rabbet cut in them. Four runners are used—two for each drawer.

Cut the runners to size and use a rabbet plane or router to cut the rabbets. The runners, since they run from the front frame to the back frame, cannot be attached until the four sides of the cupboard, but not the top, have been assembled. They are screwed to 1 x 2 blocks which are screwed to the insides of the rear and front frame. Mount the runners so the rabbeted tracks are flush with the top of horizontal member under the drawer openings. Space the runners so the drawers slide easily.

The cupboard base

The cupboard of the Welsh dresser rests on a four-sided plinth a few inches smaller than the length and width of the cupboard so the end panels and back panel can extend ½ in. down over the plinth to hold the cupboard in place. The base is not purely decorative, but acts as a scuff board.

The four pieces are fastened together with rabbeted joints. The short pieces have a tongue ⅜ in. long and ⅜ in. wide cut on their ends. When grooves of the same size have been cut in the other two pieces, glue and clamp the assembly together, making sure that it is square.

Assembly of the cupboard

The main components of the cupboard have now been constructed. After a few more steps assembly can take place. First, nail the 37 in. x 31 in. back panel to the rear frame. A ⅝ in. wide and ⅝ in. deep rabbet is cut in the inside face of the bottom piece of both the rear and front frames. This rabbet accommodates a ⅝ in. plywood panel which forms the bottom of the cupboard.

The shelf in the cupboard is supported by 1 x 2 blocks. These are nailed to the inside of the front and back frames parallel to the rails and 11½ in. from the bottom rail. The supports are fixed so that their wide surface butts against the frames.

The top is doweled to the tops of the side panels and finishing-nailed to the top of the front and rear frames. (Or it may be finishing-nailed all around.) The sides and the front and back frames are joined by the blind doweling method described in CHAPTER 15 and are glued. The front frame is set back 1⅜ in. from the front edge of the side panels to allow for the thickness of doors and hinges.

Attach the doors to the front frame with a length of brass piano hinge. They are fitted so that the bottom edge of their bottom rail is flush with the bottom edge of the bottom rail of the front frame.

Fix the blocks which carry the drawer bearers and fix the bearers to the blocks. Then fasten the top in place.

The shelf unit

The shelf unit, with the exception of its T & G paneling, is made from nominal 1 x 8. The dimensions of the unit are shown in Fig. 1. The horizontal shelves are accommodated in the vertical sides by stopped dado joints. A ½ in. thick paneling is set in a groove cut in the back of the cabinet. The unit also has a decorative curved cross-piece which butts against the bottom of the top shelf.

The first step is to cut the two sides to length and square their ends. Then mark out the positions of the stopped dado joints. Do this by laying the two pieces together with their long edges butting. Measure a distance of 13 in. up from the bottom edge of the planks. This indicates where the bottom edge of the lower shelf will be. From this point, measure and mark the thickness of the shelf and those above it.

Now cut the stopped dadoes. These are ⅜ in. deep and ⅜ in wide. Stop the dado ½ in. from the front edge of the side pieces. You can run the dado right through to the back edge of these pieces. This will leave six small gaps showing when the T & G paneling has been fitted, but they are at the back of the unit and not conspicuous.

Next cut the shelves and top piece, to length, remembering to allow for the ⅜ in. depth of dado at each end. The top is 7¼ in. wide, and the two shelves are 6¾ in. wide. This allows for the paneling, when it is fitted, to butt against the back edge of the shelves. Cut a tongue on the end of the shelves to a width of ⅜ in. and to a length of ⅜ in.

The next step is to cut the groove near the

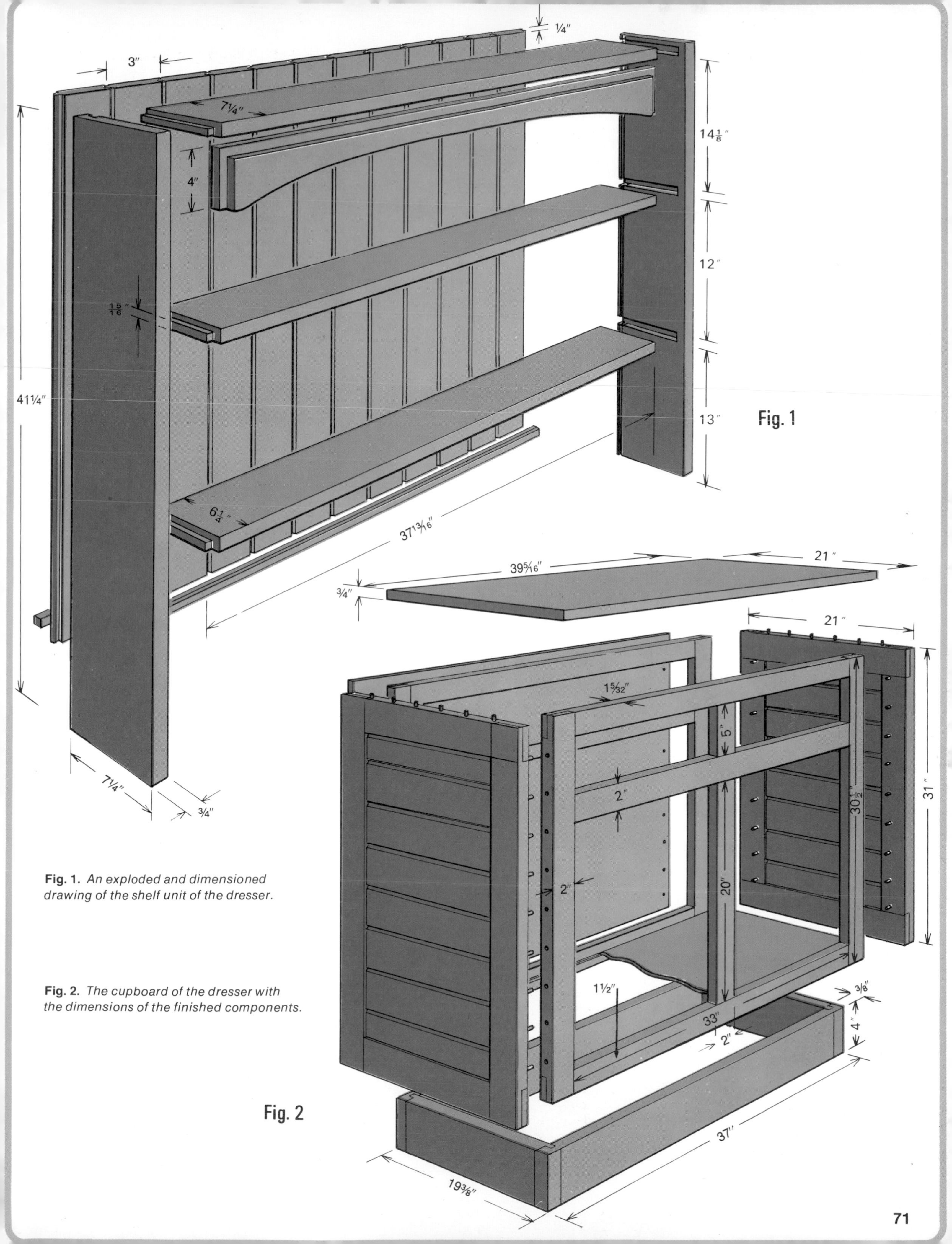

Fig. 1. *An exploded and dimensioned drawing of the shelf unit of the dresser.*

Fig. 2. *The cupboard of the dresser with the dimensions of the finished components.*

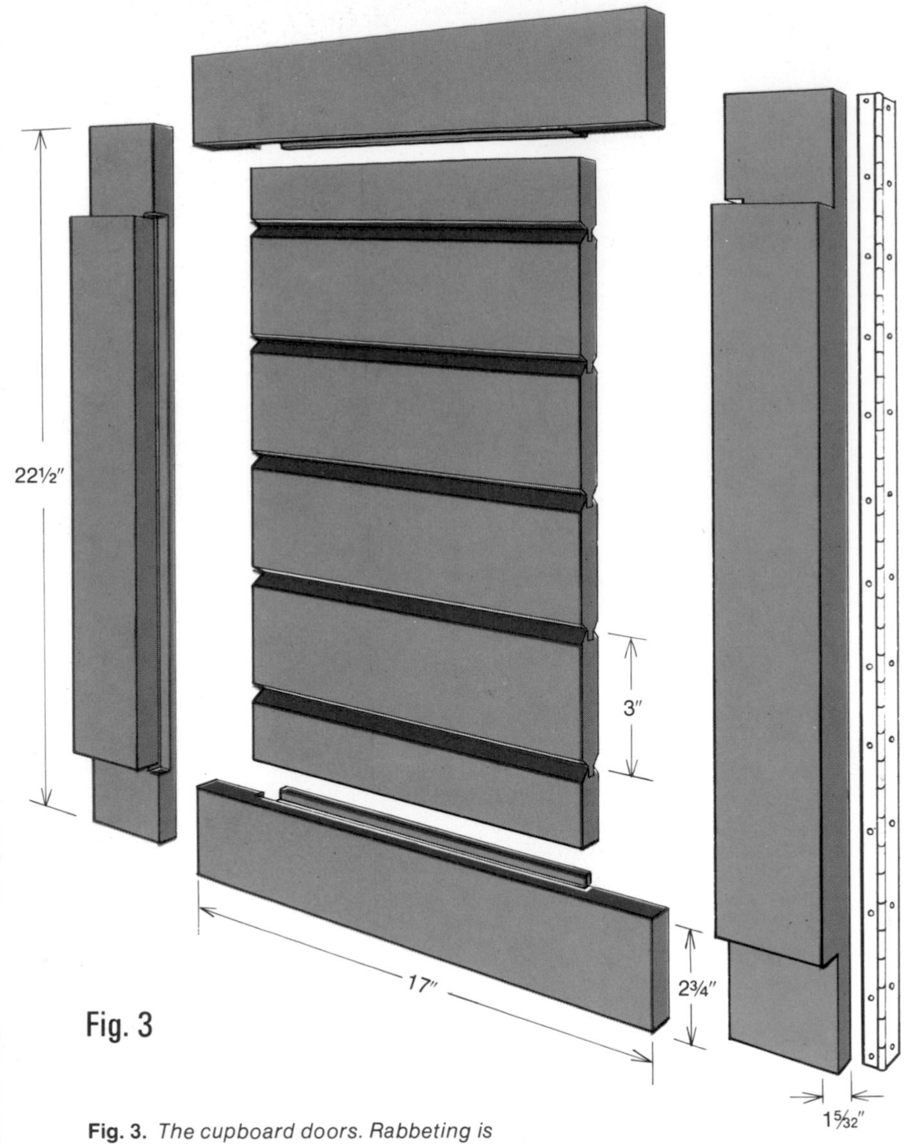

Fig. 3. *The cupboard doors. Rabbeting is used to join the stiles and rails.*

Fig. 4. *An exploded drawing of one of the cupboard drawers with the finished dimensions.*

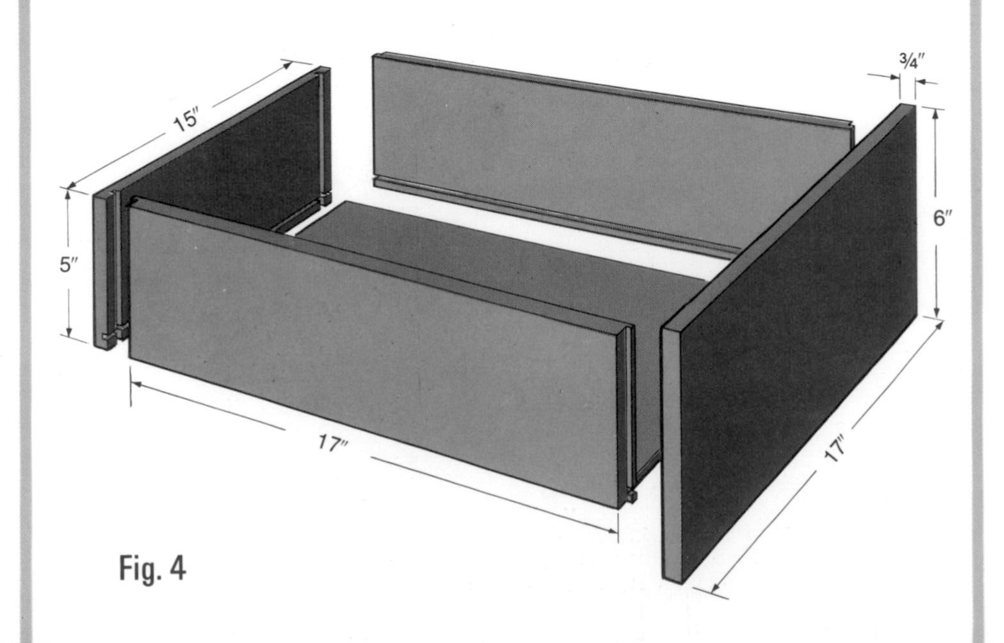

back edge of the sides and under the top shelf to take the paneling. The groove must match the paneling thickness. A tongue should be cut on the upper end of each piece of paneling to fit the groove in the underside of the top piece.

Once the grooves have been cut, the shelf unit can be assembled, glued and clamped, or use finishing nails to hold it.

In the shelf unit the paneling runs vertically unlike the horizontal boarding on the cupboard sides and doors. Assemble the sides and shelves without glue or nails for a trial fit. Cut each T & G board to length. Push the first piece (its tongue must have been removed) into the groove cut in the inner face of one of the side pieces of the shelf unit. Tap it well into the groove with a mallet and waste block used on both the bottom end and the long edge. Insert the subsequent boards in the same way.

The final few boards may have to be cut very slightly narrower and fitted together by removing the tongues or grooves (glue will hold them later). Fit them to the assembly with the outside edge of the last board just within the rabbeted groove in the shelf unit side. Push the boards flat with your hand. They should then fit tightly in place.

The cut lower ends of these boards are trimmed by two strips of ¾ in. x ½ in. molding with a beveled top facing edge. Glue one strip to the bottom of the boards at the front, and the other to the back.

The decorative curved piece that butts against the underside of the top shelf can now be cut. The piece is $37\frac{13}{16}$ in. long. Cut it to the length required. Cut a rabbet from the front face on the two ends and the top edge. This rabbet is ⅜ in. x ⅜ in. and is purely decorative. (It may be omitted.)

The curve on the bottom edge of this piece is 2 in. deep in the middle and begins ¾ in. from the ends of this piece. As the piece has no functional purpose, the curve does not have to be accurate, but it does have to be symmetrical. One way of ensuring that it is, is to make a paper template. Draw a line on a large sheet of paper. On this line mark two points, one that indicates the end of the curve and another that indicates half the length of the curve. From this second point draw a line at right angles and measure a distance of 2 in. Draw a freehand curve between the point that indicates the end of the curve and the point 2 in. above the half way mark. Then cut out the paper template and put it in place on the wood. Draw half the curve, turn the template over and draw the other half of the curve.

Cut the curve out with a compass saw. Finish it with a spokeshave and sandpaper. Glue and finishing-nail the curved piece in place.

The shelf unit is fixed in place on the cupboard by two brass mending plates with four screw holes. These are fixed so that two of the screws enter the cupboard unit and the other two into the side of the shelf unit. Set each plate about 2½ in. from the back of the cupboard.

To finish the dresser, sand all the pieces that need it to a smooth finish with sandpaper. Apply several coats of clear polyurethane varnish, rubbing down carefully between coats.

Above. *Three typical drawers, one of which is bound to be suitable for any purpose. At the top, a high-quality mahogany drawer; center, a traditional flush-fronted drawer, and at the bottom, a simple one with a laminate front.*

CHAPTER 14

Making and fitting drawers

Most homes need more storage space than they actually have; families always seem to accumulate huge amounts of possessions, equipment and junk and it has to be put somewhere. The most economical way to store it all—and the most space-saving—is to make your own built-in furniture and fit it with drawers. Here is how to make the drawers.

It is possible to buy ready-made drawer kits at a reasonable price, and they can be quite satisfactory in some situations. But the trouble with them is that they come in a limited range of sizes, and it is hard to fit them in where you want them. If you build your own drawers from scratch, you can make them exactly the size you want, both to fit the space where they are installed and to take any odd-shaped items you might want to store in them.

Basic requirements

Drawers can be built to various designs, some simple and some complicated, but all have the same basic parts and many points in common. All have a four-sided frame (unless they are being fitted into a very strange-shaped piece of furniture) and this normally has runners spaced down the sides, on which the drawer slides. A base, generally of thinner material than the frame, is fitted into it and—as well as supporting the contents—holds the frame in square.

It is essential that the front and back of the drawer are the same size, and thus that the sides are exactly parallel. Attention must be paid to this point when making the drawer, or it will not slide properly. Some types of drawer do in fact have an extra-wide front that is not the same size as the back, but this is often an attached facing and there is a conventional box-shaped drawer behind it. In any case, the sides are still parallel.

The runners must also be parallel and absolutely straight and smooth on their sliding surfaces. Some drawers use the bottom edge of the frame sides as runners, and in this case the base must be recessed a short way above these edges so as not to catch on the drawer mountings.

Drawers are held together by various joints. These must be strong enough to hold the drawer absolutely rigid; if it goes out of square it will stick. They must also be made the right

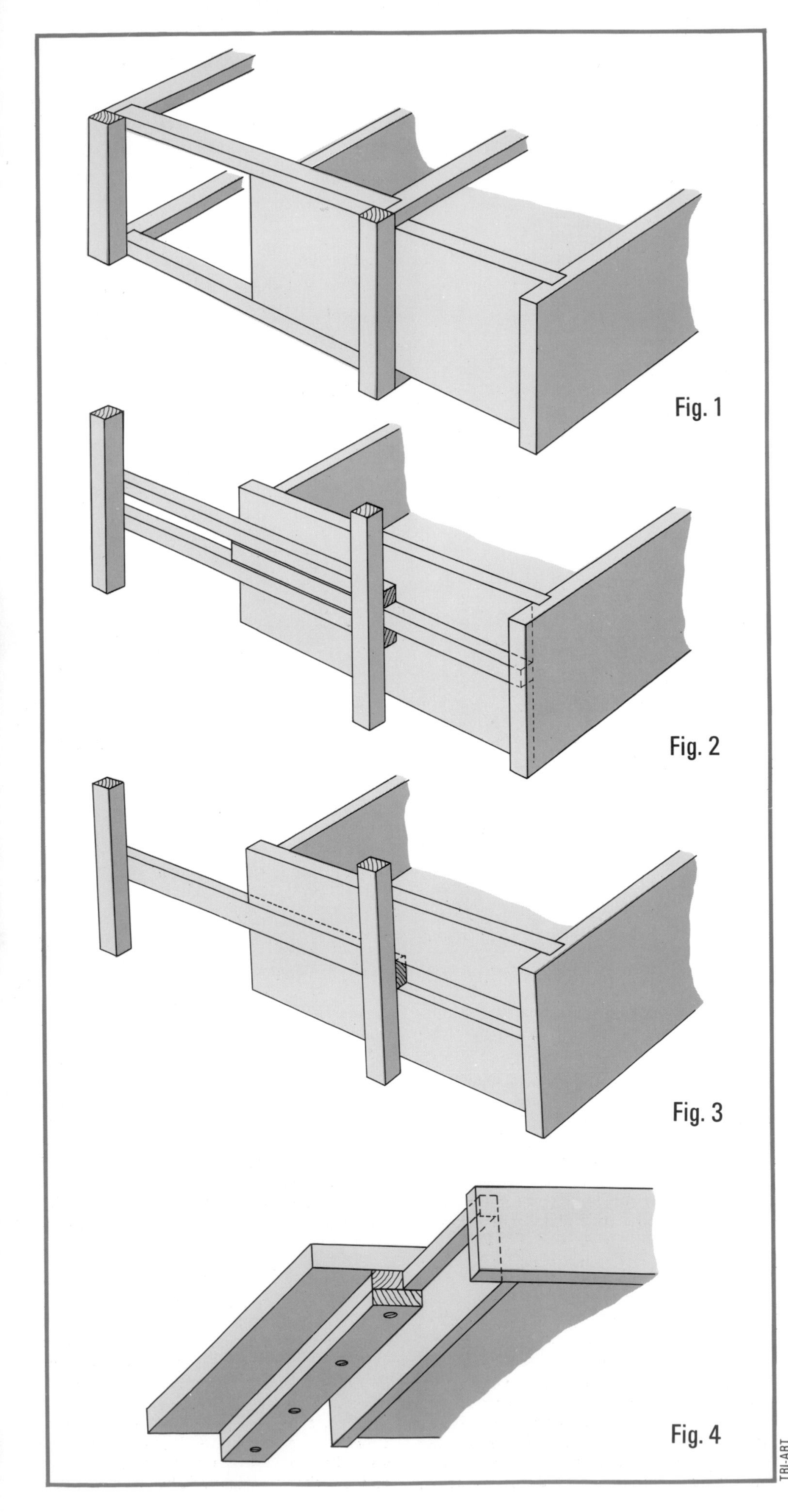

Fig. 1. *A drawer mounted on rails or battens in the traditional manner. It is supported by, and slides on, rails underneath it. Rails above stop it from tipping when open.*
Fig. 2. *A drawer mounted by the three cleat method. This is the easiest technique for mounting drawers but wastes a lot of space.*
Fig. 3. *A drawer mounted by the cleat and groove method. This does not waste space and is the normal method used in modern furniture.*
Fig. 4. *A top-hung drawer; this method is used for mounting drawers under table tops.*

way around to resist the forces acting on the drawer when it is pulled open or when heavy objects are put in it. For example, it is no use simply nailing the drawer front on from the front side, or a violent tug will pull the nails out and you will be left with a frontless drawer.

The best way of understanding the requirements of drawers in detail is to deal with the various parts one by one.

Drawer runners

The design of the runners of a drawer, and the way it slides on the supporting framework, are the most important factors affecting the design of the drawer.

The *rail technique* is the way that drawers are mounted in traditional furniture. It is shown in Fig. 1. The basic framework of the furniture is made out of wooden members, which form a frame around the front of the drawer into which it fits flush when closed. More frame members extend back horizontally from this front framing, both above and below the drawer. The battens under the drawer are used as rails to support it; there is a wide rail below each side of the drawer on which it slides, and further rails (not shown) attached to this at each side of the drawer stop it from slipping sideways.

The drawer is not fitted with separate runners, as it slides on the bottom edge of its sides. For this reason also, the base must be recessed a short way.

Drawers mounted in this way must be made very accurately, since if they are too narrow, there will be a noticeable gap at the front, and if too large, they will not go into their supporting frames.

The *three cleat technique* requires much less accuracy and is a method commonly used with modern drawers fitted with over-width fronts. It is shown in Fig. 2. A cleat, or wood strip, is fitted to each side of the drawer. Two parallel cleats are attached to each side of the space into which the drawer fits. The cleat on the drawer runs between the two cleats on the supporting framework, and in this way the drawer is held straight and level and there is no need for any other support. Sometimes there are two cleats on the drawer and one on the frame.

This method is very suitable for drawers fitted between two vertical boards, a situation often met with in built-in furniture or when fitting extra drawers to existing storage units. It is also good for fitting drawers to frames that are out of square, as the two cleats on each

Fig. 5. *A flush-fronted drawer is most easily made with end rabbets at the front, and the back dadoed into the sides.*
Fig. 6. *A drawer with a modern over-width front, and the sides dadoed into it.*
Fig. 7. *A very simple drawer with a false front screwed on invisibly from inside, and butt joints used throughout the structure.*
Fig. 8. *Joints used in drawer-making:* **a** *butt with false front,* **b** *butt with plain front,* **c** *end rabbet,* **d** *dado,* **e** *through multiple dovetail for rear,* **f** *lapped dovetail for front.*

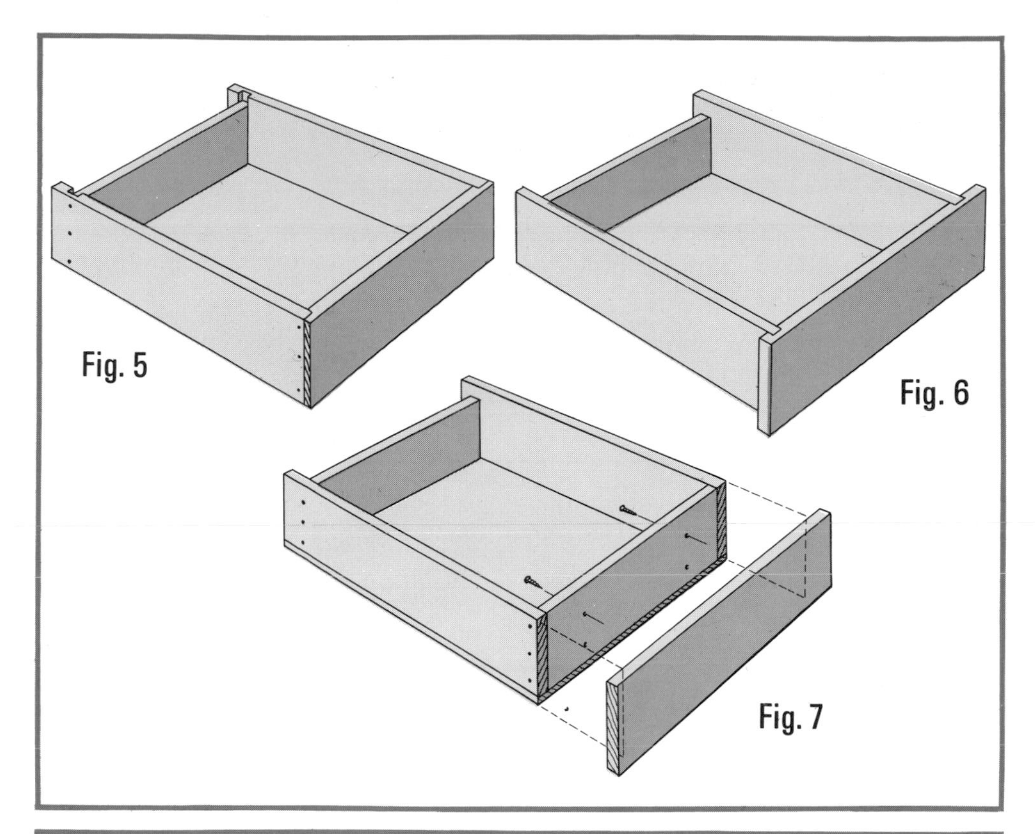

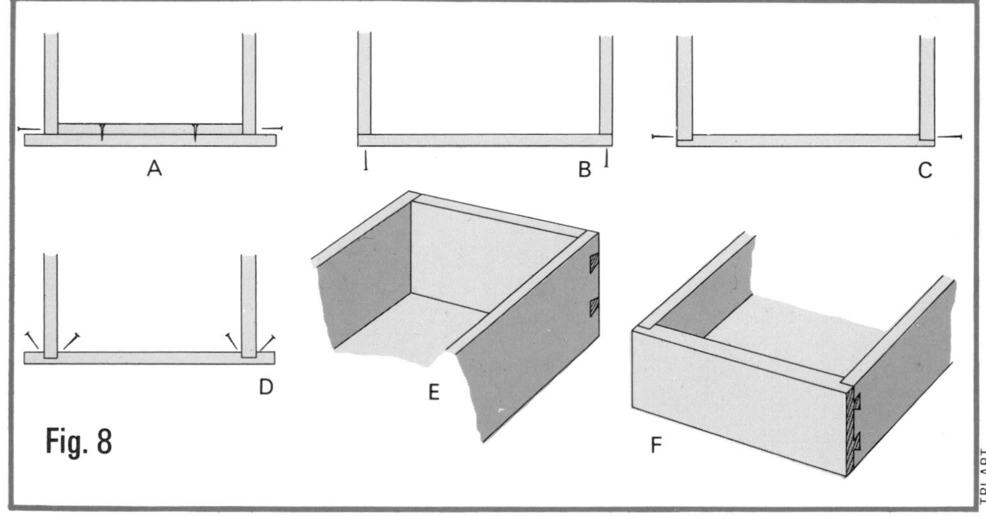

side of the frame can be shimmed out until they are parallel.

It is not, however, a good method for very heavy drawers, since the whole weight of the drawer is taken on two narrow wood strips. Furthermore, there is a large and unsightly gap down both sides of the drawer, which has to be concealed with an over-width front.

The *cleat and groove technique* is neater, though drawers using this method are harder to build. It is shown in Fig. 3. A single cleat is fixed to each side of the supporting framework, and this fits into a groove cut in the side of the drawer. If the groove is made as deep as the thickness of the cleat, the drawer can be made the full width of the space into which it fits and will not require an over-width front. The groove must stop short of the front of the drawer for the sake of appearance.

A groove cut in a drawer side must not be made too deep, or it will weaken the drawer. As a general rule, its depth should not exceed half the thickness of the wood. It must be cut with a combination plane or by some equally accurate method; if it is just chiseled out it will be too rough and the drawer will stick.

Top hanging is a special technique used for fitting drawers to the underside of a table top without any support underneath. As shown in Fig. 4, a pair of cleats with an L-shaped cross section is fixed to the underneath of the table top, and a single cleat screwed to each side of the drawer right at the top fits into, and slides in, the angle of the L.

Where two drawers are hung side by side, the cleat between them must be T-shaped, so that it can support the adjacent edges of both drawers.

This method is unsuitable for heavy drawers, which tend to tear the fastenings out of the table top. Heavy drawers should always be properly supported from underneath.

Drawer fronts

Drawer fronts can be of two shapes, flush and over-width. The choice is dictated by the desired appearance of the drawer and by the way it is mounted.

The *flush front* is the type used on traditional drawers mounted on rails, and also sometimes with the cleat and groove method. A flush-fronted drawer of a typical design is shown in Fig. 5.

The drawer is simply a plain square box, and the front does not project beyond the sides. This allows any inaccuracy in the fit of the drawer to show clearly, so it must be made carefully. It is also rather hard to attach the drawer front both strongly and in such a way that the joint does not show from the front; see below for more on this point.

The *over-width front* is the type generally used on modern furniture. An example is shown in Fig. 6. The front of the drawer projects beyond the sides, conveniently concealing the drawer runners as well as any inaccuracies in fitting. It is also very easy to conceal the way the drawer front is fixed on, because the joints are not at the corners and are therefore less visible.

Drawers with over-width fronts must be used if you want your furniture to have a modern look, with the drawer fronts butting edge to edge and no frame showing between them.

Some over-width fronts are not an integral part of the drawer frame. A plain box-shaped drawer is built, and a "false" over-width front attached to it, as shown in Fig. 7. This method is a very simple one for the beginner, as the drawer can be made and fitted in its framework, and the front attached at the last moment to get it in exactly the right place. It is also useful for drawers with fronts made out of pre-decorated materials, such as veneered or laminated particleboard. The front can be glued and screwed on working entirely from the inside, without marking the decorative covering.

Drawer bases

Drawers generally have bases made out of a single sheet of plywood of a suitable thickness. Hardboard can be used in small drawers carrying light loads; drawers in antique furniture have solid (though thin) wood bases.

Bases can be attached in various ways. The simplest, but weakest, method is suitable for drawers with false fronts. The base is cut to the width and depth of the *outside* of the basic box, and then glued and nailed firmly in place before the false front is attached. (Fig. 7).

Drawers mounted by the rail technique, and drawers that will carry heavy loads, must have the base recessed a short way up the sides to allow for a slight sagging of the base when the drawer is full. The simplest way to do this is to cut the base to the *inside* measurements of the drawer frame and support it on a strip of wood glued and nailed around the inside of the frame, near the bottom edges.

For extra strength, make the base a little larger and set it in grooves cut in the inside of the frame. This will involve you in more work than the other methods.

The friction problem

Drawer runners, whether they are cleats, grooves or the lower edges of the frame, have to be smooth so that they slide easily, and at the same time must be tough enough to resist wear. For this reason, the choice of materials is important.

The ideal material for all types of drawer runner is hardwood with the grain running horizontally. This can be sanded to a beautiful smoothness and resists wear well.

Ordinary plywood must never be used for cleats, or for drawer sides if they are grooved or mounted on traditional rails. This is because at least one veneer of the ply will present its end grain at the edge. End grain cannot be smoothed properly. Other types of manufactured board, such as particleboard, are obviously quite unsuitable.

If you are painting drawers you should never paint the sliding surfaces—although you may wish to paint the sides of the drawers. Paint on the runners prevents the drawers from opening smoothly. If any drawer sticks or becomes difficult to move; use one of the modern spray type drawer lubricants on both runners and rails.

Joints

Drawers can be assembled with many types of joint, and as a general rule you will want to choose the simplest one that will give you the necessary strength. The joints here are described in increasing order of difficulty. All are shown in Fig. 8.

The *butt joint* is the simplest type of joint to make, and quite adequate for most drawers if reinforced with glue and nails or screws. Extra strength can be added by fastening a piece of quarter-round molding down the internal corner, though this looks rather messy.

Butt joints are simple to make at the back corners of a drawer. At the front corners, however, there is a problem. If you make a butt joint where the front is the full width of the drawer (or wider) and the sides are attached to it from the back, it will be hard to nail or screw it on invisibly and at the same time give it enough strength; a firmer fastening will make a mess of the front. If, on the other hand, you sandwich the front between the sides, you will achieve the necessary strength but there will be two areas of end grain showing at the front of the drawer in a most unsightly manner. This does not, of course, matter if the drawer is to have a false front, but otherwise it looks terrible.

The *end rabbet* solves the problem neatly and without loss of strength. It is only suitable for flush-fronted drawers. The rabbet down each side of the front should not be made with a rabbet or combination plane, since it is hard to use these tools across the grain. You will find it much easier to make it with two cuts of a back saw.

The *housed dado joint* is a good way of fitting an over-width front directly to the sides of the drawer (instead of using a false front). The dadoes are cut down the rear edge of the front and the sides inserted into them. This joint does not have much resistance to pulling apart, however, and must be carefully reinforced with glue and skew-nailing from the rear, or screws passed through from the front with their heads filled over.

The *through multiple dovetail joint* is very strong and resistant to pulling apart. It is the classic joint used in drawer-making, and is nearly always found in antique furniture.

It is also very laborious to make, and there is not much point in doing it unless exceptional strength is required. Instructions on making a through dovetail are in CHAPTER 11.

Through dovetails are suitable only for the back corners of drawers, since they would be unsightly at the front. If the drawer is to have a false front, this does not matter, of course. One solution for the front corners of a flush-fronted drawer would be to use *lapped dovetail joints* (illustrated at No. 24 of the data sheet on page 41). These are not much harder to make than through multiple dovetails and have one plain face.

Sample drawers

The three drawers illustrated in Figs. 9, 10 & 11 and on the first page of this chapter are typical examples of types of drawers you might wish to construct. The first one is the easiest and simplest, and the third is the most difficult to make.

The first drawer (Fig. 9) is extremely simple but reasonably strong. It would be an ideal drawer for use in the kitchen cabinets described in CHAPTER 27, or for any other use where meticulous craftsmanship was not called for.

The basic frame of the drawer is made with butt joints; those at the front are hidden by a fascia made of plastic-laminated particleboard. The ply base is simply nailed on to the underside.

This drawer is made to be mounted on three cleats, but since it has solid wood sides of reasonable thickness, the cleat and groove technique would be just as suitable.

The second drawer (Fig. 10) is a more advanced drawer of the traditional flush-fronted type used in chests of drawers. The front is joined to the sides with rabbet joints; the back is dadoed into the sides.

This type of drawer would normally be mounted on rails, so it does not have any separate runners or grooves. The base is recessed and supported on wood strips running around the inside of the frame.

The third drawer (Fig. 11) is a high-quality drawer of a type suitable for modern furniture. The over-width front has the sides dadoed into it and fastened with glue and skew-nailing. The rear corners have dovetail joints for extra strength.

The base is recessed and supported by grooves cut into the sides and front. At the back, however, it is nailed on to the underside of the frame, which is shallower at the back than the sides. This arrangement allows the

9

10

11

Fig. 9. *Exploded view of a drawer with an over-width false front, and designed simply to be screwed together.*

Fig. 10. *The parts of a traditional flush-fronted drawer, showing the rabbeted joints at the front and the dadoes at the rear.*

Fig. 11. *Exploded view of a higher-quality mahogany drawer, with the sides dadoed into the front and dovetailed rear corners.*

base to be slid in after the frame is assembled. You have to be careful in marking out the dovetail joint to ensure the groove for the base passes through a tail and not a pin. This would weaken the joint and spoil its appearance from the side of the drawer.

Another point to watch when making this type of drawer is that you should cut the base to the width of the inside of the drawer frame plus *twice* the depth of the grooves, and to the length of the inside of the drawer frame plus the depth of *one* groove *and* the thickness of the back.

General instructions

The procedure for making different types of drawers will, of course, vary slightly, but the following general procedure should be helpful if you remember to include the modifications for your particular type of drawer.

1. Measure, cut and plane the front and back to size. Next, cut and plane the sides to length. Plane the top edges last. Check that the front and back are identical, or so constructed as to ensure parallel sides in the case of drawers with over-width fronts.

2. In making the sides remember that their length depends on the type of joints being used. Check that they are identical. Mark the sides "left" and "right" to avoid confusion.

3. Cut all necessary joints and grooves in the frame pieces.

4. Trial-assemble the frame, making sure that it fits its intended position (remember to allow for the thickness of any cleats you are using).

5. Mark out and cut the base, and check that it fits the framework. If the base is mounted in grooves, check that it will slide in without catching on the back of the frame.

6. Assemble the frame permanently with glue and nails or screws. Check it for squareness by measuring the diagonals and let the glue dry thoroughly.

7. Plane down and sand any rough spots on the framework. The best way to support it is on a wide board clamped horizontally to your work bench (this is why you have not put the base on yet).

8. Fasten the base in position.

9. Fit any necessary cleats to the piece of furniture to which the drawer will be fitted. Make sure that they are absolutely level and parallel before you insert the drawer.

Design points

As you have already gathered, there is a huge variety of ways of making drawers. By bearing the following points in mind, however, you should be able to choose the right type of drawer and mounting for your needs.

Most drawers are purely functional and only the front is ever noticed. As long as the drawer is strong enough for the job, the simpler the design the better—unless you actually *enjoy* making dovetails.

Make sure that the features you choose for your particular drawer do not clash with each other. For example, if the drawer is mounted on rails in a framework, do not fit a flush bottom, which will certainly cause the drawer to stick.

Equally, do not mount a flush-fronted drawer by the three-cleat technique, since this will leave large and unsightly gaps on either side of it.

When you are measuring the size of the space into which a drawer will fit, and working out the size of the drawer, remember to allow for the thickness of all cleats and the depth of the grooves, plus at least ⅛ in. clearance to allow the drawer to expand as the wood absorbs moisture from the atmosphere in wet weather.

There are other points to watch, of course, but most of them are common sense. In fact, drawers are easy things to make provided that you don't allow yourself to be seduced into making complicated joints that you don't need.

12

13

14

15

16

17

Fig. 12. *Cutting dadoes with a router plane, which must be sharp to cut across the grain. It should be held with both hands and pushed away from the body.*

Fig. 13. *A bevel gauge and marking knife are used to mark the slope of the dovetails at the rear corners of the mahogany drawer.*

Fig. 14. *Using a narrow bevel-edge chisel to clear the slots in the dovetail after they have been roughly sawed out. The chisel is gripped with the left index finger to brace it.*

Fig. 15. *The profile of the "tails" of the joint is marked directly on to the other half to give the size and position of the "pins."*

Fig. 16. *Skew-nailing the drawer sides into the front dadoes. The nails must be at a 45° angle, or they may come through the front. Pre-drilling the holes makes the job easier.*

Fig. 17. *Strips of wood are nailed around the inside of this drawer to support the base.*

CHAPTER 15

Wall hung kitchen cabinet

The wall hanging cupboard is one of the most useful items of furniture, creating extra storage space at a level which is easy to reach, while preserving the working area of the floor.

This unit is of comparatively simple design and can be made by anyone who has had some experience in carpentry. The main wood used is board grade pine with plywood doors faced with sheet acrylics. The thickness of the planking gives the cupboard a bold styling, although this is softened to a certain extent by the light color of the wood.

Materials and construction

The cupboard assembly, shelves and horizontal partition (those immediately over the drawers) are all made from 1 x 12 white pine. The drawers are constructed from 1 x 3 white pine. The doors are of ½ in. plywood, faced with ⅛ in. plastic laminate. A ⅜ in. plywood sheet lines the back of the shelving and the bottoms of the drawers.

The cupboard shelves are mounted with standard hardware fittings which allow them to be adjusted in height. There are many similar fittings on the market. If you want the shelves to be a permanent fixture, you can house them into the sides and vertical partitions.

The drawers, as shown in Fig. 2, are inset ½ in. at the sides as part of the design, but if you prefer, the sides can be constructed the full width of the drawer recess. In this case the guide strips along each side are not necessary, but the drawers will need to be precision-finished to avoid jamming.

Handles for the doors and drawers are cut from thick dowels—¾ in. for the drawers and 1⅜ in. closet pole for the doors. These are approximate sizes; it does not matter if they vary slightly.

The doors shown here are both fitted with brass back-flap hinges, two to each door, and closed by means of magnetic catches. If the door (or doors) opens from the side, as shown on the left-hand side of Fig. 1, no other hardware is necessary, but if you want a drop flap (Fig. 1, right side), some form of stop must be fitted to hold the door horizontal. In this case the simplest method—two lengths of chain—is used, but there are various manufactured items on the market for the purpose that you may prefer.

Four types of joints are used in the construction; *blind doweling* (Fig. 2C) which consists of five dowels, inset into the butted ends of each corner of the assembly at approximately 2 in. intervals, and at the four ends of the vertical partitions that butt on to the top and bottom; a *stopped dado* joint, which insets the horizontal partitions in the vertical partition and sides; a *mortise and tenon* for each corner of the drawers; and a *rabbet* around the rear inside edges of each side partition and the bot-

NIGEL MESSETT

Fig. 1. *The finished cupboard. Note the two different types of hardware.*

tom inside edges of the drawers to house the plywood bases.

Making a doweling template

Although a blind doweled joint is not intricate, it does require care and precision if you do not want to end up with an out-of-square cupboard. For this reason a doweling template must be made so that each dowel will locate exactly into the drill holes on each end of the board.

The dowel template is shown in Fig. 3. Note that in Fig. 3A it can be used to drill the dowel holes for both the end grain and inside edges of planking, while in Fig. 3B the template has been turned upside down and in this position it is used to drill the dowel holes in the top and bottom of the assembly (which forms a T-joint with the vertical internal partition panels).

The template is constructed from: a piece of rigid plywood (at least 3⁄16 in. thick) or hardboard, 13 in. long and about 2½ in. wide; one length of 1 x 1 or the same length as the plywood; another piece 1 in. long, to butt against one end of the long piece as in Fig. 3B; and one piece 2 in. long for the "underneath" of the template as in Fig. 3A.

The template holes must be marked out and drilled from the B side. Mark a line, exactly ⅜ in. inward from the inside of the long piece, and drill ¼ in. holes along the line at 2 in. intervals. If possible, drill all holes with a drill press.

Assembling the cupboard

The various parts, as indicated in the parts list, should be cut from standard lengths so as to minimize wastage. If you have done this, you can trim each piece as the unit is assembled so parts fit as shown in the drawing.

First cut the top, bottom, sides and internal vertical partitions to size. The sides and vertical partitions must be identical in length, so clamp these four pieces together when measuring. To ensure a perfect fit, all ends—and this applies throughout—must be absolutely square, so use the try square frequently to ensure that all cut ends are at right angles to the edges. Trial assemble the horizontal members and vertical partitions (this is quite easy with lumber of this thickness) to make sure that the fit at each butt is flush all along the joint.

Using the dowel template, drill ¼ in. dowel holes at each joint shown in Fig. 2C, *but only drill outside dowel holes* shown in Fig. 2C. Leave the other three dowel holes until later. You will now have drilled holes to take 16 dowels, leaving 24 to drill later.

Assemble the components cut so far by inserting a dowel—dry, not glued—into each recess and tapping the frame together. Use a block of wood under the hammer head when doing this so that the surface is not marked.

When the basic frame is trial-assembled, you can easily see that everything is square by measuring the diagonals.

Next mark out the dado joints for the fixed horizontal partitions shown in Fig. 2A. To do this, take one of the lengths of lumber that will be used for the drawer fronts, and lay it along one of the bottom corners as shown in Fig. 6. This will enable you to draw a line along the top of the drawer side to mark out the bottom limit of the dado joint; the pencil thickness should give you clearance between the drawer and shelf. Repeat this at all six of the bottom corners.

Gently disassemble the frame with a hammer and wood block. The "top" sides of the dado joints can now be marked. Use the direct marking system by placing the edge of one of the boards to be dadoed along the line previously drawn, then mark along the opposite side.

The stopped dado joints can now be cut and chiseled out. Avoid cutting the channel very deep—about ¼ in. is sufficient—and stop it about 1 in. from the front of the cupboard.

At this stage cut out the recesses for the back panels of the side compartments (these are the partitions which have doors). This rabbet should be cut ¼ in. along the edge of each board, and ¾ in. along the side or into the cupboard around the rear inside edges of both partitions. This will inset the back panels about ½ in. to allow room for the wall fastenings. Do not carry the rabbet where the drawer will be.

Drill out the remaining dowel holes. Put a little adhesive in each dowel hole, spread a little on each dowel, place all the dowels in their recesses along the top and bottom planks, and ease the four vertical panels into position. When the whole frame is fitted, clamp it from top to bottom, square it if necessary, and leave on a level surface to set.

When the adhesive has set on the frame, cut the three horizontal panels to slot into the dado joints. Use the direct marking system for measuring by placing each panel at the back between the two dado channels. Then mark and cut the panels to size. Slide each panel into the channel to the point where the channel "stops." Go around to the front and mark both ends of each panel where it butts against the sides, withdraw the panel, and you have accurate marks for cutting the stopped dado recesses.

Spread adhesive along the dado channels and the ends of the horizontal panels, slide into place, wipe off any excess glue and leave to set.

The doors can now be cut to size and planed to a precise fit. As the plywood for the doors has been cut slightly oversize, you can adopt direct marking to ensure an exact fit. Place one of the doors over the front of the partition to which it will eventually fit. Holding it in position, gradually lay the cupboard over on to its front; the partition edges will now be resting directly over the door. Lift the opposite end of the cupboard slightly and ease the other door underneath so that it covers the edges of the partition at the other end. When they are placed correctly, you can mark a line around the back of the doors from the "inside" of the cupboard. Saw along the waste or outside of this marked line using a fine-toothed panel saw and you have your two doors tailored for a perfect fit.

The ¼ in. plywood sheets for the back of the two side partitions can now be trimmed down so that they fit neatly into the rabbeted recesses; and secured with ¾ in. brads at 2 in. intervals.

The doors in this unit are covered with plastic laminate sheet, although there is nothing to stop you adopting some other covering such as a matching pine veneer. If you want to get a similar, but cheaper effect to the cupboard shown, sandpaper the door fronts down to a satin smooth surface and apply several coats of hard gloss paint, rubbing down with extra fine steel wool after each coat has set hard.

If you use laminate use direct marking again. Buy the laminate slightly oversize. Lay a door down over one of the laminate sheets and mark around the edges with a marking pencil. Saw carefully around the waste side of this line with a fairly fine-toothed saw such as a panel saw. If you are used to working with plastic sheeting, you might, with luck, be able to achieve a perfect fit first time. But if this is your first attempt, cutting around the waste line will leave a fraction over all around. Glue the oversize plastic sheet to the door and, when the adhesive has set, remove the oversize edges with a fine, broad flat file. For gluing, use only the proper adhesive, in this case contact cement is the usual type.

A hole will have to be drilled through the laminate to enable the screw for the handle to pass through. With patience, this can be done with a hand drill, but it is easier with a power drill. Mark the location of the hole and make a small indentation with a hand drill. Then use an electric drill, if you have one, to finish drilling through the laminate and the plywood. When the laminate has been penetrated, continue through the plywood. Countersink the hole inside the door, hold one of the door handles over the hole in the laminate, insert a wood screw from the inside or plywood side and screw into the handle. It is useful to fit the handles at this stage because they provide something for you to hold on to when you are fitting the hinges.

Cut the recesses for the back-flap hinges, and fit the hinges. If you are including a drop-flap door, fit the chain or stays, and the magnetic catches.

MATERIALS LIST

Part	Description	Quantity	Length	Nominal size
Sides	pine	2	21 in.	1 x 12
Interior verticals	pine	2	21 in.	1 x 12
Top and bottom	pine	2	6 ft.	1 x 12
Upper end-shelves	pine	2	23½ in.	1 x 10
Lower end and center shelves	pine	1	23½ in.	1 x 12
Rear panels	plywood	2	23½ in. x 17¾ in.	¼ in.
Doors	plywood	2	22 in. x 17 in.	½ in.
Door covering	acrylic sheet	2	22 in. x 17 in.	⅛ in.
Door handles	closet pole	2	2 in.	1⅜ in. diam.
Hinges	back-flap	4	—	1½ in.
Catches	magnetic	2	—	—
Chain or stays	(see text)			
Drawer fronts	pine	3	23 in.	1 x 3
Drawer sides	pine	6	9 in.	1 x 3
Drawer backs	pine	3	22 in.	1 x 3
Drawer bottoms	plywood	3	22 in. x 9 in.	¼ in. ply.
Drawer handles	dowel	3	1 in.	¾ in.
Drawer stops	softwood	6	8 in.	½ in. x ½ in.

Assembling the drawers

First cut and plane the drawer fronts so that they fit snugly into their respective compartments.

Next cut the side and back timbers for the drawers.

Plow a groove along the bottom inside edges of all the drawers. The groove should be slightly over ¼ in. deep so that the plywood drawer bottom will slide easily into place later.

Cut a tenon into both ends of each of the drawer sides. Set the marking gauge at ¼ in. and scribe a line around all four edges of the end grain. Now mark a line, ¼ in. deep around all four sides of the same piece of wood. Place one of the sides in a vise, end grain uppermost, and with a back saw cut out the tenon.

Repeat at the other end.

When all the tenons, 12 in all, have been cut, mark out the mortises on the inside faces of the front and back pieces. Place a back or front piece on edge on a level surface, butt a tenon against it where it will recess, and mark the mortise outline using the tenon as a template. Chisel out the mortises to a depth of ½ in.

Fit all the drawer handles using the same procedure as for the doors. Again, this is done for convenience in handling. Cut and fit the plywood bottoms into their grooves. Apply adhesive to the mortise recesses and smear some around the tenons, fit the joints and clamp them until the glue has set.

The last step is to fit the side battening into the lower drawer partition corners as shown in Fig. 2. The purpose of this battening is to act as a stop so that the drawers will not slide right through the partition, and also to ensure that the drawer front pushes in flush with the edges of the cabinet.

First, insert all the drawers in place and then, from the back, push a length of batten along each side to ensure that there is suf-

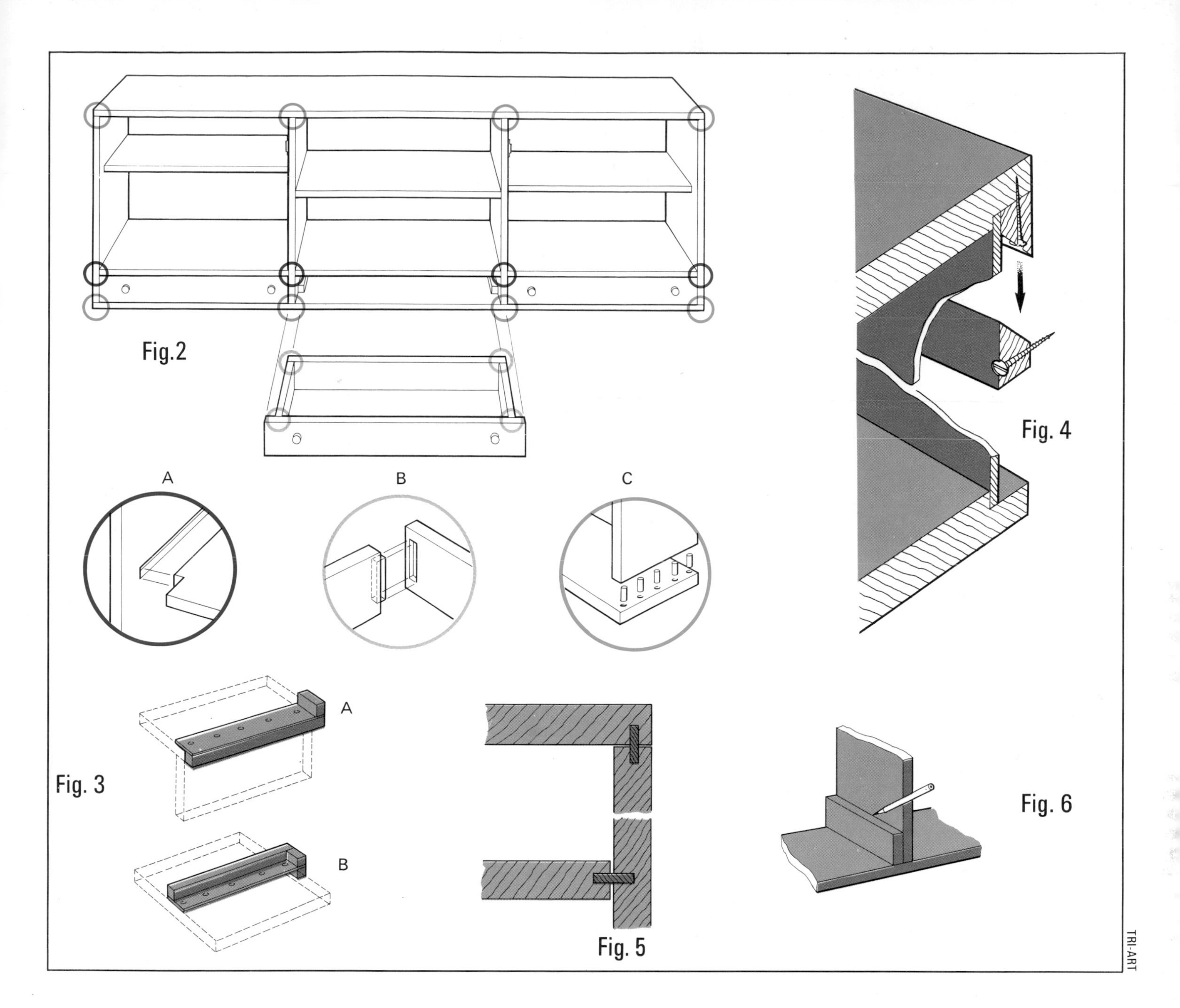

ficient space for them. If there is not, plane the strips down to size. Pull one of the drawers out, and squeeze a little adhesive into the corner where the strip will be fitted, but do not spread any glue along about 2 in. of the corners at the front of the cupboard. Press the two strips in position until they are almost, but not quite, flush with the front of the cupboard. Now ease the drawer in until it is flush with the front. This has the effect of pushing the strips back to their correct positions. Pull the drawer out gently and allow enough time for the adhesive to set. If you can manage it, drive a couple of brads through the strips into the sides of the vertical panels to "clamp" them in place while the glue is setting. (A piece of hardboard on the "floor" will prevent the hammer from marking the lumber.

Hanging the cupboard

There are many types of hardware on the market for attaching cupboards and shelves to walls, but if you do purchase one of these remember that this cupboard is very heavy and needs a heavy-duty support. It is for this reason that the back panels have been rabbeted ¾ in., so that at least ½ in. is available for attaching a support.

You can make a wall fitting yourself, and the one used in this project, consisting simply of a length of lumber cut slantwise so that one half fits the cupboard and the other fits the wall, is shown in Fig. 4. It is a very simple one and has the advantage that the cupboard can easily be removed by merely lifting it off the wall support.

Fig. 2. *The cupboard, minus doors, showing the construction. Inset A is a stopped dado joint, B is a mortise and tenon joint and C shows five blind dowels.*
Fig. 3. *The doweling template used to ensure that the doweling holes match accurately. A shows the template being used on the ends, and B, on the middle of a board to mark out the holes.*
Fig. 4. *One method of attaching the unit to the wall. This method uses a diagonally cut batten, but there is a variety of mounting hardware on the market that would do equally well.*
Fig. 5. *Side section view of the blind doweled joints at the shelf end and corner sections.*
Fig. 6. *Marking out the dado joints for the fixed horizontal partitions.*

CHAPTER 16

Hanging doors and windows

Doors and windows have two purposes: they open to provide access or let in air, and they shut to keep out rain, drafts and sound. So they must open easily and shut firmly, which means that they must be fitted very accurately into their frames.

Hanging doors and windows is not particularly difficult, and there are only a few facts and procedures you need to know about to get it right every time.

The most important part of the job is getting the door or window to fit properly into its frame. They do not fit exactly; there has to be a clearance all around to allow for expansion in wet weather and to keep the door from scraping against its frame when it is opened and shut.

Doors do not open straight out as if they were being lifted out of their frame. They swing slightly outwards as they pivot on their hinges, and the lock side of the door, opposite the hinge, has to be cut away to allow for this movement. Different kinds of hinges cause the door to move in different ways, and you have to know how a particular type of hinge will make the door behave.

The hinges themselves have to be installed strongly but accurately, so that the door does not sag or hang at an angle. It is not hard to put them in correctly when you know how.

Fitting doors and windows

These instructions apply to doors and windows alike, except where stated.

Doors (and wood-framed windows) invariably expand slightly when the humidity rises. If they are painted, or sealed in some other way, it slows down the rate at which the humidity affects them, but they still change size.

Outside doors are obviously more affected than inside ones, but even so, inside doors in a house that is centrally heated during the winter or air-conditioned during the summer, may change their size quite noticeably from season to season, perhaps by as much as 1⁄16 in.

Paneled doors made of solid lumber move more than modern ones made from manmade boards. Softwood ones move more than hardwood ones. And of course, wide doors move more than narrow ones (in actual distance, not in proportion to their width).

One way to estimate the right clearance is to leave a minimum of 1⁄16 in. around the top three sides of the door, and at least 1⁄8 in. at the bottom (but more about clearance *under* doors later). Then you can add to this figure 1⁄32 in. for each factor that might make the door expand an extra amount—for example, if it is an outside door in a centrally heated house, or if it is made of solid lumber panels. The average clearance for an inside door is about 1⁄8 in. on the top three sides, and 1⁄4 in. at the bottom.

There used to be a joiner's rule of thumb that you should be able to get an old penny (just over 1⁄16 in. thick) between a door and its frame all around. This was before central heating, however, and the gap is a bit small by today's standards. But it is a good way of measuring if you can find a coin of the right thickness for the clearance you need.

The method for fitting a standard-sized door into a standard-sized frame is as follows: first buy a door of the correct size and leave it in the room where it is to be fitted for a couple of days to let it adjust to the prevailing humidity. New doors with solid lumber rails and stiles (this includes many flush doors) sometimes have the top and bottom ends of the stiles (vertical side pieces) left uncut, and these projecting *horns,* as they are called, should be sawed off and planed smooth.

When the door has acclimatized itself, hold it up to the hinge side of the door-frame and examine the edge of door and frame to see if they are parallel. Many frames are leaning or curved, even in new houses. If the frame is really irregular you may have to scribe the door to it, but otherwise an ordinary plane and your own judgment should be enough to make it fit. Make sure that you don't remove so much more wood from one end than the other that the door is no longer vertical; an occasional check with a spirit level should guard against this.

If the door is not too tall to fit into the door frame at this stage, hold it up with a couple of wedges underneath it while you check it against the frame. If it is too tall, rest it on the floor and put wedges under it only after it has been trimmed to the right size and shape (see Fig. 1).

After the hanging edge of the door is trimmed to fit, do the same with the top, then the bottom. Then raise it to its correct height on wedges, and trim the last side, which is the side with the lock.

Finally, run a coin around the top three edges of the door to check that there are no tight spots.

Non-standard sizes

Most door frames in houses less than about a century old are in standard sizes, and you can buy doors that only need a little planing to fit into them. If, however, you are unlucky enough to have a door frame in a non-standard size, you are going to have to alter your new door more drastically than by merely planing it.

Nearly all doors, including flush (flat-surfaced) modern doors, have some kind of supporting framework of stiles and rails to strengthen them and keep them from sagging or warping. This is held together by mortise-and-tenon or doweled joints. If you remove more than, say, 1⁄2 in. from any side of the door, there is a danger of cutting too deeply into one of these joints, and this could seriously weaken it. Check on the trimming limits where you buy the door.

Another way around this problem is to order a specially-made door, but this might turn out to be expensive. Ask a building supplier and see if you think the job is worth the money. If not, you could just buy a large piece of 3⁄4 in. blockboard, apply an edging strip all around and use that as an interior door. It might, however, warp a bit because of not having a proper frame.

The best solution is to dismantle a door and re-cut the joints to alter the size—if you can be bothered to do it. The technique for an ordinary paneled door is described fully in CHAPTER 26. It is a laborious job, however, for door manufacturers use good glue and strong joints deliberately to stop their doors from coming apart.

Altering the size of window frames in this way is nearly impossible for the amateur. They will almost certainly have to be made to measure by a professional.

Fitting the hinges

Nearly all doors and casement windows are hung on butt hinges (see Fig. 2). These come in a good range of sizes from 1 in. long, used for cupboard doors, to 6 in. long, used for heavy front doors. Most ordinary-sized doors use 4 in. hinges. For really large doors—the front doors of some Victorian houses are a case in point—you can use three hinges instead of two, with the third hinge halfway between the other two. This also helps to prevent warping.

To attach a pair of hinges, first position the door in its frame in the exact position it will occupy, propping it on wedges to held it steady. Then make a mark on both door and frame (and at the same level on each) 6 in. from the top of the door and 9 in. from the bottom. For small casement windows, halve these measurements.

Then take down the door and draw around one flap of the butt hinge with a marking knife to mark its position on both door and frame. Top and bottom hinges should be positioned *inside* the lines you have already marked. The hinges should be set so that the pivot is just clear of door and frame.

When you are satisfied that the position of the hinges is correctly marked, set a marking gauge to the thickness of the hinge flap and mark the front surface of door and frame to show how deep the cutout for the hinge is to be. Be very careful not to mark it too deep.

Some marking gauges will not adjust far enough to make such a shallow mark. One solution to this is to put in a new metal spike at the other end of the arm of the gauge in such a position that the sliding part of the gauge can move right up to it. You could make a good spike from a small finishing nail sharpened with a file.

Now chisel along all the marked lines to ensure that the wood will be removed cleanly (see Fig. 3). Then turn the chisel bevel-edge down and make some diagonal cuts to the correct depth to make it easier to remove the wood (see Fig. 3 again). Finish the cutout neatly by slicing out the raised wood chips with the chisel held right way up. Using a broad chisel improves accuracy.

Place a hinge leaf in each of the cutouts to ensure that it fits with its upper surface flush with the edge of the door or frame. If the cutout is too shallow, and the hinge stands too high, remove more wood. If it is too deep, you will have to pack it out with a piece of veneer or ply—but this is *not* recommended and it is much better to cut too shallow and work down.

When all the hinge leaves are set in properly, drill *one* hole through each side of each hinge into the door and frame. Drill through the center hole of each hinge leaf, using an awl on small hinges, a drill of the correct size for the mounting screws on large hinges. Don't drill the other two holes in each side yet.

Prop the door up on its wedges and mount it temporarily on its hinges with one screw per hinge leaf. Then open and shut it—gently, so as not to tear the screws out—to make sure that it is not catching on anything. You will almost certainly find it does catch on the lock side, because the projecting hinge pivot makes it swing slightly wide. The remedy is to bevel the edge of the door slightly with a plane (see Fig. 4). Nearly all doors and wood-framed windows are beveled in this way.

Open the door as far as it will go to make sure that it does not catch on the floor. If it does, you will have to plane a bit more off the bottom edge.

The door should fit flat into its frame and stay there without having to be held. If it sticks and has to be forced, or swings open of its own accord, the screw holes are wrongly sited. Take the door off and plug the misplaced holes with glued-in dowels. Let the glue dry and try again.

When you are satisfied with the fit, take the door down, drill the remaining holes and refit it with all its screws. Make absolutely sure that all the countersunk screw heads are in the whole way, or the hinge will not fold flat. If they stick out, file the heads flat—this is cheating, but it works. The installation is now completed except for any locks or bolts. These come in a multitude of shapes, sizes and types, but any good-quality lock comes with instructions.

Other types of hinges

Special problems in fitting doors and windows may call for special types of hinges. For

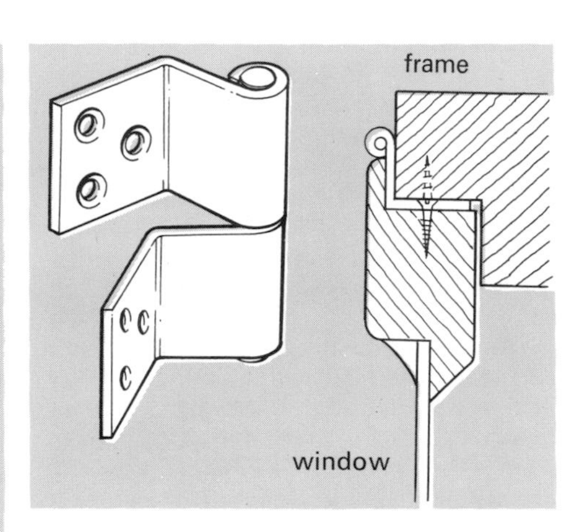

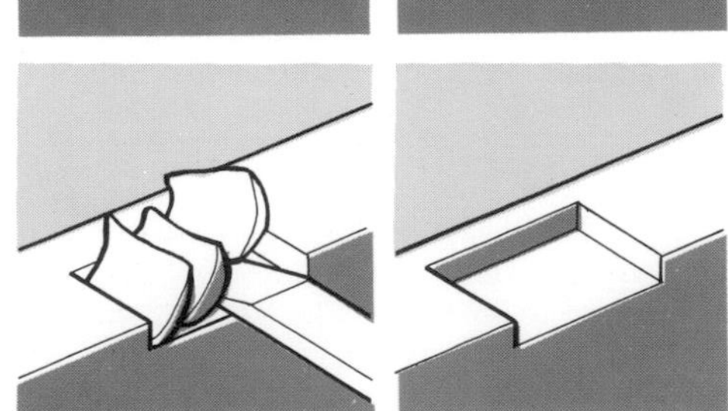

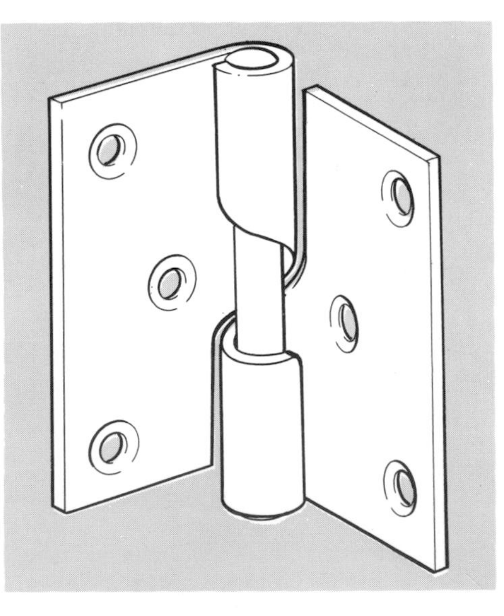

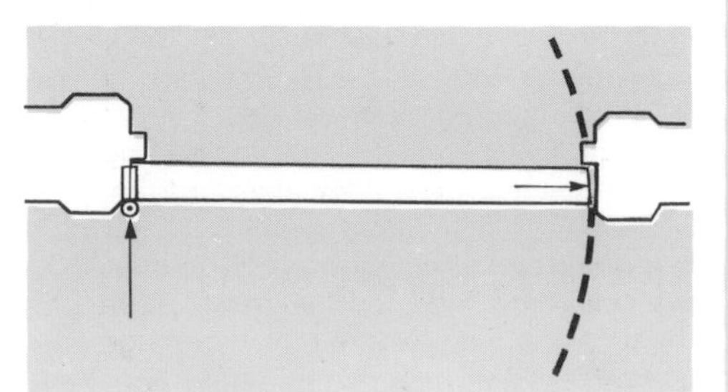

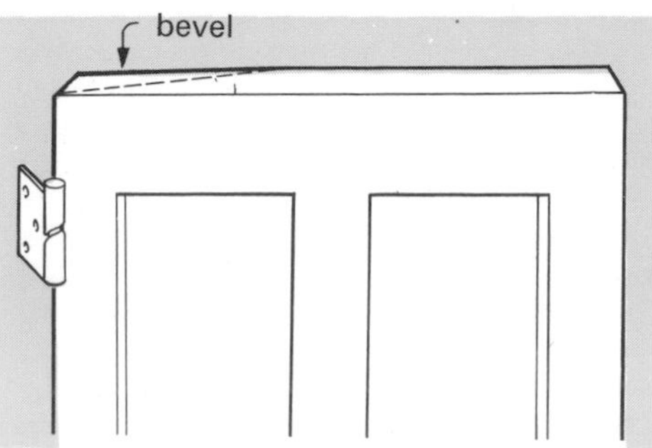

Fig. 1. *A door propped up on wedges to bring it up to the right height while it is being fitted into its frame.*
Fig. 2. *A typical butt hinge. The screw holes are offset to keep the screws from splitting the wood of the door and frame.*
Fig. 3. *Cutting in a butt hinge. First the outline is cut, then the wood is chiseled out to leave a flat bottom.*
Fig. 4. *The edge of a door must be beveled, because the position of the hinge pivot makes it swing outward.*
Fig. 5. *A stormproof hinge, which fits into the L-shaped space between some types of casement window and their frames.*
Fig. 6. *A rising butt hinge, which raises a door a short way as it is opened.*
Fig. 7. *The top of a door on rising butts must be beveled to allow it to clear the frame as it rises during opening.*

TRI-ART

example, a door or window that has to fold flat against a wall needs a special type of hinge, known as a "parliament" hinge. This has an offset pivot that moves the door well away from the frame as it opens, allowing it to be opened much farther than normal. It makes the door swing very wide in the first few inches of opening, so that the lock side needs a heavily angled bevel cut on it.

Casement windows often have a rabbeted edge and frame to keep the rain out. These must have a special L-shaped hinge sometimes called a "stormproof" hinge (see Fig. 5). Unlike butt hinges, these come in left- and right-handed versions, depending on which way the window opens (though of course, they are generally bought two pairs at a time for paired windows). They are installed in the same way as butt hinges, but the window frame can be a looser fit than usual, since the rabbeted front seals it.

In houses with irregular floors or thick carpets, there is often a problem with the door catching on the floor as it opens. If enough wood is taken off the door to clear it, the wind whistles through the huge gap underneath. The solution is to use "rising butt" hinges, which raise the door as it opens (see Fig. 6). To stop the top of the door from catching on the frame as it rises, a special tapered bevel has to be cut along one-third of the length of the top edge of the door at the hinge end (see Fig. 7).

The only way to get the shape of the bevel right is by trial and error, removing a very little wood at a time. The length, angle and depth of the bevel vary with particular installations. The hinge is installed like a normal butt hinge, so if you hang the door temporarily on single screws and keep taking it off, planing a bit more wood off and replacing it, you should soon get the shape of the top edge right.

Certain types of hinge are installed in plain view on the front (or back) of the door. These include the long sheet-metal hinges found on shed and outbuilding doors, and special self-closing spring hinges, which have such huge pivots that it would be impossible to hide them. Both types are very easy indeed to install, because the screws are exposed and you don't have to keep taking the door off to get them positioned correctly.

Above. *These two doors must be hung at exactly the same height to avoid giving the installation an untidy appearance.*

CHAPTER 17

Louvered doors and shutters

If you are looking for a way of brightening up your doors and cupboards, why not try louvers? Don't be put off by their complicated appearance. There is no reason why you shouldn't make a whole roomful of louvered doors in the space of a single weekend.

Louvers are an elegant feature in almost any room. They can be used in inside doors or cupboard doors, and are most attractive in shutters. The loudspeaker cabinets of stereo systems can also be faced with louvered panels, which let the sound through but hide the uninteresting cloth front that most cabinets have. The only place where louvers are not suitable is outside doors (for obvious reasons), but even then you could draftproof the door with a glass panel.

It is, of course, possible to buy ready-made louvered doors and shutters, but these are fairly expensive and come in a limited range of standard sizes. Making a louvered panel is not at all difficult, and although it can be a long job setting in each slat, this article will show you several short cuts that reduce construction time to a minimum.

Strictly speaking, a louvered door should be made by setting each slat individually into a pair of stopped dadoes cut at an angle into the inside edge of the door stiles (see Fig.1). You can imagine the time it would take to cut out each dado separately and at the right angle. Furthermore, if you cut even one in the wrong place, you would have to throw away the whole door stile, an expensive piece of lumber on which you would already have spent a lot of effort.

The way to avoid this laborious task is to make up a light rectangular frame separate from the door and set the louvers in that. Then the frame can be set in the door or shutter. Doors and shutters, incidentally, are made in exactly the same way the only difference between them is one of size.

DRALON

It is much easier to attach the slats in this kind of frame than in a door stile. They can just be nailed to it through the frame's outer edge. You can't do this with a door stile because it is 3 in.–4 in. wide.

Shapes, sizes and spacings

The appearance of a louvered door can be altered quite a lot by changing the size and angle of the louvers and the way the louvered panel is set in the door frame. Before you start work, it is a good idea to make a scale drawing of the whole door, making it as realistic as possible, to ensure that you are satisfied with the proportions.

The possibilities are wide. For example, the front edge of each slat may be planed off parallel with the face of the door, which gives the louvered panel a smooth appearance (see Fig. 2). Or it may be left at its original right angle to the face of the slat, which intensifies the contrast between light and shade in the panel and makes it look more serrated.

The louvered panel can be set in the door so that its surface is flush with the rest of the door, or it can be set in or made to project. Several possibilities are shown in Fig. 3; the first one, where the slats are flush but their light frame is set back a short way, is probably the neatest as well as the easiest to build. It is proposed to set the slats in housed dadoes to save trouble, so the louver frame should not project beyond the front of the slats or the ugly ends of the dado slots will be visible. This could, however, be overcome by edging the frame with a piece of molding.

It is normal to set louvers with the lower edge of each slat toward you, which makes the panel opaque when seen from above (unless the slats are spaced wide apart). Shutters, however, should be the other way around if they are used to keep out direct sunlight or rain. In this way you will be able to look down through the slats into the garden or street, but people at the same level as you or above will not be able to look into the room. The spacing of the slats can be altered to make them more or less "peep-proof"—see below for details.

The angle of the slats also makes a difference, both to their appearance and to how easy it is to see through them. The nearer they are to being horizontal the easier they will be to see through. An angle of 45° is a good choice, not just because it looks right but because it allows you to mark out the dadoes with an ordinary miter try square. Angles nearer the vertical are often found in ready-made louvered doors, because they save wood. This is not really a problem for the amateur, however, since the saving is only a few cents per door.

If you want slats at an unusual angle for some reason, use an adjustable bevel gauge for the marking out, but otherwise proceed normally.

Opposite page. *An unusual and very successful effect is created by mounting a pair of louvered shutters inside a window frame, where they can be seen when open.*

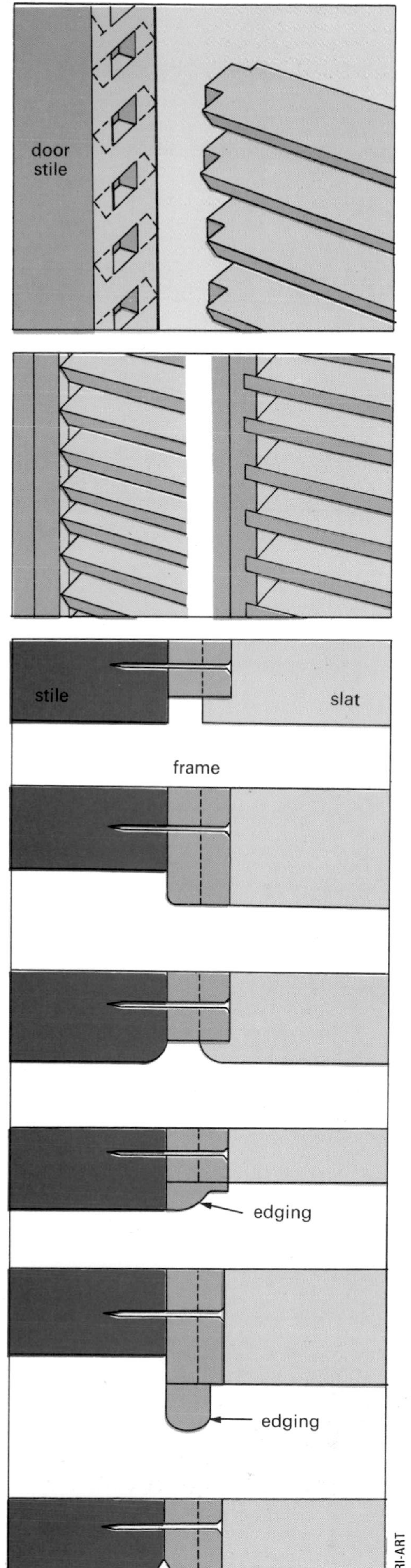

Fig. 1 *(left). Louvers should ideally be made by setting slats in stopped dadoes cut directly into the door stile. Of course, nobody bothers to do this today.*
Fig. 2. *Two styles of louver slats. The one on the left has its frame trimmed with an edging strip; the ends of the slats are planed flat to accommodate the strip.*
Fig. 3. *Slats and their frames can be mounted flush or set forward or back.*

The size of lumber required for the components varies with the use the door or shutter will be put to. A light wardrobe door could have a main door frame with the stiles and top rail made of lumber as light as 1 x 2 in; the bottom rail, which should always be deeper to give the door good proportions, should then be of 1 x 4 lumber. The slats should be of lumber ¼ in. thick—this is a minimum thickness for any application—and 1½ in.–2 in. wide, depending on the design. Material of this thickness, called "lattice strip," is available at lumberyards.

Shutters should have a heavier frame. A hinged, useable shutter might have an outer frame of ¾ x 4 in., including the bottom rail, and ½ in. thick slats, which will be more resistant to warping and accidental damage. The slats should be 1½ in.–2 in. wide, as before.

If you want shutters for decoration only, there is no point in making them as strong as this, or in mounting them on real hinges. It is much simpler and cheaper to make them out of lighter materials, such as those for the wardrobe door, and screw them to the wall on either side of the window. Normally, open shutters would have the upper edge of each slat outermost, but a fake shutter set directly against the wall should be the other way around, both for appearance's sake and to stop it from channeling rain to the surface of the wall and making it damp.

Room doors should be heavier than shutters. Nominal 2 x 4 lumber is about right for the outer frame, with a 2 x 6 bottom rail. The slats should again be ½ in. thick and as wide as is appropriate for the design.

In all these applications, the thickness of the lumber making up the light inner frame that holds the slats should be the same: ½ in. It does not need to be strong, but it will have ¼ in. deep dadoes cut out of it to take the slats, so this is a minimum thickness. The width of the inner frame will vary with the style of louvered door you have chosen.

If you are replacing an existing paneled door with a louvered door, and the old door is in good condition, you might consider replacing the panels in it with louvered panels instead of replacing the whole door. This will give you a smart and unusual louvered door with four panels instead of one or two, and will also save you the trouble and expense of building a whole new door from scratch. If you decide to do this, the best way to remove the old panels is to smash them up with a mallet and chisel and pull the pieces out of the grooves in which they are set. The smashing-up should be done as gently as possible so as not to damage the frame of the door. The grooves where the panels fitted will be covered by the light framing of the new louvered panels, so there is no need to fill them.

The outer frame

If you are making a complete new door from scratch, the first thing to make is the main outer frame. This is a solid but simple structure with fewer pieces than a conventional four-paneled door (see Fig. 4). It has five pieces: left and right stiles and top, center and bottom rails. The bottom rail should be 1½–2 times as wide as the other pieces, as already mentioned.

The four joints at the corner of the frame are haunched mortise-and-tenon joints. This joint is shown on page 39 in the DATA SHEET on joints. The procedure described there, however, is for the slightly angled joint needed for a stepladder frame. Here, a plain right-angled joint is wanted, so no complex marking-out procedure is needed and you can do everything with a normal right-angled try square.

The two joints where the center rail joins the stiles are ordinary through mortise-and-tenon joints, which are described fully in CHAPTER 23. They should be absolutely plain sailing if you follow the instructions there.

When making the door frame, pay particular attention to the squareness and exact size of the two rectangular spaces in the middle of it. You can always plane bits off the *outside* of the door if it doesn't fit its site. But if the louver frame has to be made deliberately out-of-true to fit into a crooked rectangular space, you will have a lot of trouble.

The louver frame

A typical louver frame, consisting of a light ½ in. thick framework and slats of the same thickness, is shown in Fig. 5. The corners of the frame do not need to be strong, so they are just butt-jointed, glued and nailed. The only exception to this would be where a frame of

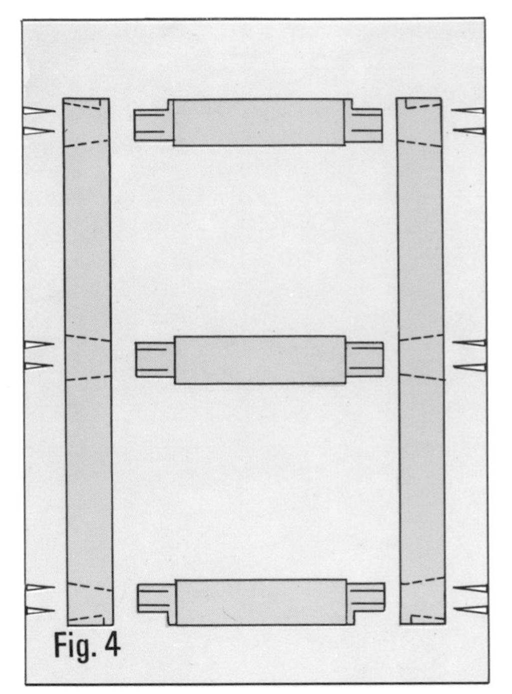

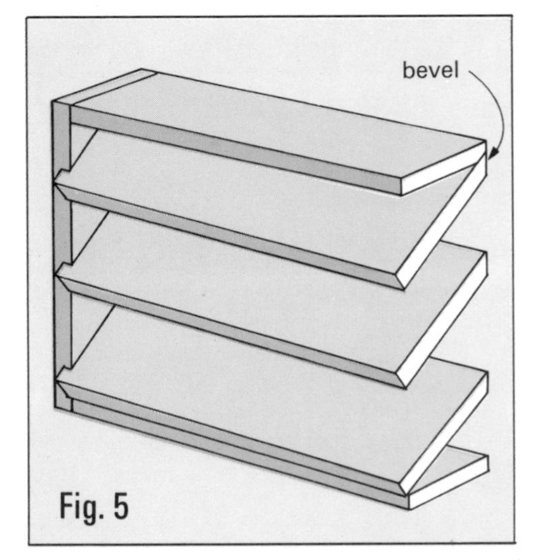

this type has to stand on its own, as it might if it was used for a loudspeaker grille. Then, if you can be bothered to do them, tiny dovetails should be made to hold the frame together more positively (see the detailed instructions given in CHAPTER 11). The tails of the asymmetrical joint should be on the top and bottom of the frame, so that assembling it locks the slats into their dadoes.

Do not make a frame of this type with mitered corners. They are just not strong enough.

The first stage in making a louver frame is to make a really detailed scale drawing, showing all the measurements. This will both tell you how much wood to buy and save you from tiresome and expensive mistakes later.

There is a special technique for marking out louvers to fit into a space of a certain size. It is only suitable for calculation in *decimals,* however, not ordinary fractions. This is because if you start doing the calculations using sixteenths and thirty-seconds of an inch, you will need the mathematical ability of an Einstein to get through them without at least one mistake—and even one will spoil the whole procedure. Furthermore, even if you do avoid making a mistake, the result may well be in sixths or ninths of an inch, which are not marked on an ordinary ruler.

This doesn't mean that you have to work in millimeters if you are unused to them. Use tenths of an inch. Ordinary carpenter's rules don't have tenths of an inch marked on them, but architects' rulers do. They are very cheap and obtainable at art suppliers.

Before outlining the marking-out procedure, three points should be noted. First, the top rail of the louver frame should be beveled at its rear edge (see Fig. 5) so that the top slat can overlap it by at least ²⁄₁₀ in. or ⅕ in. Then, if either piece shrinks or warps, no ugly gap will be visible. If the front of the slats are flush with their frame, there should be a bevel at the bottom, too. All measurements should be made

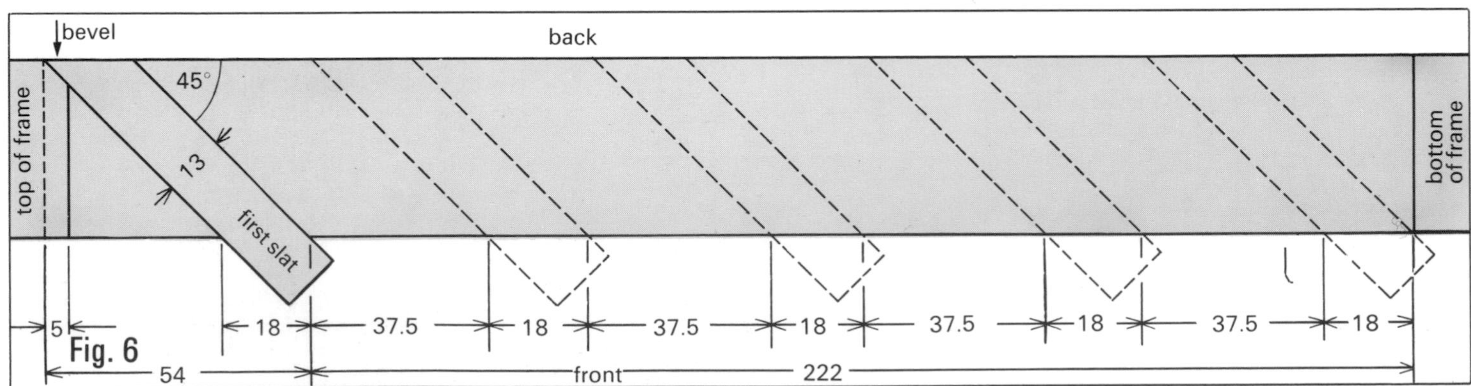

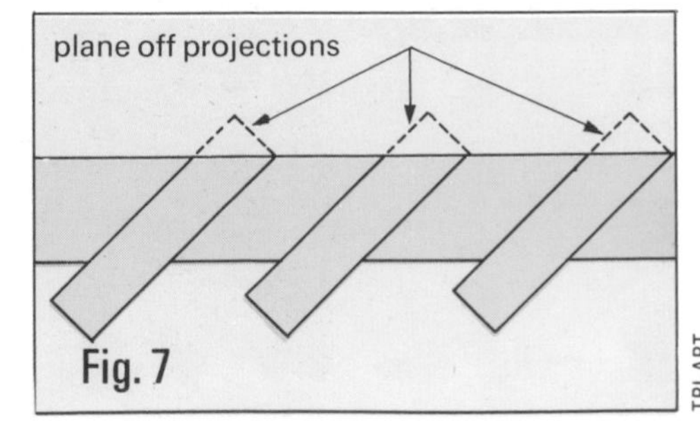

Fig. 4 *(top of page). The simplified door frame in which the louvered panels are set. It has haunched mortise-and-tenon joints at each corner and normal mortise-and-tenon joints at the ends of the middle rail.*
Fig. 5. *A typical louver frame. Note the bevel at the top, which prevents a gap from forming.*
Fig. 6. *A marking-out diagram; its use is described on page 89 (opposite).*
Fig. 7. *The rear edge of each slat should be planed flat after the frame is assembled.*

TRI-ART

from the far side of the bevel, not from the flat inner face of the frame.

Second, a louver frame, unlike any other frame with bars, has the same number of spaces as slats. A glance at Fig. 5 will show you why.

Third, the number of slats needed to fill the frame (this number is given at step 3 of the procedure below) is the minimum number that will make the louvered panel opaque when viewed from straight in front. If you want to make it more peep-proof, add at least one slat to every six that the calculations tell you you will need, and carry on with the subsequent steps using the increased number.

Marking-out formula

1. Make a scale drawing of the frame from the side, using Fig. 6 as an example but altering it to your own measurements. Draw in the first slat, set at 45° or whatever angle you have chosen. Measure its total width along the frame—this is 54 mm. in the example, *including* the 5 mm. overlap where the frame top is beveled, but *excluding* the part of each slat that projects out of the frame at the front.

2. Measure the remaining space from the bottom of the first slat to the other end of the frame (in this case, 222 mm.).

3. Divide the remaining space by the width of the first slat (222 ÷ 54 = 4.11). The result, taken to the nearest whole number, gives you the *minimum* number of slats you will need in addition to the first one (4).

4. Measure the diagonal thickness of a slat by holding your ruler parallel to the frame edge. A slat of 13 mm. (about ½ in.) will measure 18 mm. in this direction if set at 45°.

5. Multiply this measurement by the number of slats still to be drawn in (18 × 4 = 72).

6. Subtract the result from the remaining space (222 − 72 = 150).

7. Divide this figure by the number of slats still to go in to give the width of the spaces between them (150 ÷ 4 = 37.5).

8. Now you know the width of each slat measured along the frame (18 mm.) and of each space between two slats (37.5 mm.). Mark these two measurements alternately right down your drawing to see if the slats fit exactly. If they don't, you have made a mistake. Transfer all these markings to the real frame sides, marking the two sides together to keep the slats absolutely level.

Above. *This door is given a touch of distinction by its polished wood louver panels, which stand out from the contrasting white-painted frame.*

Assembly of the louvers

Now that you know how to set out louvers, everything else is easy. Cut the dadoes as marked, using a back saw and a chisel of the same width as the thickness of the slats. Don't make each dado separately. First make *all* the back saw cuts to a depth of ¼ in. Then chisel the wood out of all the dadoes. This "production-line" technique saves a lot of time.

Fit the frame together dry to make sure that it is (a) square, (b) the same width at top and bottom, (c) the right size to fit in the door or shutter. Measure the internal width between the pairs of dadoes to find the right length for the slats.

Now make up a rough jig to enable you to cut all the slats the same length. A typical jig would be a box like a miter box, whose bottom is made of the same lumber as a slat, so that it is the right width internally. It has a stop at one end and a slot to guide the saw at the other. Cut all the slats to length in this box. Do not

Below. *How to add interest to a hallway: fit louvered doors and panels all around it. The mirror on the right gives the impression that there are even more louvers.*

Below. *The gentle, diffused light from a set of unpainted pine shutters softens the angular lines of the stark modern furniture in this living room.*

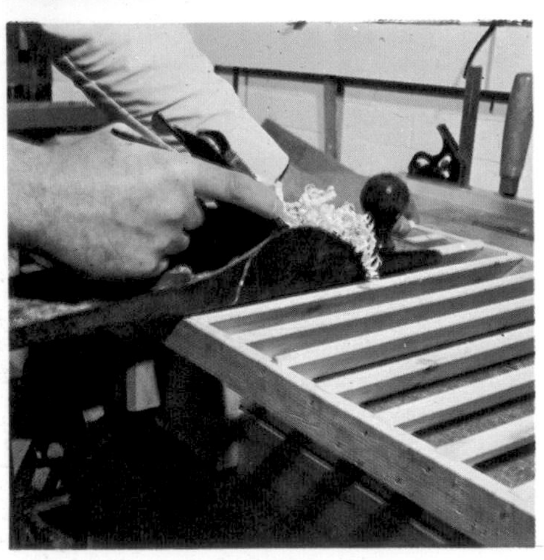

1. *Marking out the dadoes for the slats with a miter try square and dividers.*
2. *Marking the depth of the dadoes with an ordinary marking gauge.*
3. *Sawing the sides of the dadoes. Do the whole lot in one step for accuracy.*
4. *Chiseling out the dadoes with a chisel held horizontally, bevel side up.*
5. *The box-shaped jig used to ensure that all the slats are cut to the same length.*
6. *Nailing the slats to one side of the frame; use one or two nails per slat.*
7. *Planing the back of the slats off flush with the frame. You can plane the front too.*
8. *A few nails are enough to hold the louver panel in its door; don't glue it in place.*
9. *The completed door fitted to a cupboard unit, which will be installed later.*

plane their edges at an angle, even if they are going to be angled later.

Glue and nail all the slats into their dadoes on one side of the frame. If the rear of the panel is to be planed flat, the slats should still project slightly beyond the rear of the frame so that none of the joints is visible (see Fig. 7). Then fasten on the other side of the frame in the same way. One nail per slat is enough.

Slide the top and bottom of the frame into place between the sides. Glue them to the sides and the top and bottom slats, and nail them through the sides to hold them while the glue dries. Waterproof glue should be used for work that is to go outside.

Test each completed panel for square and leave it to dry. Then do any necessary planing on the slats and drop the panel into position in the rectangular hole in the main frame of the door or shutter. The outside of the frame might need a bit of planing before it will go in, in which case you will have to punch all the nail heads below the surface to avoid ruining the blade of the plane.

When the panel fits, glue it in and secure it with a *few* nails—you might have to take it out later for repairs.

Finally, clean everything up with fine sandpaper, then prime, undercoat and paint it, or give it three coats of polyurethane varnish. Shutters have a great deal of wind and rain to contend with, and the better the paint you put on the longer they will last.

You may find the instructions on hanging doors, listed in full in CHAPTER 16, useful for installing your new louvered door or shutter.

CHAPTER 18

Modern lines of a boxed-in bathtub

The days are gone when bathtubs were tall cast-iron structures with ornate, exposed legs. The simple lines of a modern bathroom require the sides of a bathtub to be boxed in.

Boxing-in a tub is a comparatively simple job and can be done with a wide range of surfaces—ceramic tiles, plastic laminate, tongue-and-groove boarding, or even gloss-painted plywood. None of these presents any serious technical problems.

All these surfacing materials can be mounted on a simple but robust wooden frame fastened to the floor around the edge of the tub. The installation, however, has to meet various requirements.

First, the frame must be strong enough not

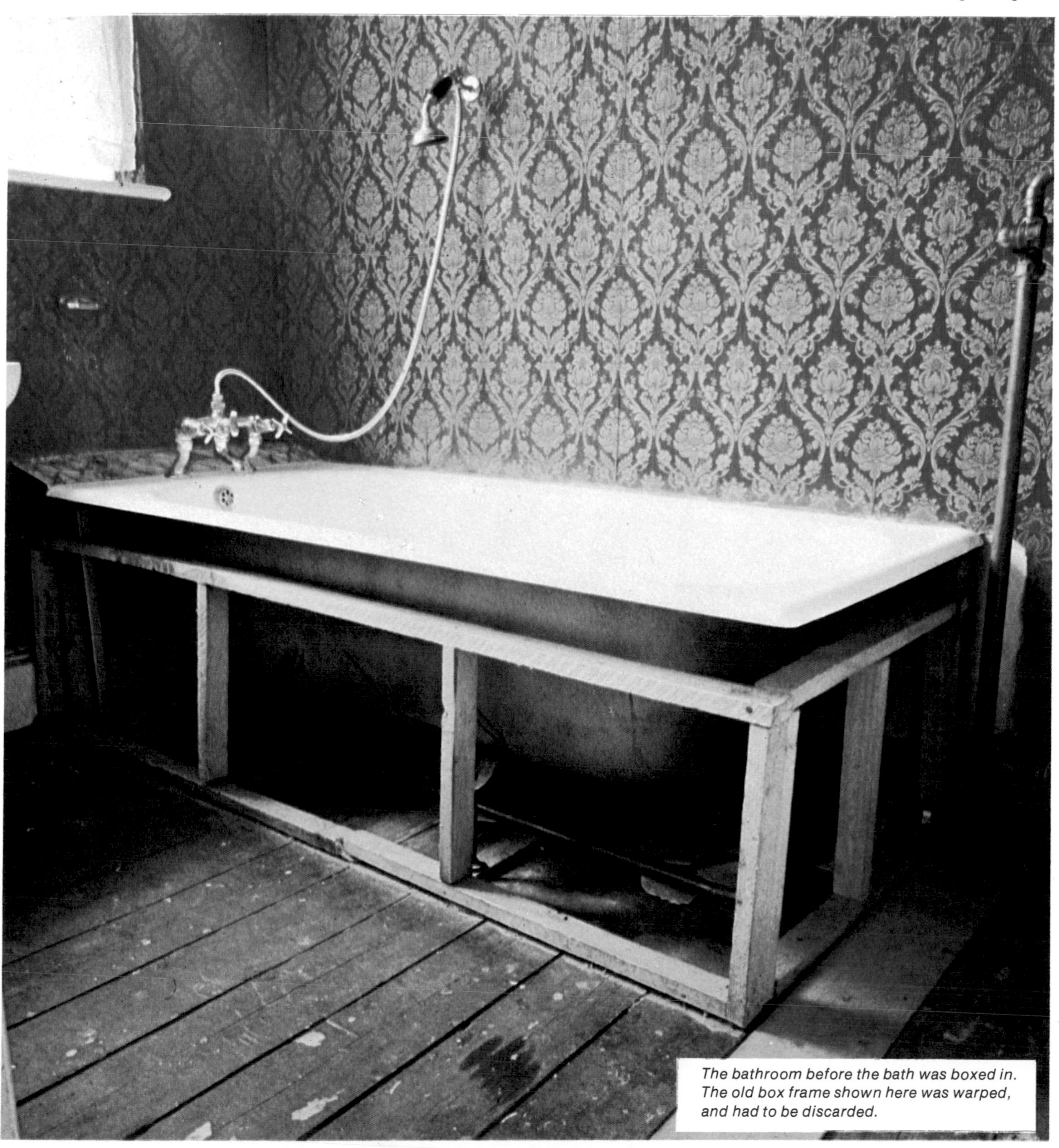

The bathroom before the bath was boxed in. The old box frame shown here was warped, and had to be discarded.

to warp—this is a serious problem in the steamy atmosphere of bathrooms. The tendency of wood to warp in damp air can be reduced by painting all the parts with waterproof paint or varnish on all sides, so that the humidity of the wood remains constant.

Second, it must be properly fastened to the surfaces it touches, so that it will resist kicking, blows from mops when cleaning the floor, and so on. The frame can be nailed to a wooden floor or it can be fastened to a concrete floor with Rawl plugs and screws, or with masonry nails. Nearly all bathtubs have at least one side or end against a wall, and the ends of the frame can be plugged and screwed to this too.

Third, the outer surface has to be reasonably watertight, so that water does not seep down behind it and under the tub where it cannot be mopped up. This is simply solved by setting the panel about ⅛ in. back from the outer lip of the tub, so that any drips from it run down the front of the panel instead of seeping through its back and rotting the frame. At the same time, the joint between the tub and the wall should be sealed, either with a "bath trim kit" consisting of narrow tiles with an L-shaped cross-section or (more simply and cheaply) with the white tub sealant obtainable from any hardware store.

Fourth, the pipework under the tub must be accessible for maintenance. This includes not just the water pipes but also the trap in the waste pipe, which has to be cleaned out occasionally. The best solution here is to screw on the panels with chrome screws, which have decorative heads so that you don't have to disguise them. Don't use too many screws; six or eight is ample for each panel if it has to be removed from time to time. (It is a good idea to inspect the floorboards under a bathtub for rot every few months.)

Materials

The framing should be made of nominal 1 x 2 (actually ¾ in. x 1½ in.)softwood throughout. The amount you will need of this, and all the other materials, depends on your particular bath and bathroom, so you should make a scale drawing of everything you are going to make before you buy anything.

The paneling material depends on what you are going to surface it with. An ideal material for tiling, painting or laminating is ½ in. exterior plywood which will last for many years. Particleboard and hardboard are cheaper but less durable. If you do use them, choose exterior grade board or tempered hardboard, both of which are more waterproof than the ordinary kind.

Other materials you can use are a good-quality ready-laminated particleboard if available (the edges should be given several coats of polyurethane varnish to make them waterproof) or any kind of tongue-and-groove boarding (this should be thoroughly varnished on both sides—do the back and ends before you put it on).

You will also need some chromed or stainless angle strip to neaten and protect the corners of the box, unless it is tiled or the bath

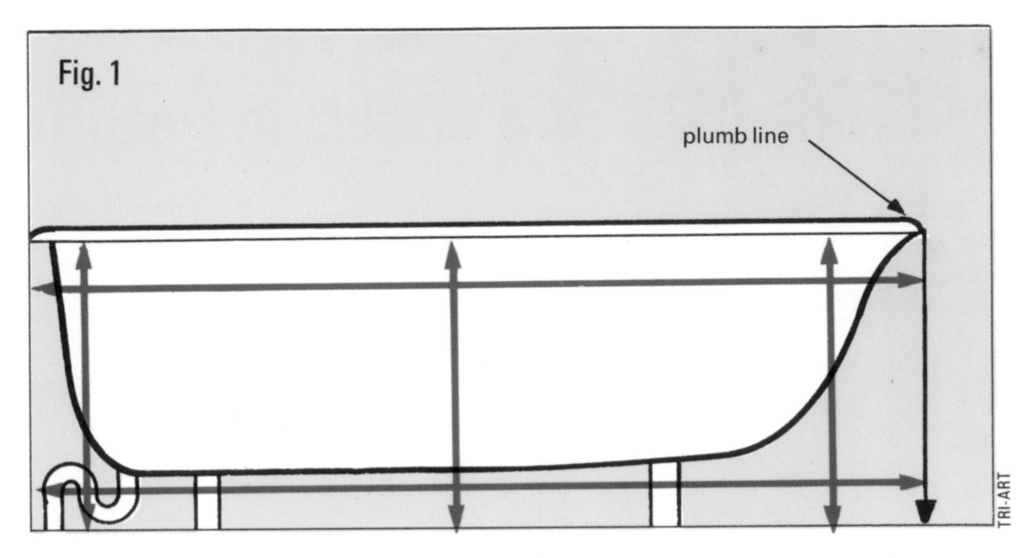

Fig. 1 *(above). The vertical distance from the tub rim to the floor, and the horizontal distance along the tub, should be measured in several places (marked with red arrows) to ensure complete accuracy.*

Fig. 2 *(below). The basic frame for the tub. All types of paneling fit this frame. It can, of course, be built with as many side and end panels as you need to suit the layout of your own particular bathroom.*

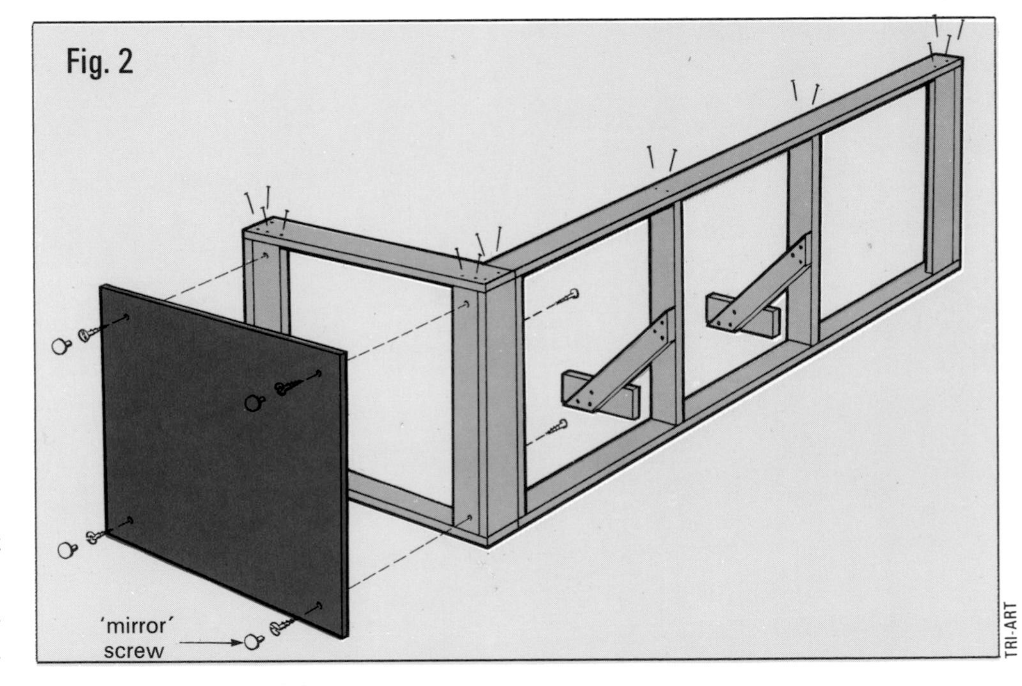

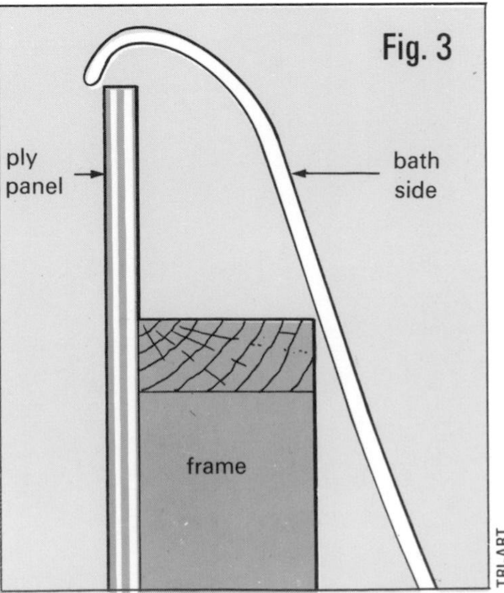

Fig. 3 *(left). A bathtub with sloping sides and a narrow rim may need a shallower frame than normal to fit into the limited space.*

Opposite page: Fig. 4 *(top). The frame should be braced upright with a diagonal strut fastened to a block nailed to the floor. The nails can be driven in diagonally as shown; neatness is not important.*
Fig. 5 *(second from top). Tongue-and-groove boarding should be made up into a single panel by nailing it to battens. This panel can then be removed in one piece when you want to inspect the pipework.*
Fig. 6 *(bottom). Ceramic tiles are heavy, and a tiled panel running along the whole of one side of the bathtub would be very difficult to remove and replace. The solution is to make a smaller, removable inspection panel, and attach the rest to the panel permanently to the box framework.*

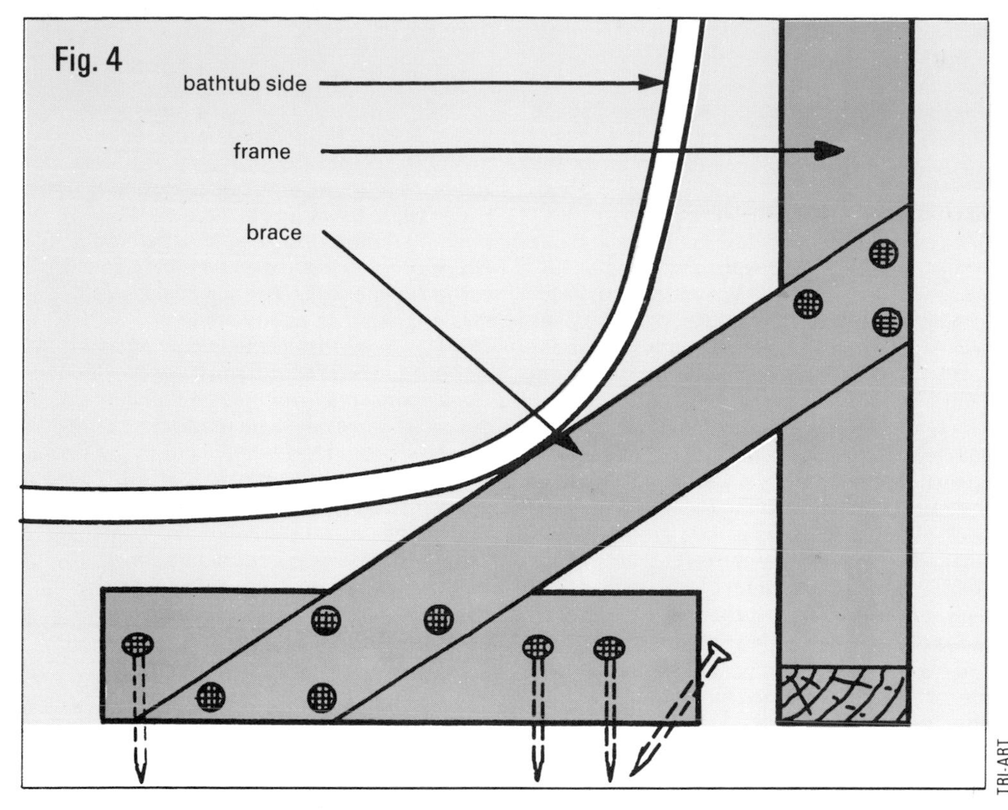

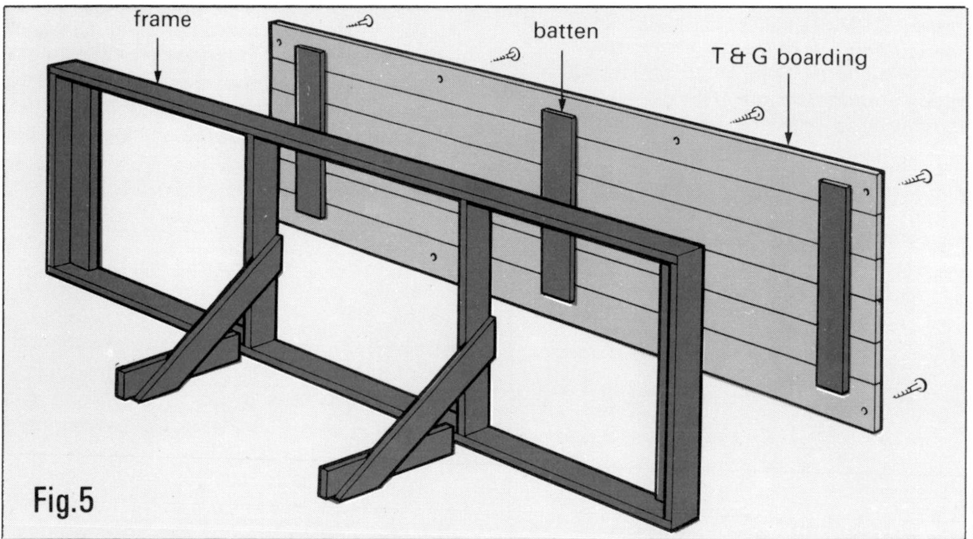

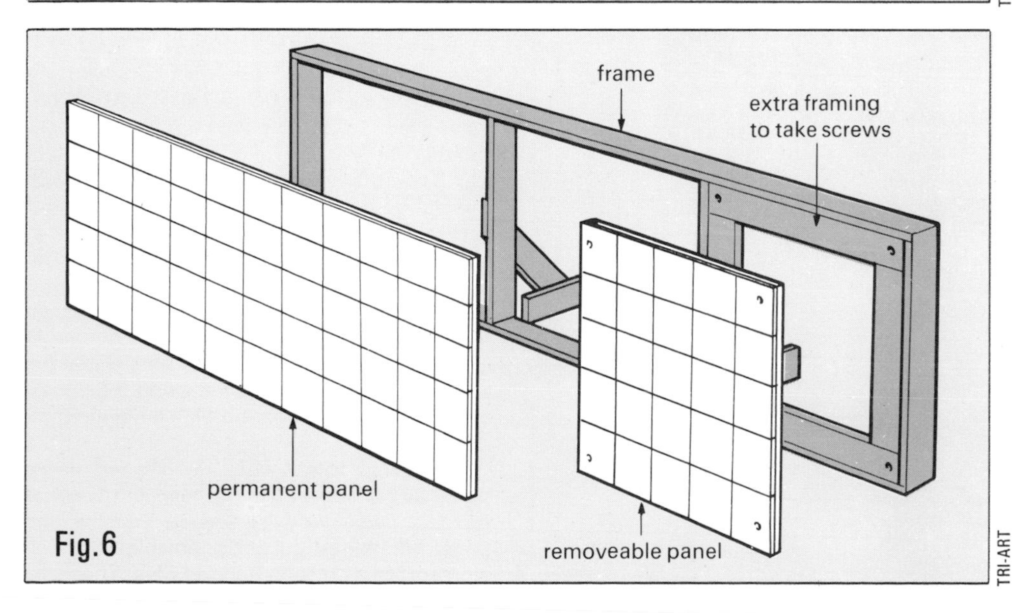

has a wall at each end. Other necessities are surfacing material for the panels, if they are to be surfaced, and plenty of 2 in. wire nails, 2 in. No. 10 screws, and 1½ in. chrome screws.

Measuring

The first thing to do—before you buy anything—is to measure the bath carefully. The space into which each frame must fit should be measured at several points, to allow for the floor not being quite even and the walls being slightly out of true (as they almost certainly will be, in an old house).

Measure the height of the underside of the rim of the tub above the floor in at least three places (see Fig. 1). Then measure the length of the outer edge of the bathtub rim. If the bathtub is set between two walls, measure this distance from wall to wall; if there is a wall at one end only, measure from that wall to a plumbline (or a small weight on a string) hung over the outside corner. If there is no wall at either end, use two plumblines. Any length beginning or ending at a wall should be measured both at floor level and at bathtub rim level to see if the wall slopes.

A typical frame for a box around a tub is shown in Fig. 2. To find the length that each side of the frame should have, take the *shortest* measurement of the length of the tub, i. e., the horizontal distance measured at the height where the tub bulges out most. Subtract from this ⅛ in. or the thickness of the rim of the tub, whichever is greater; the thickness of the plywood (or T & G boarding) that you are going to cover the frame with; the thickness of the tiles or laminate (if any); and the width of the frame members (at each outside corner of the frame not touching a wall).

This last measurement should be subtracted on *one* side of each corner only, since it allows for the overlap of the frames where they meet at each corner. Note that the frame members are nailed together on edge and not flat (for extra rigidity) so that a frame of nominal 1 x 2 is 2¼ in. wide, seen from the end.

To find the height of the frame, take the shortest measurement from the floor to the lower edge of the tub rim and subtract ⅜ in.—this will make the frame easier to insert. The panel over it must, of course, be the full height and project above the frame a short way.

Measure also the horizontal distance from the inner edge of the tub rim to a point at the same level on the outer wall of the tub. This is to check that the frame will be narrow enough to fit into that space without the tub's outer wall pushing it beyond the outer rim of the tub. On most tubs, there should be enough room. If there isn't, you can make the top of the frame lower (see Fig. 3) so that it will fit—but don't make it more than 3 in. lower if you are paneling with ½ in. ply or particleboard, or about half that if you are using T & G boarding or hardboard. Otherwise, the paneling may curl at the top. If you cannot fit the frame in and still comply with these limits, you might cut away part of its rear edge to make it fit.

When you have sorted out the size of everything, make scale drawings and from them work out how much wood and other materials you will need. Allow a reasonable amount for waste, particularly with large tiles if you are buying any, because they often break when you are trying to cut them.

Order of work

The frames are so simple to make that no detailed instructions are required. There are no fancy joints, because they would be out of sight anyway, and because the frame is not required to be very strong. A few points to watch: the pieces should be skew-nailed, i. e., the nails should be put in slightly crooked in opposed pairs. This makes them less likely to come apart. All the frame members are set on edge except for the inserts at each end, which should be set flat and flush with the front. The purpose of these inserts is to provide something to screw the panels to that is not too near the end of the panels, and so likely to split them.

When the frames are made, mark the floor where they are to go by dropping a plumb line from the rim of the tub at various points. This will ensure that they are vertical—but remember that the frames are set back from the place where the plumb line hangs, so that there is room for the thickness of the paneling and the slight projection of the rim of the bathbub.

Fasten the frames to the floor in a suitable manner, and brace them upright with battens running diagonally back from as high up the frame as possible to blocks fastened to the floor under the bath (see Fig. 4). The blocks should be made of nominal 2 x 2 and fastened to the floor by nails driven diagonally through the sides, or small angle brackets or angle strips screwed to both block and floor, if space permits.

You will probably find it easier to screw the corners of the frame together than to nail them, because it is difficult to hammer a nail sideways into a slightly flexible frame in close proximity to a tub. The ends of the frame must be screwed to the wall with toggle bolts or "mollies". If the wall is irregular, pack the gaps between it and the frame with scrap plywood.

When the frame is completely installed, cut each piece of ½ in. ply, if that is what you are using, to slightly more than the correct length and height. Then get a helper to hold it steady against the frame with its top edge 2 in. above the bottom of the rim of the bathtub—mark the back of the panel to help you to locate it correctly. Set a pair of compasses to 2 in. and use them to scribe the bottom of the panel to the floor, or use a 2 in. block and a pencil instead of the compasses.

If the wall is irregular, you will have to scribe the panel to that, too. This should be done after the panel is scribed to the floor, removing as little wood as possible. Trim the other end to length afterwards.

Fig. 7 *(below). An old-fashioned bathtub with round corners can be put into a square box by bridging the gap at the corners with a double-round-edged "corner" tile.*

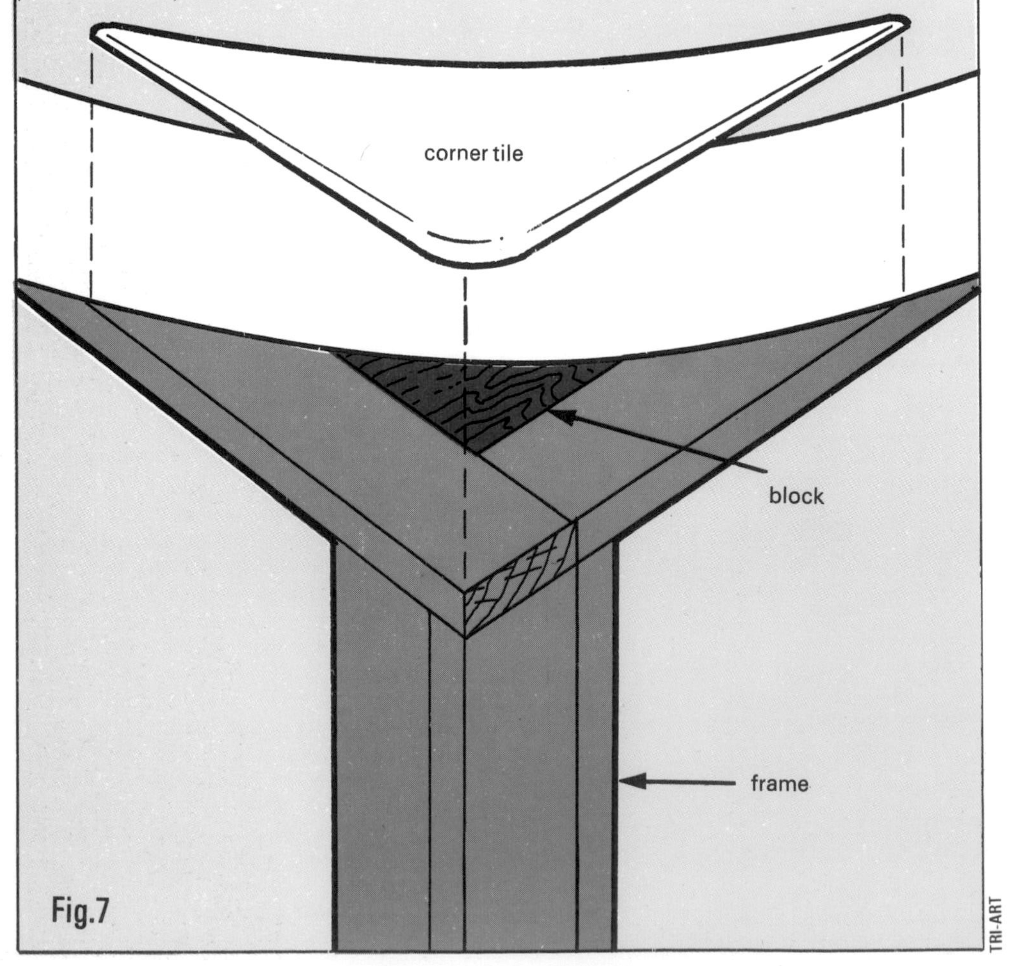

Opposite page: Fig. 8. *Skew-nailing the frame together. Nails are quite strong enough for this type of construction; there is no point in wasting time on complex joints.*
Fig. 9. *Nailing the diagonal struts to the frame. This frame has been set a short way above the floor to provide a toe recess underneath, which gives a little extra standing room to this very small bathroom.*
Fig. 10. *Fitting the frames together at the corners of the bath. They can be nailed or screwed together, as you prefer.*
Fig. 11. *Check that all the frame parts are straight and level before you go any further. The sloping panel mounted behind the taps drains splashes back into the bath.*
Fig. 12. *The finished frame, a sturdy job that will stand up to years of hard use and constant dampness.*
Fig. 13. *Fastening a laminated particleboard panel to the frame with chrome screws.*
Fig. 14. *The completed job. The bathroom has also been improved by changing the wallpaper and laying a carpet.*

When the panels are the right shape, they can be painted or covered with plastic laminate, glued in place with a contact adhesive. Then they can be installed with four, six, or eight chrome screws, depending on the length of the panel. Finish the corners with decorative angle strip and the job is complete. If, however, you are using T & G boards or tiles, read on.

Special panels

Tongue-and-groove boards can be nailed direct to the frame if you like. Instructions on how to make them fit are given fully in CHAPTER 6. But if you have to remove them to inspect or repair something under the bath, it will be a great bore to pry them off one at a time. (All T & G boards should be mounted with tongued edge down.)

A better idea is to nail them to vertical battens set a short way in from each end, and not quite reaching to the top and bottom of the "panel" of boards. When the panel is laid against the frame, these battens will fit into the spaces between the frame members, and the boards themselves will lie flat against the frame (see Fig. 5). The whole made-up panel can then be fastened on by screws in the usual way. The outside corners can be finished with a metal angle strip or wood molding to conceal the end grain of the boards, or beveled at 45° to create a neat mitered corner (this is laborious, but probably worth it).

If you plan to apply tiles to the bath paneling, the tiles should be applied to a normal ½ in. ply panel. The only difficulty here is that the weight of the tiles makes removing the panels difficult, and you might break tiles in attempting it.

If you are content with only one removable panel at the end of the tub, and the tiles are not too thick, then go ahead. Tile the end panel before you put it up, finishing the outside corner with round-edged tiles. Then drill

8

9

10

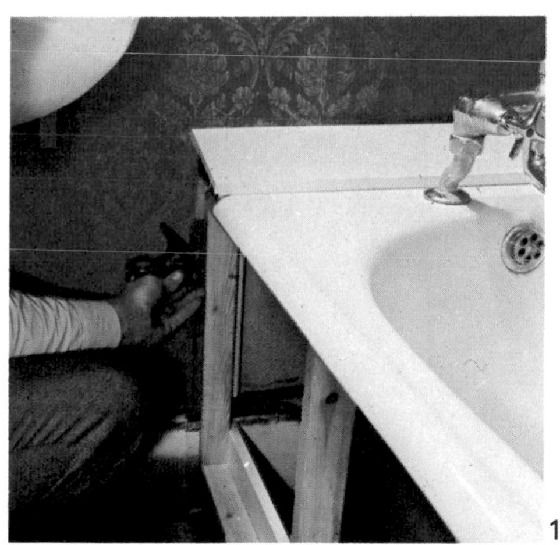
11

12

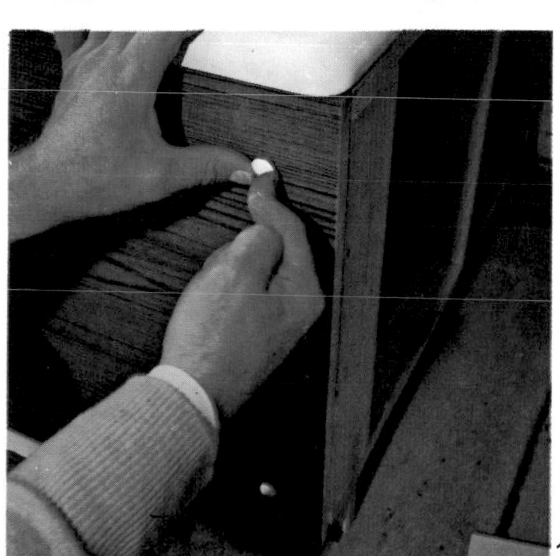
13

holes (slightly oversize) for screws, using a masonry drill, and put the panel up in the normal way.

If, however, the shape of the bathroom requires you to remove a long side panel to get at the pipes, it is a better idea to have only part of the side panel removable, as shown in Fig. 6. Fasten battens all around the inside edge of one bay of the frame, flush with the front, to provide something to fasten the removable panel to. Install the larger, permanent part of the panel with ordinary countersunk screws, and tile and grout it in the usual way. Tile the removable part of the panel before you put it up, making sure that the arrangement of the rows matches that of the permanent part. Then drill oversize holes in the corners and put the panel on with screws. You can wipe grout into the joint between the two panels if you like, but it will fall out the first time you remove the panel.

The only other problem you are likely to have is if you are boxing in a very old-fashioned tub with round corners. The best way to deal with this is to treat the bathtub as if it had square corners. Make the box the same length and width as the tub, so that it projects beyond the bathtub rim at the corners. Then insert a block behind the corners of the frame, flush with its top edge (see Fig. 7) to bridge the triangular gap. Finish the projecting top corner of the box with a white corner tile with two round edges, and its back cut to fit the curve (shown in Fig. 7). These baths are nearly always white, so a white tile blends in and makes it look like a modern squared-off bath.

14

CHAPTER 19

Versatile room dividers

Room dividers have many uses—when two children share the same room, for example, or when one room must serve two different purposes. And they can be both elegant and easy to make.

Very large rooms can often benefit from a partition. It may range from an elaborate wall structure to a simple screen, but the function of all partitions is essentially the same—to separate expansive areas into smaller, more contained ones.

Simple wall screens are especially useful for dividing a room into smaller areas that do not require absolute privacy. Frames may be constructed from stock lumber and enclosed with anything from cloth to decorative plastic paneling.

Estimating the size

The room partitions featured here are designed to fit floor-to-ceiling heights; this is both for strength and visual appearance. The frames were made from nominal 1 x 2 softwood and separated into panels 2 ft. wide by 2 ft. high. They were made to fit an area 6 ft. wide by 8 ft. high. If your room is larger or smaller than this you will simply need to make additional panels—or fewer ones—according to the same directions. Widths have been calculated to fit even distances. If the height of your ceiling is uneven—i.e., if it cannot be divided into 2 ft. areas—the frame should be constructed according to the alternative method suggested below.

In addition to the lumber you will need a number of 1½ in. and ¾ in. nails, ½ in. quarter round molding, 3 in. No. 8 screws and plugs, and white glue.

Building the frame

When working with nominal size lumber, the dimensions of the wood are somewhat different from the sizes quoted. 1 x 2 will, in fact, be ¾ in. x 1½ in. To allow the inner area of each square to be exactly 2 ft. sq.—a size which cuts exactly out of most standard sheet materials—the actual dimensions for a "6 ft. x 8 ft." frame will be about 6 ft. x 3 in. x 8 ft. 3⅜ in.

It is probable that your ceiling will not be perfectly level so, when making your frame, you will find it easier to stop it about ½ in. below the ceiling. The gap can then be filled with a shimming strip of wood slightly narrower than the frame. From eye level, this gap will not normally be noticeable, especially if cove or quarter round molding (Fig. 13) is used to hide it.

The lumber frame

As can be seen in Fig. 15, the frame consists of top and bottom *plates,* vertical members, called *studs,* which run between the two plates, and horizontal members, or *rails,* which run between the vertical members.

Begin constructing the frame by establishing the correct height for the vertical members. Square off the end of one of the vertical members using a try square and marking knife, and cutting with a back saw or fine-toothed panel saw. Stand this piece in position next to the wall against which you are building the divider, keeping it as close to the baseboard as possible. Mark the upright so that it

NIGELL MESSETT

will fit ½ in. below the ceiling. Cut the remaining vertical studs to the same height, making sure that the ends are all square.

Hold the first upright against the baseboard and scribe off the necessary amount so that the upright will fit flush against the wall. Cut out the marked area with a coping saw.

If the baseboard is too thick you will not be able to scribe and cut the upright to fit over it. The way to get around this is to use a filler strip of wood from the top of the baseboard to the top of the upright. Then, when the partition is erected it will stand against both the baseboard and the wall.

Next mark and cut the top and bottom plates to length (6 ft. 3 in. in this case). If a filler strip is used, as described, add its thickness to the length of the top plate. Join these four outer pieces of the frame together at the corners with *corner lap joints* (see Fig. 10). The principles for making all the lap joints used to construct this frame are simple and are fully described in CHAPTER 5. Glue and nail these four joints firmly together using 1½ in. nails.

Once this has been completed and you have checked to see that the outer members are square, you may fix the vertical studs in position. These should be 2 ft. apart, measuring from surface to surface (Fig. 15). Attach these vertical pieces to the top and bottom plates using end lap joints as shown in Fig. 11, 1½ in. nails, and adhesive.

Now, cut the remaining horizontal members to the same size as the top and bottom plates. Mark out 2 ft. distances on these from the innermost edges of each of the vertical members. Do this at the bottom of the vertical members, not midway. Measuring and cutting out the horizontal members in this way will help to straighten out any bows in the uprights. You will need to use a try square to see that all the angles formed between the cross members and the vertical ones are right angles.

Each cross rail should be fixed in position with lap joints (Fig. 12). The outer joints will be middle lap joints, and the internal ones will be cross lap joints. Check at all times that the angles are right angles by measuring the diagonals of each panel area. Glue and nail these cross rails into place, using 1½ in. nails. Allow a day for the adhesive to set.

Erecting the frame

Once the frame has been constructed it must be hoisted up into position; you will need at least two people to do this safely. First check the ceiling for the position of the joists. You can locate them by gently tapping a hammer along the ceiling. Areas along which joists run will emit a much less resonant sound than other areas. Make a final check with an awl. Remember that the joists may run either along the room or across it, and that they are usually at 16 in. centers. If the joists run in the same direction as the eventual position of the partition, you must attach the partition to one of them or fasten it through the ceiling wallboard or plaster with toggle bolts. If the joists run at right angles to the partition, there will be no problem except, of course, in locating them and marking their position so that you will be able to place the fastenings correctly.

Stand the frame upright in its correct position, so that the cut out portion or the filler strip of the first upright fits over the baseboard. Hold the frame in its exact position, against the wall, making sure that it is square, and mark out four positions for drilling screw holes into the frame and the wall. These positions should be at the top and bottom of the upright, near the outer plates, and at equidistant positions moving toward the center. Someone must hold the frame for you while you drill. Use a power drill with a masonry bit set at slow speed and work carefully. Drill a

Opposite page. *Lightweight plastic paneling available in several colors and textures, is one attractive way of finishing your screen.*
Center *and* **below.** *Fabric-covered softboard is another—and offers enormous variety.*

RAPHO

NIGEL MESSETT

hole into the wall to accommodate a nail into a stud or a toggle bolt through the plaster if the frame is not at a stud location.

Now fit a shim strip of softwood, about ½ in. thick between the top rail and the ceiling, to make a snug, even fit. Once this is done, check with a plumbob that your frame is upright, and adjust it at the base if it is "out." Then drill through the top rail and the ½ in. shim strip into the ceiling at the points previously marked to indicate the position of the joist(s) or use toggles if not at joist locations. The bottom plate can be screwed directly into the floorboards. Attach molding around the top, if necessary (Fig. 13), mitering it at the corners and then finish the frame with paint or varnish.

Alternative frame method

If your ceiling is unusually high—12 ft.–14 ft.—or is extremely uneven, or is an odd height, you may not wish to construct a frame to the full ceiling height. If this is the case, it would be possible to make the frame in the same manner as described, but you would have to allow for the outermost upright (the one farthest away from the wall) to extend to the full height of the ceiling and to be fixed to the ceiling joist by skew nailing (this is illustrated in CHAPTER 2). In this case, you will need an outer post of nominal 2 x 2 for stability, and a matching inner one for appearance. The sides of the frame should be fastened to the wall as described before.

Filling the panels

The choice of insert is mainly a matter of personal taste, but it should be something which is reasonably lightweight or the frame will not be able to support it. Another thing to consider is that the divider will be seen from both sides, so the panel inserts will either need to be suitably finished on both front and back or some method of double covering will be necessary.

Cloth-covered panels

Any of the wide variety of fabrics that are suitable for use as wallcoverings can be used on ¼ in. plywood panel inserts but it is best to use a lightweight, fairly stiff, simply designed fabric. Dark fabrics or fairly heavy ones, such as felt or burlap, can be used to fill in some, but not all of the panel areas. If used exclusively, these fabrics can produce too dense an effect.

All fabric coverings will need to be tacked to a firm backing of some sort, such as light plywood.

The primary reason for using plywood is that both sides of the panel will display the fabric covering. To fill a panel, use a back saw to cut a piece of the plywood to fit into the interior area of the panel (in this instance 2 ft.

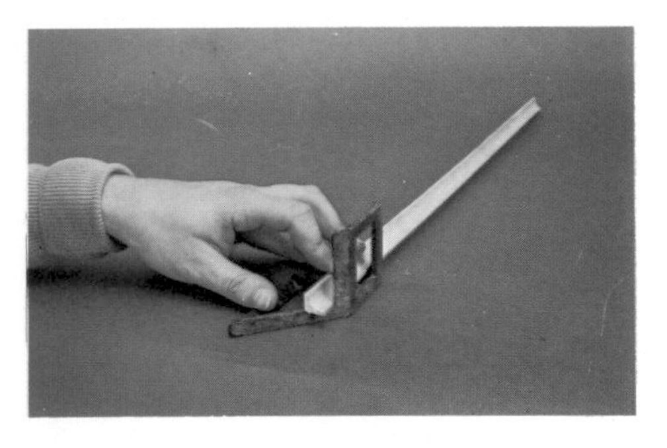

Fig. 1. *Using a miter iron to cut cove molding. A try and miter square may be used for this.*

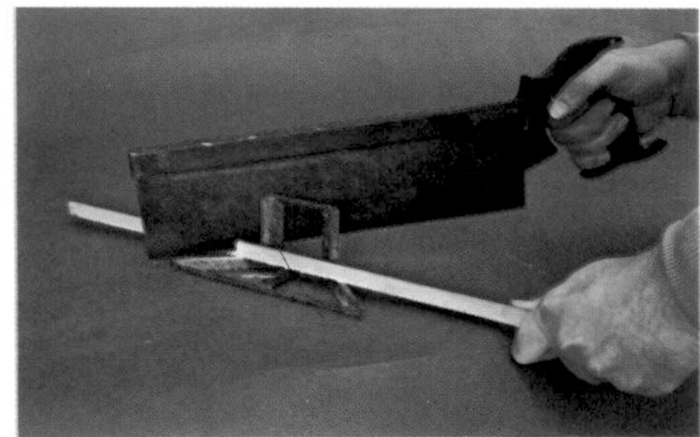

Fig. 2. *Next, a back saw issued to trim the wood. Cut from the good side toward the back, so that any splintering is hidden.*

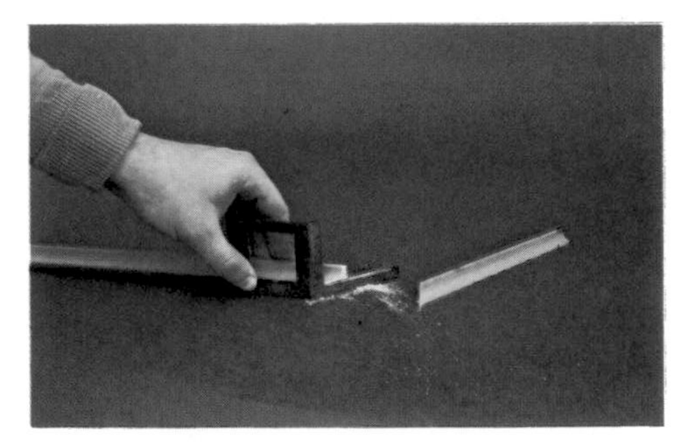

Fig. 3. *With the offcut severed, you can check that the angle is correctly sawed—and not upside down as for an external corner.*

Fig. 4. *A trial assembly of the first corner will show whether you are cutting correctly. Here, a sharper angle will be needed.*

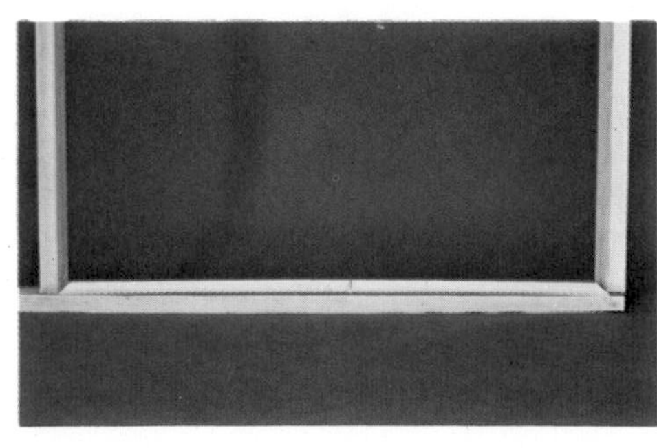

Fig. 5. *For internal angles, as in this project, cut cove molding a shade over-length and "spring" it in to ensure a tight corner fit.*

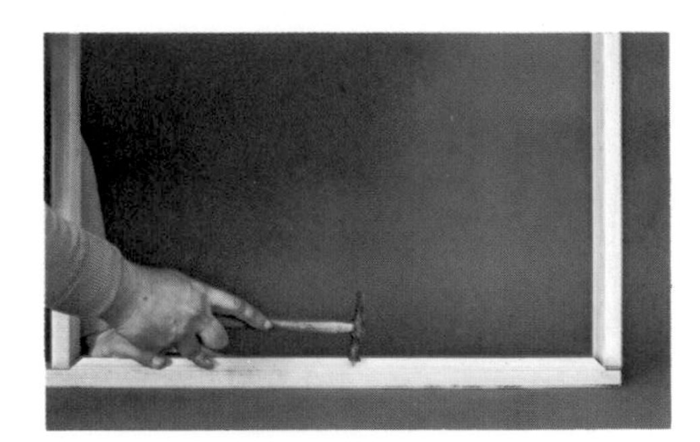

Fig. 6. *Initially, tack the first length of cove molding in the middle only. This allows for adjustment when fitting the other pieces.*

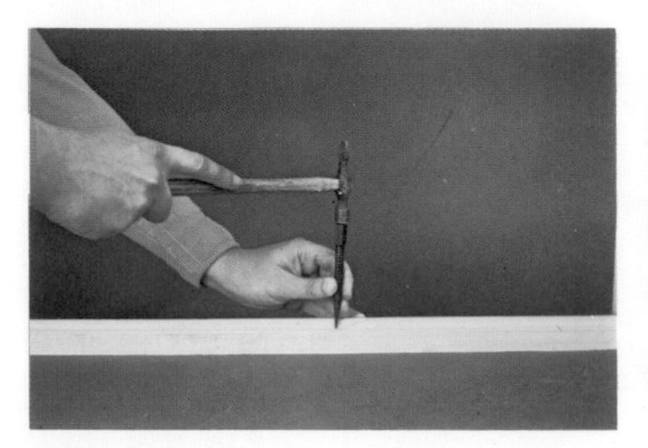

Fig. 7. *All nails should be punched, and the holes filled with wood filler. After filler hardens, sand smooth.*

Fig. 8. *A brad pushing tool is handy for fixing moldings. By pushing rather than driving the brad, it helps prevent the wood splitting.*

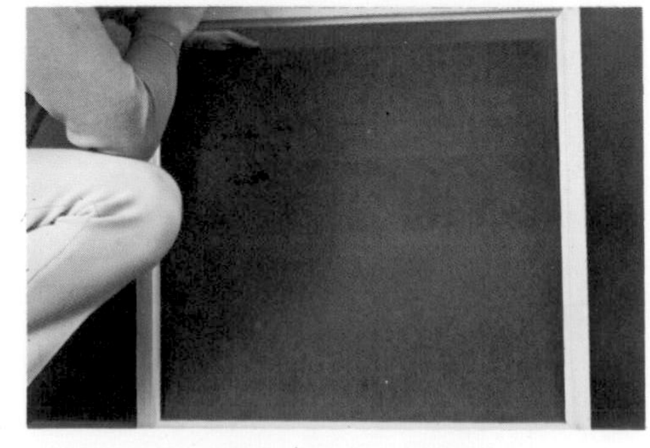

Fig. 9. *When all the moldings are fitted, test-fit the sections of paneling to make sure there will be clearance for the fabric.*

x 2 ft.), but first mark out the area on the plywood and check to see that it is in square by measuring the diagonals. Test the fit by inserting the plywood into the frame—it should fit snugly, but should not have to be forced into place. Trim off any uneven sections if necessary.

The covering should now be cut to fit this plywood—two pieces for each panel—with ¼ in. extra left on all sides to overlap the edges. Make sure that the surface of the board is free from dirt and bumps, and then spread a piece of the material across, smoothing out any wrinkles with your hands. Keep the material centered across the board so that the overlap on each edge is equal and, using a staple gun, tack the material to one edge of the plywood. Staple it securely across the edge about six to eight times, keeping the material taut; do not, however, stretch the cloth. Continue stapling the material around the remaining edges, making certain that you do not distort the weave of the cloth. Now trim off the overhang on each edge. Edges do not have to be perfectly neat, since they will be covered by the molding.

Do exactly the same with the second piece of cloth on the opposite side. Once you are certain that the material is securely fastened, you can position it in the panel.

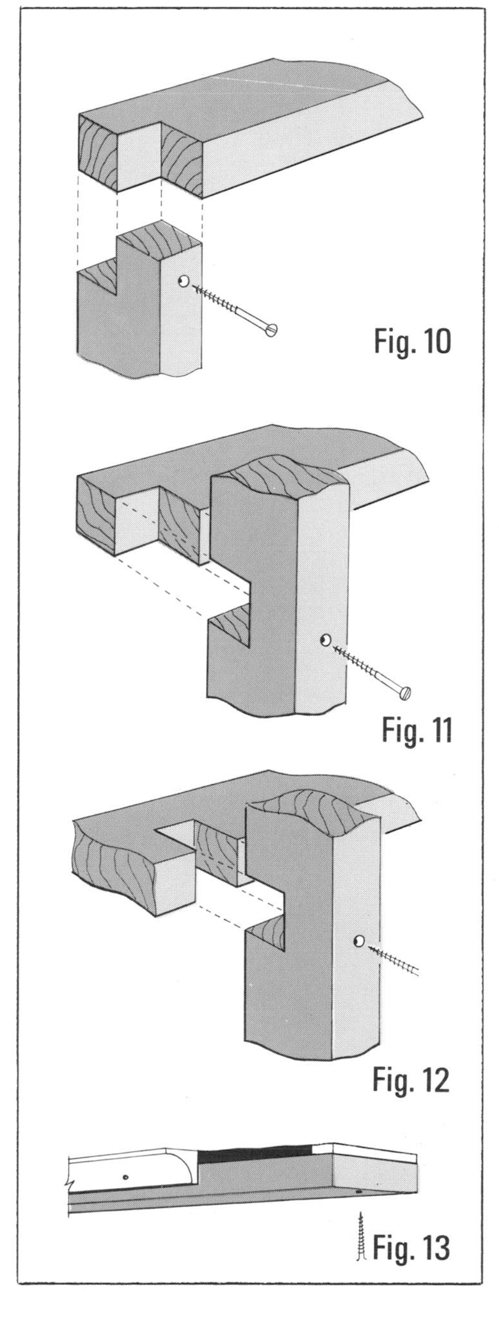

Left *(from top). Details of the lap joint at the extreme corners of the frame; on an outside edge; and where two rails cross. The bottom illustration shows how shim strips are used to fix the screen to an uneven ceiling, and a molding used to cover the gap.*

Mitered molding

In order to hold the panel material securely inside the frame—whether it be cloth, or special paneling pieces—molding will need to be cut and mitered to fit the inner panel area. Each panel will need eight pieces of molding, four to run around each side. Cove or quarter round molding ½ in. thick should be mitered to the correct size (follow the directions in Figs. 1 to 9) and fixed to the edges of each panel with 1 in. brads. Cut each piece of molding slightly oversize so that it can be sprung into position. The molding should be finished to match the frame before it is installed.

First, attach the molding to the outer edges of each panel area that you are covering, tapping the brads in just below the surface of the molding. Insert the covered board *behind* the molding and nail it lightly to the top and bottom of the molding to hold it in place. Now attach the molding around the edges of the panel area on the opposite side, so that the covered board is fixed squarely and securely *between* the lengths of molding. The molding will form a kind of rabbet for the insert and will cover the edges of the covered board by ½ in. all around.

Alternative fillings

Cloth-covered panels are only one of the many materials that may be used to fill the panel areas. You can also use hardboard to fill the gaps. Locally-available ranges vary somewhat, but you may find both boards with an attractive perforated pattern and another kind with a textured surface. This needs only to be cut into squares to fit the panels. The squares should be set in place between the molding as directed for the cloth-covered plywood.

If desired, these hardboard panels may be used in long sections rather than in square sections. To do this, construct the frame with the outer members and inner vertical members only—no cross members are needed. It is helpful, however, if the inner area of the frame—from upper rail to bottom rail—is exactly 8 ft. or less in length, since the hardboard comes in a standard height of 8 ft. Cut and miter the molding to fit the dimensions of 2 ft. x 8 ft., remembering to cut each piece slightly oversize so that it can be sprung into place. Fix the molding in place as described earlier, using 1 in. brads, and position the paneling between the sections of molding as described above. The hardboard can be painted or finished as desired.

Modern plastic materials are also available for use in room dividers. Most lumberyards carry them in various forms. This material is usually translucent and comes in a variety of colors, textures and designs. It can be cut with any type of small-toothed hand saw.

These panels can be fixed in place using moldings as for the other materials. If you wish to seal them into place to prevent rattling, use a mastic sealing strip or even self-adhesive foam strip around the edges. Usually such plastic-type panels do not come in sections larger than 2 ft. x 6 ft. Besides being lighter than glass, they are much less liable to breakage.

Fiberglass and acrylic plastic are other possibilities. Alternatively, you may decorate plain cloth panels with variously designed appliqués, or decorate only some of the panel areas.

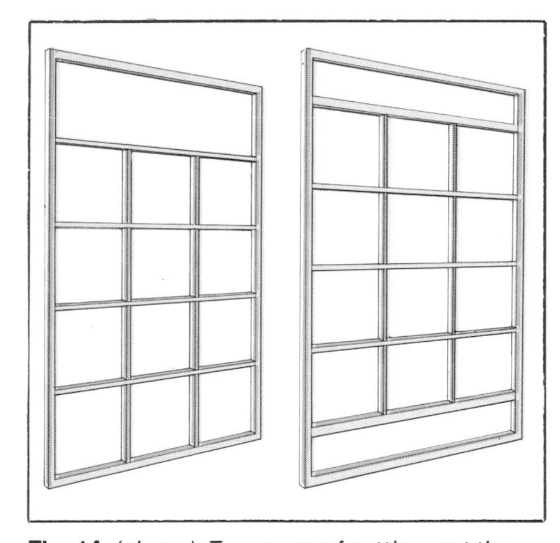

Fig. 14 *(above). Two ways of setting out the frame for ceilings whose heights do not match the standard module of the paneling material.*

Fig. 15 *(below). The horizontal and vertical members are all cut in continuous lengths, then cross-lapped to fit over one another.*

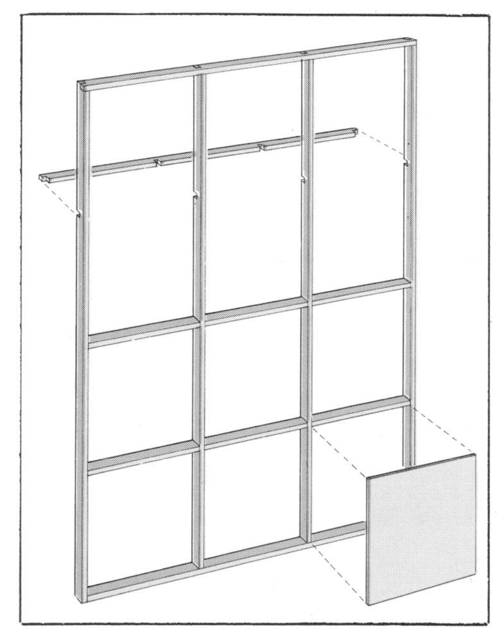

CHAPTER 20

Building a flat roof —English style

This is the first of three roof types described in this book for amateur or professional builders who seek to add a distinctive touch to their work with styling details from other parts of the world. The photos and drawings of this roof and those that follow it are of homes actually built in Britain. As the styling details often result from features of the structure itself, the terminology and the lumber sizes are also those used in Britain. You can, however, duplicate the styling with American lumber and materials, sometimes with a few design modifications to suit your local building code.

Of all the common types of roofs—flat, lean-to, gabled, hipped and mansard—the flat roof is the simplest. And, since the same lengths of timber serve as both rafters and ceiling joists, it is also the least expensive.

Pitching a flat roof

Flat roofs are seldom truly flat. If they were, rainwater would not drain off and, sooner or later, it would penetrate the roof covering. And there are few more dramatic—or disheartening—events in housebuilding than what happens when a large volume of water, filtering through a leaky roof into a "pond" between the ceiling joists, suddenly bursts through the ceiling and floods the room below!

In Britain, the minimum *pitch,* or slope, permitted is 1 in. in 5 ft. But a steeper pitch, around 1 in. in 3 ft., is better anywhere, and essential in areas subject to heavy rain or to gale-force winds, which can blow water *up* a slight slope.

The lean-to roof

In Britain, the building regulations state that a roof can slope by as much as 10°—about 1 in. in 5½ in.—and still qualify as a "flat" roof.

Above that pitch, it becomes a *lean-to* roof. On garages and workshops, where a steeply-sloping ceiling is acceptable, this is built in the same way as a flat roof. But on living accommodation, where a flat ceiling is desirable, both ceiling joists and rafters are needed. So the lean-to roof is then built by the same methods as those for a gabled roof (CHAPTER 34 describes this) except that an extra set of wall plates (see below) is necessary where the joists and rafters meet the wall of the existing house.

Parts of a flat roof

The main components of a flat roof are:

Wall plates. These two heavy pieces of timber are fixed to opposite walls of the structure to support the ends of the rafters, which are nailed to them. (In some countries, a wall plate bolted to the side of an existing wall is known as a *ledger.*)

Rafters/ceiling joists. These are also heavy pieces of timber, running from one wall plate to the other and forming the main support for the roof covering. The bigger the span—distance from one wall plate to the other—the bigger the rafters used. The higher end of a rafter is called the *head;* the lower end the *foot.*

Noggings. These are short lengths of timber nailed between the rafters to stiffen them and stop them twisting. Their depth matches the depth of the rafters.

Strutting. An alternative to nogging, this consists of lighter pieces of timber fixed between the rafters in a herringbone pattern; that is, each pair of struts forms an "X" shape between two rafters.

Soffits. These are the coverings—of timber, plywood or asbestos-cement—fixed to the underside of the rafters where these protrude beyond the walls of the building. They neaten the exterior appearance, help form a watertight seal around the ends of the rafters and—not least important—stop birds from nesting under the eaves.

Fascias. These are wide boards fixed around the outside edge of the roof to neaten its appearance and to provide "nailing" for the roof guttering.

Sarking. This name is sometimes used for the continuous boarding which covers the rafters, and to which the roof covering (roofing felt, lead, etc.) is fixed.

Planning your roof

Rafters in a flat roof usually cover the area below across its shorter dimension. On an extension measuring 12 ft. by 8 ft. for example, they normally run from one of the 12 ft. walls to the other. This means that there are two quite different methods of constructing the roof.

When the rafters run in the same direction as the slope of the roof:

In this case, you must decide whether you want a dead-flat ceiling or one with a slight slope.

If you want a flat ceiling, the wall plates are fixed in place so that both are level, and the rafters are also installed flat. The slope on the roof is imparted by fixing *firring pieces* on top of the rafters (Fig.1).

If a sloping ceiling will do, on the other hand, one wall plate is fixed higher than the other so that the rafters slope automatically (Fig. 2).

When the rafters run at right angles to the slope of the roof:

In this case, the roof slope is imparted by fixing the firring pieces, not on top of the rafters, but on top of the wall plates (Fig. 3). This means that each successive rafter is automatically a little higher than the one before it—and also, of course, that a flat ceiling is impossible.

So the first thing you must decide is which way you want the rafters to run. And if, for example, your 12 ft. x 8 ft. extension is joined to the house across its 8 ft. wall, and you want a flat ceiling, you will find you have no option but to run the rafters across the 12 ft. span. This will mean larger, and somewhat more expensive, rafters.

Finishes for flat roofs

At the planning stage, you must also decide what type of finish your roof will have, since the kind of fascia/soffit combination you use will affect the placing of the outside rafters. The most common types of finish are:

The flush finish

For this finish (Fig. 4), the outside pair of rafters are kept flush with the side walls, and the ends of the rafters are also trimmed off flush. The fascia boards are fixed direct to the sides and ends of the rafters, and must be deep enough to cover both the rafters and the top ½ in. or so of the brickwork or wall cladding.

For this finish, it is usual to "sink" the wall plates in the top course of brickwork, as shown in the diagram; otherwise, the fascia would have to be very deep indeed and it would look rather ugly.

The boxed-in finish

For this finish (Fig. 5), the outside pair of rafters are laid just *inside* the inner walls. Then short pieces of rafter, called *stub noggings,* are cut, laid across the walls at right angles to the rafters, butted against the rafters and skew nailed in position. The stub noggings provide support for the soffits and the fascia, while the outside pair of rafters—being just inside the walls—are handily placed to receive the ceiling lining.

Rafter sizes

Local authorities in different countries differ as to the safe dimensions for a rafter spanning a given length. But the British sizes given below are close enough for you to use on your preliminary drawings—and if a larger size is required the building inspector will soon tell you!

Sizes of rafters for a flat roof, if the rafters are spaced at 16 in. centers:

Up to 7 ft. 9 in. span, use 4 in. x 2 in.
Up to 9 ft. 8 in. span, use 5 in. x 2 in.
Up to 11 ft. 4 in. span, use 6 in. x 2 in.
Up to 13 ft. 4 in. span, use 7 in. x 2 in.
Up to 15 ft. 3 in. span, use 8 in. x 2 in.
Up to 17 ft. 1 in. span, use 9 in. x 2 in.

These sizes are for roofs to which only limited access—for maintenance and repair—is required. If you want to use your roof as, say, a sun-deck, you will generally need rafters one size larger than those given.

Fixing the wall plates

Wall plates (except in timber-frame houses, where they form part of the wall) are usually of 4 in. x 3 in. timber, so that their cross-section matches one course of bricks. They are fixed in place by rag bolts mortared into the brickwork.

The first job is to fix in place the plate or plates which are set into the inside skin of the cavity wall. Begin by cutting the timber to

An attractive finish for a flat roof. The rafter ends are left uncovered to reveal the neatly laid continuous boarding, or sarking, which covers the rafters and supplies a fixing to the roof covering.

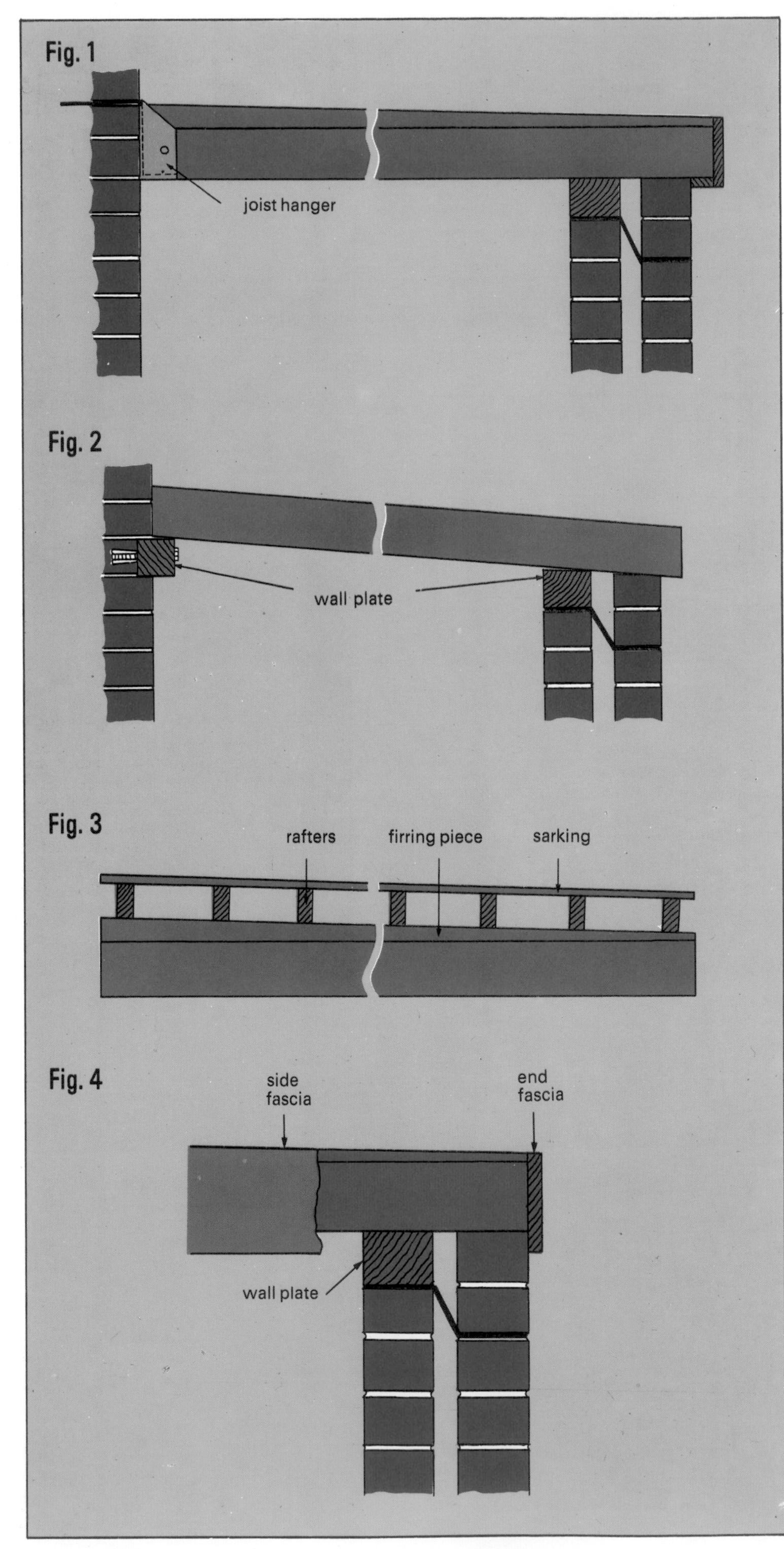

length and then, unless you are using timber which has been rot-proofed, apply a lavish coating of wood preservative such as one of the penta types. Put both plates side by side while you mark on them the positions that the rafters will occupy; this job is most easily done at ground level!

Stand the plate on edge beside the rag bolts and mark off on the plate the positions of the bolts. Drill holes at these points in the plate, ease the plate in position over the bolts, and tighten firmly.

If one of your plates has to be fixed to the side of an existing wall, its strength will be greatly increased if it is partly embedded in a brick course as well as being bolted in place. First, measure the height at which you want the plate, and mark the corresponding course of bricks with chalk. Using a bolster and club hammer, chip away no more than one-third of the thickness of the bricks—that is, about 1½ in. from a standard brick.

Next, drill No. 10 holes in the plate at 1 ft. intervals along its length. Hold the plate in position (you will need a helper) while you check with a spirit level that the channel you have cut in the brickwork is level. If necessary, insert temporary packing pieces in the brickwork to level the plate. Then mark the positions on the wall where the screws will enter, drill for 4 in. No. 10 screws, insert fiber or plastic wall plugs, and screw the plate home. Fill with mortar any gaps above and below the plate and, when the mortar has hardened, remove the packing pieces.

The firring pieces

If your firring pieces are to go on top of the plates, fix these next. Otherwise, fix your rafters first and then the firring pieces.

Since it is difficult to cut a firring piece to a sharp taper, make its thinner edge about ½ in. thick and increase the other end correspondingly. For example, an 8 ft. long firring piece, cut 2½ in. thick at one end and ½ in. thick at the other, will give you a slope of 1 in. in 4 ft.

Cutting firring pieces by hand is a long, laborious job. If you cannot buy them pre-cut from a timber yard—and British timber yards usually carry them in stock—you would be wise to hire a hand-held power-saw, which, incidentally, will also make short work of cutting all the other heavy timber in this job.

Fig. 1. *Setting wall plates level and using firring pieces on all the rafters gives the necessary slope and a flat ceiling. An alternative to the wall plate at the house wall is joist hangers—square-sectioned metal supports which are bedded in a mortar course to form a box-like support for each rafter.*

Fig. 2. *If you do not require a flat ceiling, the slope of the roof can be achieved by setting one wall plate higher than the other.*

Fig. 3. *If the rafters are to run across the slope of the roof, fix a firring piece on each wall plate.*

Fig. 4. *The flush finish. Here the end rafters are flush with the side walls and the feet of the rafters are trimmed off flush.*

The firring pieces are fixed in place simply by nailing to the plates or rafter tops, as appropriate.

Cutting the rafters

The next step is to cut the rafters to size.

If you are going to use a *flush* finish, begin by squaring the end of one piece of timber, laying it in position, carefully marking the right length, marking a vertical line with a pencil and spirit level, and cutting it to size. Then cut all the remaining rafters from this first one. Do not measure the third rafter from the second, fourth from the third, and so on—you may make a progressive error and be well off line by the time you cut the last one. If you are going to use a *boxed* finish, square one end only of all the rafters, and leave the other end over-size. They will be cut to length later.

Mark on the rafters the positions that the nogging will occupy. Then place the rafters on the marks you have ready on the plates, and skew nail them through both sides with two 3 in. nails.

Next, if your rafters are overhanging, run a builder's line (a length of nylon fishing line is fine for this) across each row of rafters, and mark where each has to be trimmed off. Use a builder's level, upright, to ensure that your trimming cuts will be vertical. Trim off all the unwanted ends. The reason for doing this, instead of pre-cutting the rafters to length, is that you will automatically correct any variation in the overhang caused by, for example, one of your walls having a slight "wander" in it.

Fixing the nogging

The nogging is cut from short lengths of timber the same depth as the rafters, and spaced at 2 ft. centers.

To save time, cut out together, and to the same length, all the nogging except those pieces which will go between the last two rafters. In this case, since the rafters are at 16 in. centers, the nogging will be 14 in. long, or slightly longer if you are using planed or dressed timber.

Fix in place all these identically-sized pieces of nogging, skew nailing them into the rafters through the *sides* of the nogging. Then measure the lengths for the pieces in the last row, cut them to size, and fix them individually.

Fixing the boarding

The main objective with the boarding on top of the rafters is to get as smooth a surface as possible, so that puddles of water cannot collect on your roof. Flat-sawed timber 1 in. thick will do, but tongue-and-groove boards ¾ in. thick, or a ⅞ in. thick high-quality impregnated chipboard, would be better.

Lengths of timber can be fixed at right angles to the rafters but, if you care to take the trouble, fixing them diagonally is better. This braces the roof by creating a series of triangles, and also means that, should the boards begin to "cup" with time, the ridges thus created are running down, and not just across, the slope of the roof.

Fixing the fascia and soffit

Fascias can be made from random-length boards, but any joints should be over a rafter or stud-nogging end. Joints should be splay-cut—that is, cut at an angle so that one overlaps the other—so that any shrinkage in the boards will not create an ugly gap.

Paint all sides of the fascia boards with primer before you fix them on, using plenty of paint at joints and on end-grain. Fix with galvanized nails.

If you are using tongue-and-groove boards for your soffits, fix these using the methods given in CHAPTER 6.

If you are using asbestos-cement sheets, cut them to size by scoring them deeply with an old file or chisel, and then snapping the sheet over the edge of a length of timber.

Different methods of fixing are used for asbestos-cement sheets in different countries, because the density of the sheets varies. In Britain for example, you must drill holes in the sheets with a masonry drill and fix them with screws; in Australia, on the other hand, the sheets are softer and can be nailed in place with galvanized flat-headed nails. If in any doubt, ask the merchant from whom you buy your sheets.

Finishing off

There are several types of roof covering suitable for a flat roof. Transparent plastic sheets, usually corrugated, and bituminous roofing felt are the most common. Whichever one you choose however, you must be careful that, where the roof abuts an existing wall, you provide adequate *flashing* between the old wall and the new roof.

This is a sheet of roofing felt or metal, let into a mortar course immediately above the roof where it joins the house wall, providing a runoff for rainwater.

Select a joint about 6 in. above the roof, and with a cold chisel and hammer rake out some mortar to a depth of 2 in. Insert the flashing, wedging it with small pieces of brick if necessary, and run the flashing down so that it covers at least 6 in. of roof. Fill the joint with a good bricklaying mortar and leave for two or three days to set.

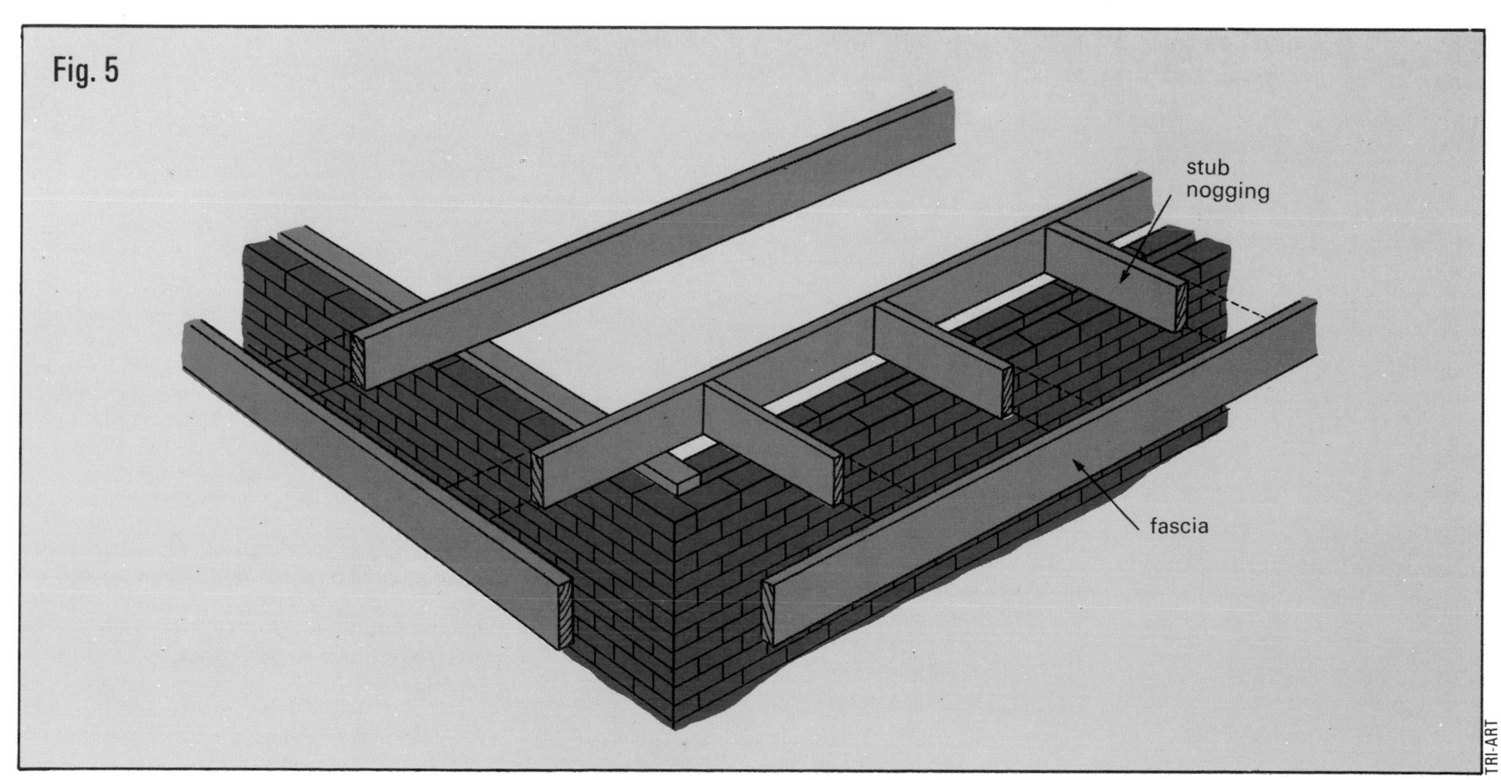

Fig. 5. *The boxed-in finish. The end rafters are laid inside the walls and stub noggings provide fixing for the fascias and soffits.*

CHAPTER 21

Sprucing up old floors

Floors in old houses tend to be covered with stains, dents, cracks and layers of old finish. Yet underneath all these surface defects, the planks are generally sound, and the floor can be made to look as good as new.

The best way to restore a floor like this is to sand the surface flat. The majority of the defects are removed, the sound wood underneath is revealed, and at the same time it is smoothed so that it can be refinished.

Floor sanders shaped like vacuum cleaners, and which can be operated from a standing position, can be rented from tool rental shops or paint dealers. The daily rate is not expensive, and sanding sheets and other necessities can be bought at the same time. Many firms will, upon request, deliver and pick up the machine.

It is essential to wear very old clothes when using a large sander, since it throws up a good deal of dust and dirt. It is also a good idea to wear a face mask, to avoid inhaling dust.

Before you arrange to rent the machine, check that your floor is not suffering from dry rot or any other condition that will involve you in replacing all or some of the boards. Check also that the boards are still thick enough to sand—they may be worn very thin around doors or in other heavily-used areas.

The minimum acceptable thickness depends on the type of boards. Square-edged boards (as in century-old homes) must be thicker than tongue-and-groove ones, which interlock and to a certain extent hold each other even.

If any of the boards are too thin, they will have to be replaced. Try not to replace them with new boards; even after old and new have been sanded together, there will still be a marked difference in color between them. You will very often be able to get matching old boards from a demolition contractor. If this fails, you could take boards from a carpeted room in your house and fill the gaps there with new boards. All this applies to very old flooring.

Preparation

Floorboards in old houses are often so badly shrunken that there are large gaps between them. This can also result from long periods of excessive dampness. This causes the boards to swell so much that the edge fibers are crushed. Then, when the boards dry out a permanent gap (called "set") results. If there are gaps all over the floor, running the full length of each board, probably the only thing to do is to take up the boards one by one and move them sideways. This will create a large gap at one side of the room which can be filled with an extra board, sometimes ripsawed to less than full width. Floorboards, whether square-edged or tongue-and-groove, should be pressed hard against their neighbors and held there while being nailed down. This can be done most easily with rented floor-nailing tools. An alternative method is to protect the edge of the board with a strip of wood the same thickness as the board (and grooved if it is a tongued board), and then to lever the board into place with a stout chisel (see the illustration in CHAPTER 6).

Smaller gaps can be filled by making appropriately-sized hardwood fillets with a wedge-shaped cross-section, and forcing them into the cracks with a mallet.

Some boards may be split. If the split is really bad, the board may need replacing. Moderately bad splits can be repaired with wood filler of matching color.

If some of the floorboards have been frequently taken up and put back (to reach pipes or wiring, for example), they may be full of messy-looking jagged or split nail holes. These can be repaired with wood-filler. Another method for very large holes is to drill them out and insert a piece of dowel.

On heavily-worn floorboards the knots, which are harder than the surrounding wood, may be sticking out. You can cut them down level with a "Surform" or power sander, starting with coarse abrasive, and ending with fine.

It is essential before any sanding is done to punch *all* the nails in the floor below the surface. This is a boring task, but not particularly strenuous. There won't be many nails in the surface of the usual edge-nailed modern floor, however. If a carpet has ever been tacked to the floor, there will probably be some tacks left in it. They can easily be removed with an old chisel and mallet or a pair of pliers. After you think you have removed or punched down everything, double-check the whole floor area. One projecting nail will rip up a sanding sheet, or may even ruin a very expensive rented sander.

If the floor is painted, the sander will remove paint perfectly well, even if it is thick. Unfortunately, the machine cannot sand right to the edge of the floor (see below). If there is only one coat of paint on the floor (and you plan to paint it again), you do not have to bother about those edges the machine cannot reach. If there are several coats (which is much more likely), you will have to use an edger-sander, which you can rent, to get the paint off. This is a disk type sander, smaller than the regular floor sander.

Parquet floors

Parquet floors have their own preparation problems. Damaged blocks may have to be replaced (matching blocks can generally be obtained from a demolition contractor) and loose but sound ones stuck back. Parquet blocks are hardly ever nailed down. They are stuck down with mastic, or sometimes asphalt, either direct to the floor or to a suitable underlay. In any case, the blocks should be stuck back with what was used to fix them originally.

If the parquet was not laid when the house was originally built, it may have been laid with a wood strip extending from it under the baseboard, with a gap between it and the wall to allow it to expand with changes in humidity. This strip may need replacing. Where original parquet extends under the skirting, no strip is needed.

There is a modern type of parquet floor which comes in large pieces, each appearing to consist of many individual blocks. In point of fact, the blocks are not blocks at all, but pieces of veneer stuck to a plywood backing with tongue-and-groove edges. This type of floor can be sanded, but it must be done gently. If you "sand through" the damage can be repaired only by replacing an entire plywood "tile," though you can patch the veneer if the damage is small enough in area and you work carefully.

Types of sanders

The floor sanding machines you can rent will sand up to within a few inches of the baseboard but no nearer. You will have to sand the edges and any awkward corners in some other way. You could do it by hand in a small room, but it would be exhausting. A disk sander of the type that fits on to a power drill will do it quickly and thoroughly, but may leave circular "swirl marks," which you may consider unsightly. Special flexibly-mounted disks are made that reduce swirl marks, but probably the best solution is to use an edger-sander, which can be rented from the same firm that supplies the floor sander.

The floor sander you rent should be the standard size even if you live in a mansion, as these machines work fast. The rental firm will normally supply you with plenty of coarse, medium and fine sanding sheets to fit the machine; you pay only for the ones you use. They will also give you advice on how to use the machine, but basically it is no harder to operate than an upright vacuum cleaner. Like a vacuum, it has a dust bag that collects most of the wood it removes from the floor. The only point to watch when using it is that the cord should be kept well away from the machine or you might sand through it. It is often a good idea to hang the cord over your shoulder while working.

Getting down to it

Start sanding with coarse abrasive sheets, working along the length of the boards, and not across them. The sheets must be fastened securely to the revolving drum on the machine or they will come to pieces. Continue with coarse sheets until the worst of the dirt and old finish has come off the floor, then switch to medium to smooth it off. Do not over-sand the floor, or the boards will become dangerously thin. And always keep the sander moving. If it remains in one spot, even briefly, it may sand a hollow in the floor.

Once the floor is in a reasonably clean condition, go over the edges with the smaller sander until they match the middle. Then, if you want a polished floor, put a fine sheet on the big sander and go over the floor repeatedly with it. If you persist, it will become smooth and shiny. Finishing the edges to the same standard should not be too hard with an orbital sander fitted with a fine sheet. Vacuum the floor carefully afterwards to remove all dust.

To protect the floor, it should be coated only with a finish made for floors. Follow the manufacturer's instructions as to the sealer coat (if recommended) and the number of finish coats. Avoid mixing brands, as many finishes are formulated as part of a "system" and may not be compatible with other brands.

Remember to start painting on the side away from the door and work toward it—this may sound like an unnecessary piece of advice, but it is surprising how many people forget and have to climb out of the window.

Warning

One final point: the wood dust from the dust bag can be put on the garden only if the lumber has not been treated with a preservative. And **NEVER PUT IT ON THE FIRE** or even let it get near one. Fine wood dust is *highly* flammable and can be explosive. It should be thoroughly wetted before disposal and put where there is no possibility of it catching fire.

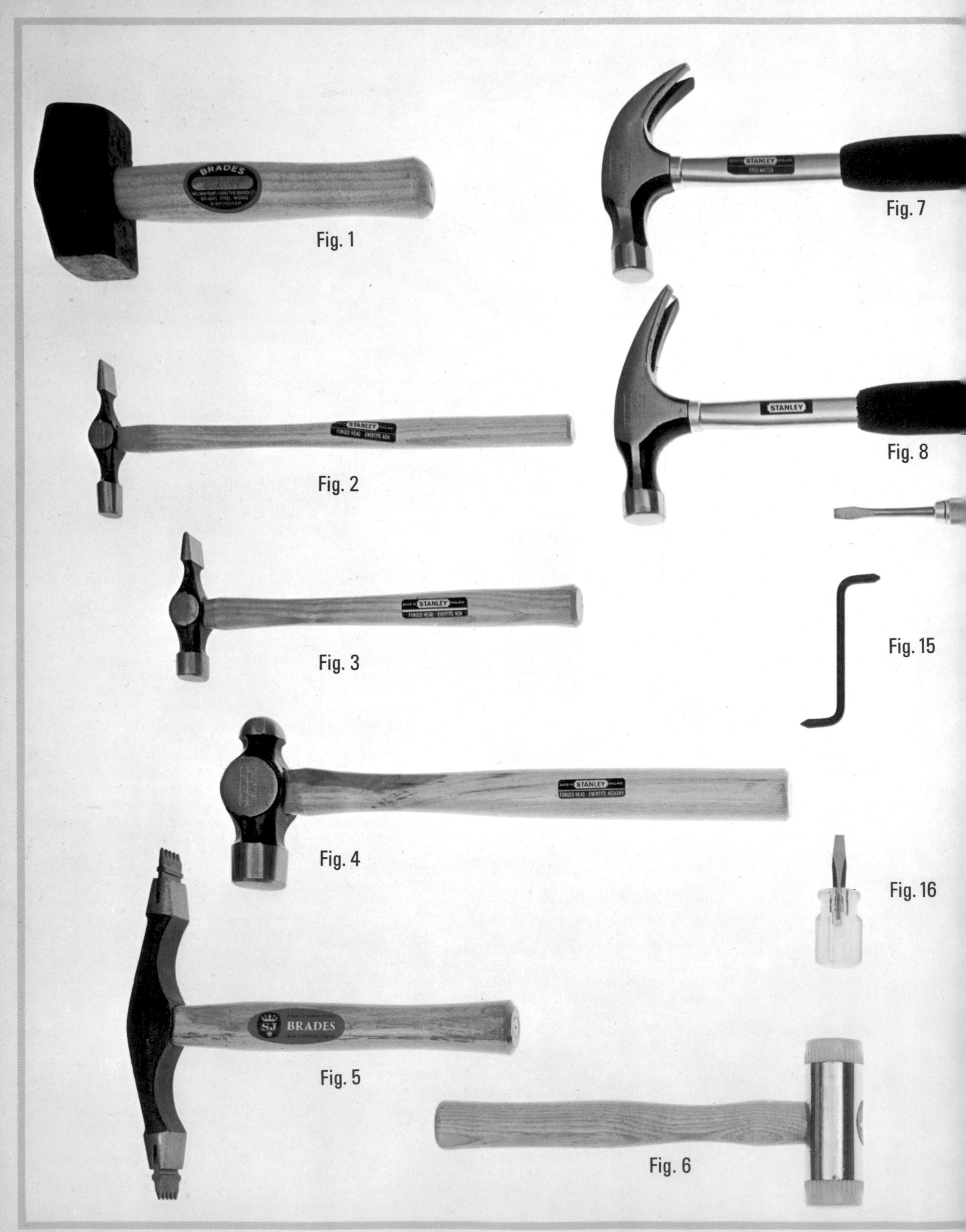

Fig. 1

Fig. 2

Fig. 3

Fig. 4

Fig. 5

Fig. 6

Fig. 7

Fig. 8

Fig. 15

Fig. 16

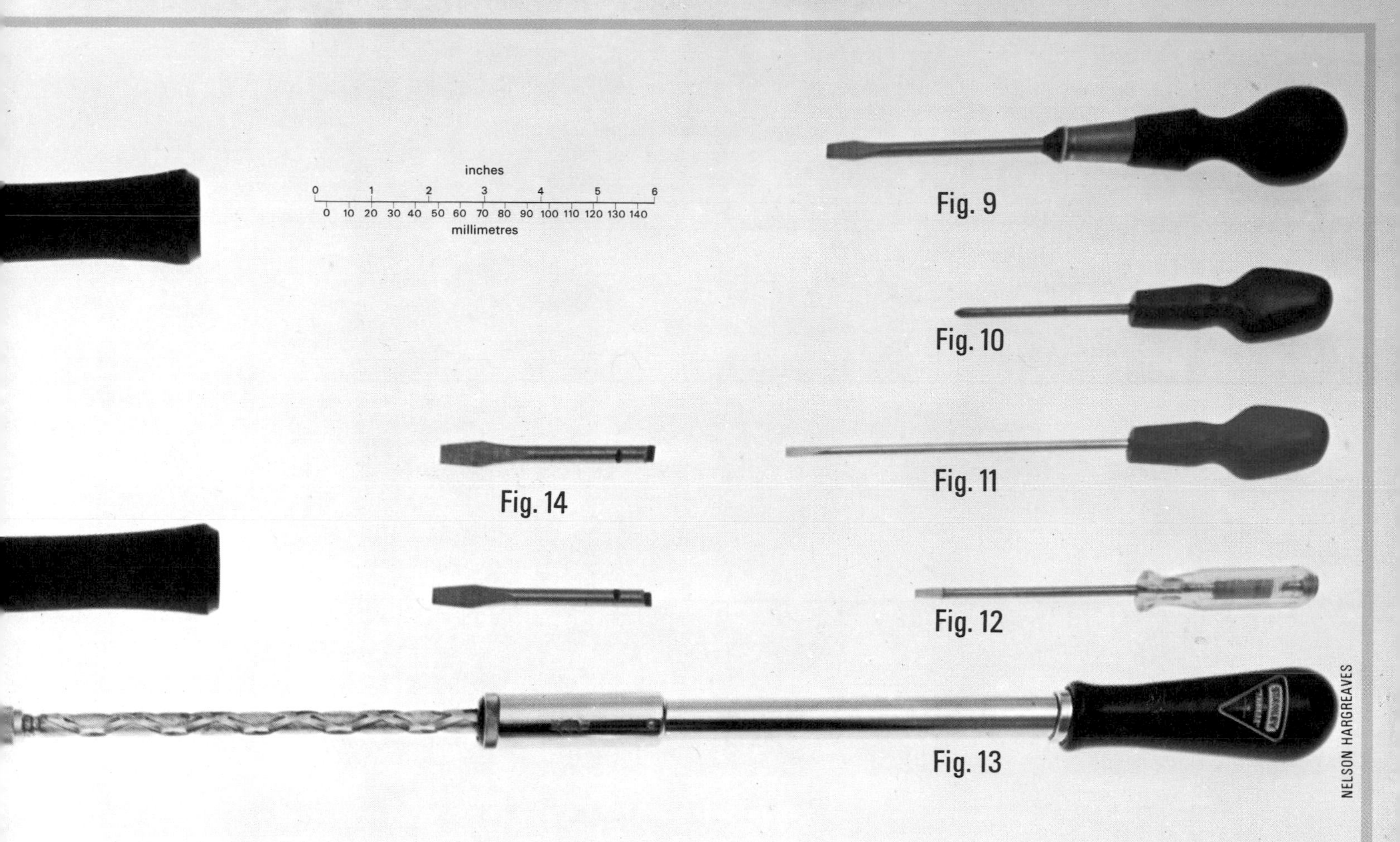

DATA SHEET

Hammers and screwdrivers

The right tool for the job makes all the difference—saving time, tempers and often money as well. The range of choice for even simple tools such as hammers and screwdrivers is extremely wide, and each one is designed with a particular task in mind. This DATA SHEET will help you match tool to task.

Hammers range in function from those used for heavy demolition work to those appropriate to fine nailing. In addition, each type of hammer comes in various weights.

Fig. 1. Stone cutter's hammer. 2½-4 lbs. Used for general heavy hammering, particularly in building and demolition work. In conjunction with a stone chisel it is used for cutting bricks, shaping paving stones, knocking through brickwork and so on. The *sledgehammer* (not shown) has a straight-sided head, a 3 ft. handle, weighs between 4 and 20 lbs., and is used for driving metal stakes, demolition work and so on.

Fig. 2. Long-handled cross pein hammer. 3½–4 oz. Used for tacking and light metal work. Wedge-shaped pein end spreads metal, as in lengthening a flat strip.

Fig. 3. Heavy cross pein hammer. 6–16 oz. Can be used for nailing. Designed for metal work, as above.

Fig. 4. Ball pein hammer. 4–40 oz. Used for metal working. The round end is used for starting rivets, also metal shaping, for example.

Fig. 5. Scutch or comb hammer. Used for trimming and shaping common or hard bricks. The combs can be replaced after wear. A special order item.

Fig. 6. Soft face hammer. Used in metal beating and in general work where it is important not to damage a surface. The soft head also avoids the possibility of a spark causing an explosion. Replaceable heads may be made of plastic, lead, copper or leather.

Fig. 7. Heavy curved claw hammer. 16–24 oz. Used for heavy work, as in framing.

Fig. 8. Ripping claw hammer. Used for "ripping" work, as in opening crates. Straighter claws make it easier. Can also drive nails.

Screwdrivers, too, have a range of functions and sizes. Match the screwdriver tip as closely as possible to the screw slot to prevent damage to either.

Fig. 9. Standard screwdriver. Used for general screwdriving of single-slotted screws.

Fig. 10. Crosshead screwdriver (Pozidriv, Reed & Prince, or Phillips). Used with cross-slotted screws to provide greater purchase and positive location.

Fig. 11. Cabinet screwdriver. Used in engineering, and otherwise, when the screw is inside a deep recess.

Fig. 12. Electrical screwdriver. The insulated handle increases safety in electrical work.

Fig. 13. Spiral ratchet screwdriver. Used for general purpose screwdriving. Pushing the handle home automatically drives or removes screws. Set slide button for driving or removing.

Fig. 14. The chuck can take blades of different widths, and even drill bits.

Fig. 15. Offset screwdriver. Used for driving awkwardly-placed screws, and for increased leverage.

Fig. 16. Stub screwdriver. Used in confined spaces. You can grip the square shank with a monkey wrench to give greater purchase.

CHAPTER 22

Repairs to sash and casement windows

If your house is an older one there's a good chance that the windows will need repairs from time to time. Although the details of their construction vary considerably, the repair methods described in the following pages can serve as a general guide. Simply adapt your fix-it procedures to the type of windows you have, and the end result should keep them trouble free for years to come.

Lack of maintenance is the usual cause of deterioration in softwood windows, causing them to jam and refuse to open and close properly. If paint is not renewed, it ceases, sooner or later, to fulfill its essential protective function and will flake and deteriorate, allowing the lumber to become saturated and swell in wet or damp weather. Tenons may rot and break, and glue joints come apart. In severe cases of deterioration, the only solution is to discard the window sash and fit a new one, since rot may have irreparably harmed the lumber fibers.

Ideally, repairs should be tackled during dry weather. In damp conditions, apart from your

ALAN DUNS

discomfort in working and letting cold into the house, lumber will remain swollen and the trouble difficult to rectify.

Where a number of windows require attention, remove one window at a time and carry out inspection, repair and replacement. A piece of polyethylene sheeting can be tacked in place across the window gap as a temporary protection. Wrap the polyethylene around thin battens, or hardboard, to prevent it from pulling away from the nails.

With care, it may be possible to remove upstairs windows from inside, and also fasten polyethylene, without using a ladder.

Where a ladder is needed, check that it is sound. Stand it at an angle of about 30° and on a firm base. Ensure that all rungs are firm. Steady the ladder firmly by sash lines fixed to it, and secured to a No. 14 screw eye driven into a convenient point on a soffit board, or into other firm woodwork (Fig. 1).

Lightweight scaffolds are a useful aid for this type of job. These can be rented and can be quickly assembled by slotting the frames together. Some can be moved around on wheels to the desired position, where the wheels can be locked.

Types of windows

The two basic types of windows are the *casement* window and the double-hung *sash* window—although the latter term is misleading, since the movable part of any window is called the "sash." Deterioration in a casement window is apparent because it will not close fully. Sash windows will stick and not slide properly and inspection will normally show distortion, sometimes causing windows to crack or break; this is often accompanied by deterioration of the sash cords, which have to be replaced.

The main operations are removing the windows; removing the glass; dismantling the frames; cleaning up the joints; gluing and repegging the joints; checking for squareness and general alignment; removing old paint from the frames; repriming and repainting; reglazing; and, finally, rehanging. If necessary, broken or rotted tenons may have to be remade.

Removing casement windows

If you are removing an upstairs window single handed, support the window by a sash cord or a strong clothesline before you loosen the hinges. The cord may be tied to a screw-eye, handy drainpipe, or a sound gutter support. This will help to prevent hazards such as windows falling out and becoming wrecked—or even hitting someone below.

Remove the screws holding the hinge to the frame. Those holding the hinge to the window are more easily removed once the window is taken out.

Before attempting to remove the screws holding the hinges, remove all paint from the grooves in the ends of the screws. This ensures that the screwdriver can get a good grip in the screw, avoiding slipping and damage to the screw head, which may then jam and have to be drilled and filed out. Clearing the screw head is best done with a spiked tool, such as an awl or icepick.

If the hinge screws prove difficult to remove, try tightening them first before undoing them. If this fails, a few sharp taps with a mallet on the end of the screwdriver might do the trick, or a nailpunch may be driven at an angle against the outer end of the screw groove to free it. Penetrating oil left to soak in will help to ease rusted screws.

Once you have removed the screws, place a pad on the ground below the window and lower the window on to this. The best pad is a sack filled with straw, rags or similar material. Remember, a window sash may be heavier than you think, so have some help handy, just in case.

Removing sash windows

Sash windows, correctly known as *double-hung sashes* or *box windows,* operate with cords, pulleys and weights, which counterbalance both inner and outer sash while they slide up and down. One end of the cord is nailed to a groove in the side of the sash, and the other is attached to a weight hidden in the frame.

The pulley wheels are attached to the *pulley stiles*—the upright sides of the frame which hide the weights. Part of each pulley stile consists of a removable piece of wood known as a "pocket," which fits flush with the stiles and serves as an access hatch to the weights. These pockets may be screwed in.

Carefully remove the glazing molding around the inside edge of the window frame. Start in the middle of a long run by gently prying it away from the main frame by about an inch. You can use an old chisel for this. Next, tap the molding smartly back into position; the nails holding it should pop up through the surface of the wood, to be removed with pliers. (A piece of thin ply between the pliers and the wood surface will prevent damage to the finish.) If this does not work, drive a wedge in the middle and use the chisel to lever progressively toward the ends of the molding.

Next, remove the parting strip (if present) between the sashes by using an old chisel to ease it out of its groove.

Now you can take out the lower sash and rest it on the window sill. Before going any further, mark the front of the sash with a pencil to show where the ends of the sash cords come. Make a corresponding mark on the frame.

Next, remove with a pair of pliers the fastenings holding the sash cords—but keep hold of the cords so that the weights at the other end do not crash down behind the stile boards. The inner sash can now be removed and stood aside. Repeat the marking procedure for the outer sash.

Finally, unscrew the pocket covers in the frame (or lever them out if they are not screwed, but wedged in place); remove the covers and take out the weights by pulling them through the pocket openings. The window detail is shown in Fig. 2.

Removing cracked glass

The most useful tool here is a beer can opener for removing the old putty if it is loose. For hard, firm putty use an old screwdriver or chisel for removal. When you replace the glass drive the glazing points below the surface of the wood with a small hammer. (These are small triangles of metal that hold the glass.)

First, chip out the old putty from in front of the glass, then remove the old glazing points. Lift out the glass (you will need old gloves to handle it). Then clear out the ⅛ in. or so of putty around and behind the glass, so that you are down to bare wood.

Old glass which is brittle and needs replacing can be broken out into a large bag or a paper sack. Use pliers to remove jagged pieces from around the edges.

Dismantling and repairing windows

First remove all window hardware—catches and so on—following the procedure above for removing tough screws if necessary. Then remove the glass.

For tenon joints, remove the wedges in the joints by drilling a bore hole down the middle of the wedge and then using a slim chisel to pry out the bits (Fig. 3 shows details of this type of joint). Not all windows have it.

For doweled tenon joints, remove the dowels by drilling them out, using a drill of the same size as the dowels.

Next, mark the lumber on each side of each joint with a letter or number. This will make it easy to identify the correct section when reassembling the sections.

Tap the joints apart by holding a block of wood against the frame and tapping them with a mallet. This must be done carefully in order to avoid damage.

Remove all the old glue from the joints (if glued) by using an old chisel or scraper or by brushing out loose particles. Clean the joint with fine steel wool.

Repairing tenons

Broken or damaged tenons should be cut off and replaced by a new tongue. This is done by replacing the tenon with a hardwood fillet, half of which becomes the new tenon, and the other half of which is inset into the rail (horizontal member of the window sash).

First, measure the length, depth and thickness of the old tongue. Next, mark out a piece of hardwood whose depth and thickness is the same as the old tenon's, but whose length is twice that of the old tenon plus ¼ in. For example, if the old tenon is 2 in. long, the fillet should be 4¼ in. long. The extra ¼ in. waste is allowed so that, when you reassemble the sash, it projects right through the mortise hole and is sawed and planed off to make a flush, flat surface.

Next, set a mortise gauge to the width of the tenon and mark it back along the rail. Cut off the old tenon. Continue the tenon lines across the end grain. Measure out where the fillet will be fitted into the rail, as in Fig. 4, using a rule. Place the rail in a vise and, using the techniques described in CHAPTER 5 for making the

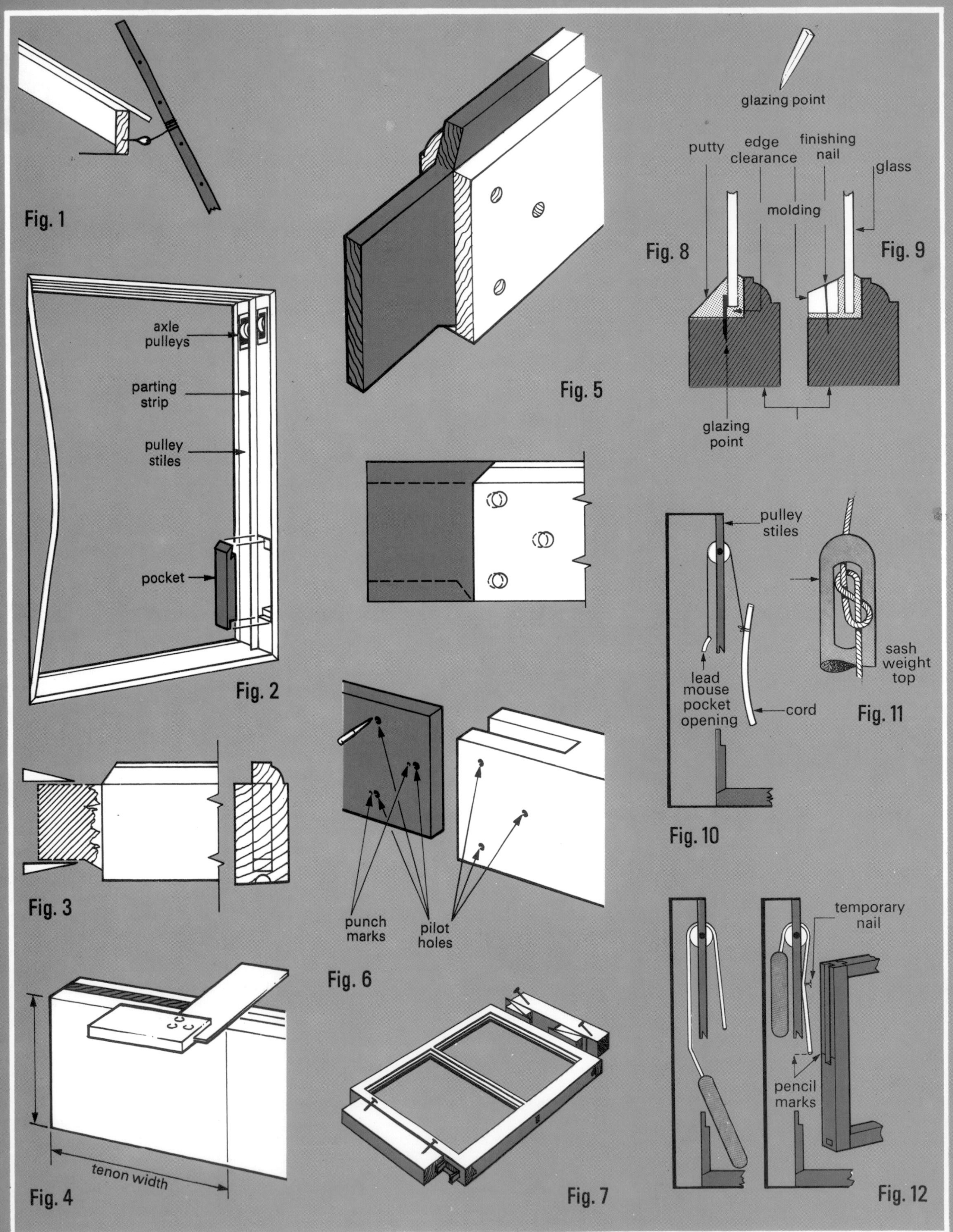
Fig. 1
axle pulleys
parting strip
pulley stiles
pocket
Fig. 2
Fig. 3
tenon width
Fig. 4
Fig. 5
punch marks
pilot holes
Fig. 6
Fig. 7
glazing point
putty
edge clearance
finishing nail
glass
molding
Fig. 8
Fig. 9
glazing point
pulley stiles
lead mouse pocket opening
cord
Fig. 10
sash weight top
Fig. 11
temporary nail
pencil marks
Fig. 12
TRI-ART

lap joint, cut out the unwanted section. Remember to saw on the waste side of the line. Now remove the waste lumber in the middle with a coping or bow saw and finish off with a firmer chisel of the appropriate width.

The new tongue is fixed into the rail by three ⅜ in. dowels and waterproof glue (Fig. 5). First, drill three pilot holes through the rail so that these just mark the hardwood tongue. Then remove the fillet from its position and, using a nail punch, mark new points about 1/32 in. on the outside of the three marks (Fig. 6). (Drilling the holes off-line in this way will help tighten the joints.) Now enlarge to ⅜ in. the holes in the rail, taking them right through to the other side, and drill matching ⅜ in. holes through the fillet. Slightly point three oversize lengths of dowel. Coat the end of the tongue, and the dowels, with an exterior grade of glue, tap the tongue into place and firmly hammer in the dowels. After the glue has set, cut off the excess dowel and plane smooth with the surrounding surface. The new tenon may now be haunched (cut back) as necessary to fit the mortise (Figs. 3 and 5).

Reassembling the sashes

The sashes can now be reassembled. Coat the tenons with adhesive and slide them into position in the mortises. Great care is necessary to see that the sashes are square. You can check the corners with a try-square, but a more accurate method is to measure from corner to corner across the diagonals; if the sash is square, the diagonals should measure exactly the same.

Once all four sides of the sash have been assembled, cut new hardwood wedges the same thickness as the tenon, glue-coat them, and drive them in from the outside, above and below each tenon (Fig. 3). To hold the work steady, and square, while you do this use sash clamps if you have them. If not, you can improvise clamps by nailing blocks of wood to a floor, as in Fig. 7, and using wedges to tighten the joints. Where large frames are being clamped, use wedges at both ends. Blocks and wedges should be clear of joints to allow for movement.

Or you can make a "tourniquet" of rope, tied right around the sash and twisted tight with a piece of wood. Before you leave the adhesive to dry, check again that the sash is square, and remove excess glue.

Sometimes excessive paint causes sashes to stick, and this is a good opportunity to remove and renew it. Strip the old paint, reprime and undercoat, and apply two top coats of paint.

Replacing the glass

Avoid using old or weathered glass for replacement glazing, as it may be difficult to cut—it usually gives a ragged cut edge. Some glass becomes more brittle as it ages.

You may wish to cut your own glass, but usually a glass supplier or handyman's shop can cut it for you. Most windows need single thickness sheet glass (24 oz.). For larger windows, double thickness glass may be needed.

Measure carefully for the new glass, using a steel tape. The measurement is made from the inside edge of each rabbet to the inside edge of the opposite one; then you subtract 3/16 in. to give an adequate clearance. Too tight a fit may make the glass crack.

A steel glass cutter is satisfactory for most glass cutting and works out cheaper than the traditional glass cutter's diamond. You need a large, flat surface to cut glass on. A felt-tipped pen can be used to mark guide lines on the glass, and a long straight edge or yardstick, or a home-made T-square, is needed to guide the cutter accurately.

First clean the glass. Then lubricate the line of the cut with turpentine. To cut, use a firm stroke, holding the cutter vertically. Never back-track, since the glass is then unlikely to break along the cut line. After the surface has been scratched, put a strip of wood, or the yardstick, beneath the glass under the score line.

Place your fingertips a few inches from the line and press down slowly and firmly on both sides. This should give you a clean break. If you have to trim any surplus from the glass, scratch a further line and gently break off the waste in small bites with a pair of pliers, or with the notches in the glass cutter end.

It is best to hold the cut sheet in a fold of newspaper or rag. Never stand it on or against a rough surface such as concrete, without protecting the edges from becoming chipped.

Before glazing, apply linseed oil or primer to all the rabbets to protect the wood and prevent oil in the putty from being sucked out.

As a rough guide, allow about 4–5 oz. of putty for each foot of frame, but always have an extra 1 lb. handy for contingencies. In some putties, there is a tendency towards excess oiliness and over-softness of the material, particularly if it is bought in a polyethylene or plastic wrapper. To remove the excess, the putty should be wrapped in newspaper which will absorb the oil.

The putty should be rolled in the hand until malleable. Use linseed oil to soften it if necessary. Next, line the rabbet of the frame with bedding putty to a depth of about ⅛ in. by "rolling" it from a ball of putty in the hand.

Place the pane gently into the frame, bottom end first, and holding it by the edges. Press it in at the edges; *never* from the center. This pressure will squeeze out surplus putty, leaving a bed to a depth of 1/16 in. or so. With heavier panes, you may need to put a couple of pieces of matchstick at the bottom to keep the pane from sliding down and displacing the putty.

Secure the pane by tapping the glazing points into the side of the sash, hard against the glass, at about 4 in. intervals. Keep the hammer sliding on the glass as you use it, so that you do not knock the glass and break it.

Now that the glass is in place, with putty behind and all around it, you need a strip of weathering putty around the outside. Feed it from a ball of putty in your left hand while you use a putty knife in your right hand to spread it, at an angle of about 45°. Fig. 8 shows the finished sections.

If the putty shows signs of sticking to the putty knife, keep the knife moist with water.

Attaching glazing molding

If you are using wood glazing molding instead of exterior putty, use rather less putty to line the rabbets. Prime the inside surface of the glazing molding and let it dry. Use only 1/16 in. or so of putty between the strips and the glass, and nail the molding to the sash with finishing nails. Then trim the putty at a slight angle to allow water to run off (Fig. 9).

Finishing off

On the inside of the pane, trim off surplus putty by cutting at a slight angle. This helps to stop condensation from collecting.

Both inside and outside, finish off by brushing over the putty with a soft paint brush. Clean any smears from the glass or surrounding woodwork with methylated spirits.

Putty should not be painted for about four weeks after it has been applied. When painting the frame, carry the paint line just beyond the putty and on to the glass to seal the joint and prevent water from getting between the putty and the glass. You can remove excess paint with a scraper after it dries.

Replacing sash cords

If the sash cords need replacing, buy only sash-type cord; otherwise you will have to allow about an inch for stretching in use.

Begin with the outer (upper) sash. As well as the cord, you need a length of string and a flat piece of lead made into a "mouse." Roll the lead around the end of a 5 ft. or 6 ft. length of string at about the thickness of a cigarette but half as long. Bend the lead mouse slightly in the middle and feed it over the groove of the outer pulley wheel until the mouse falls down behind the stile. Tie the new sash cord to the other end of the string. The cord can now be pulled over the wheel and out through the pocket opening, and the mouse removed (Fig. 10).

Tie the sash cords to the top of the weights. Either use a flat-finish knot, or bind the loose end of cord, so there is no lump to interfere with the window-opening action (Fig.11).

Pull the weights up about 2 in. from the bottom, and half drive a nail through each cord into the pulley stile to hold the weights temporarily in position. Cut each cord level with the pencil marks on the stile (Fig. 12). Be sure weights will have full range of travel as sash moves up and down.

Now position the outer sash so that you can fit one of the cords into its groove. Align the end of the cord with the pencil mark on the sash edge, and fasten it with four or five nails, starting at this point (Fig. 12) if sash is of the type shown.

Once both cords are fastened, take the temporary nails out of the cords and stiles, and lift the sash into position. Give a trial run by sliding the sash vertically.

The two weights for the inner (lower) sash are corded in the same manner.

The pockets can then be replaced, the parting strip sprung back, and the inner sash lifted back. So that the sash slides smoothly, run spray lubricant in the two channels and on all sliding surfaces.

TUBBY

Left. *A Dutch door from the kitchen to the garden prevents children tramping in and out on a wet day and yet provides extra ventilation.*

CHAPTER 23

Dutch doors for a farmhouse look

Dutch doors, traditionally seen in farmhouses and rustic country cottages, can be both an attractive and functional addition to various rooms in the more conventional home.

Dutch doors are doors that are separated into two pieces across their width. This enables one half to be opened independently of the other. Traditionally Dutch doors were ledged and braced but any design, providing it incorporates a middle horizontal rail, can be copied as a Dutch door.

Standard doors are 6 ft. 8 in. high in different widths, but there are many variations. The size of the door you build will obviously depend on the size of the door you intend to replace. The construction involves conventional joining techniques and rabbeting.

Construction

A Dutch door consists of a frame composed of four horizontal members, or rails, and two vertical pieces, or stiles, which run the whole length of the outsides of the door though cut at the door's separation point. The top rail of each part of the door is made from lumber of the same size. For standard doors and all but very large doors pieces of 4 in. x 1½ in. (cut from nominal 2 x 6) lumber are ideal. The bottom rail of the top part of the door is slightly wider than the rail it meets to allow for a ½ in. matching rabbet to be cut. 4½ in. x 1½ in. is ideal for this rail. The bottom rail of the finished doors should equal the finished width of the two middle rails—8½ in. x 1½ in. being about the right size. The bottom rail is double haunched mortise-and-tenoned into the stiles, and the other three rails are single haunched mortise-and-tenoned (Figs. 3 to 4).

The space between the rails and the stiles is filled with tongue-and-groove boards, or "T & G." These are nailed to rabbets cut into the inside edges of the stiles and rails. Dutch doors are usually made with braces—diagonal pieces of lumber the same width as the stiles, placed on each half of the door along the T & G material. The braces run parallel to each other, are fastened to what will be the back of the door, and their top faces, when fastened, are flush with the top faces of the stiles and rails. Braces are not essential but they do give extra strength—provided they run *upward* from the hinge side.

A matching rabbet is cut into the two middle rails, and across the stiles that meet them, to provide a weather-tight joint where the two halves of the door meet.

Setting out

If the two halves of a Dutch door were made separately, they would almost certainly not be parallel when they were hung. So you make the whole door in one piece, using stiles which are over-length, and cut it in two when construction is nearly complete. This means that you must set out the height of each half by measuring over the *rails,* not by measuring down the stiles. Remember to allow an extra ½ in. for the overlap in the middle; your doors will shrink by this amount when you rabbet them together.

The stiles should be cut about 2 in. too long. This will allow for ½ in. waste at each end and 1 in. in the middle. The inch in the middle allows adequate room for waste. The gap in the middle of the door will resemble a narrow mail slot when the door is assembled. The rails must also be cut slightly over-size; the waste is removed after the mortise and tenon joints are completed.

At this stage the areas of the joints in the stiles and rails can be marked. Lay the two stiles together with their long edges butting. Measure and mark a point in the middle of the stiles' length. On each side of this point measure and mark ½ in. toward the ends of the stiles. Square lines through these points.

All measurements for the areas of the joints must be made from these squared lines. For the area of the joints in the upper part of the door mark from the top squared line. To mark the area of the joints in the lower part of the door measure from the other squared line.

From the respective squared lines measure and mark half the length of the finished door toward the end of the stiles. From this point marked in the top part of the door measure and mark 4 in.—the width of the top rail—back toward the center of the stiles. From the points marked near the ends in the bottom part of the door measure and mark 8 in. back toward the center of the stiles. This indicates the width of the bottom rail. Returning to the central squared lines, mark off toward the ends of the stiles the width of the two middle rails—4½ in. for the rail in the top part of the door and 4 in. for the rail in the lower part of the door.

Working from the middle, mark out each rail using the same method as for the stiles. The first pair of squared lines should mark the length of each rail between the stiles; what is left over (4 in. and a bit) is for the tenons, which run right through the stiles and protrude slightly on the other side.

Now lay the rails and the stiles together in the positions they will occupy in the finished door. Mark the top surfaces and the edges that are to be rabbeted. This will avoid the danger of cutting joints or rabbets the wrong way around.

Cutting the mortises

The next step is to cut the mortises in the stiles. Fig. 4 shows the shape of the haunched tenon which will later be cut on the ends of every rail (except the bottom rail) in the finished door. The mortises should be cut to accommodate this shape. The tenon is one third the thickness of the rail—in this case ½ in. The tenon also has a stepped shoulder which is equal to the depth of the rabbet which will later be cut on the inside edge of the stiles and rails. The depth of the rabbet, given the measurements stated earlier, is ½ in. The haunch in the tenon is about one third the length of the tenon—which equals the width of the stile. So the mortise has three depths, the main part of the tenon going right through the stile, the stepped shoulder entering to a depth of ½ in., and the haunch entering to a depth of about 1⅓ in.

Carefully mark out the mortises on the ends of each stile, referring to the data in Figs. 3 and 4. In the case of this complicated joint, do not cut the whole mortise yet. Cut the part that accommodates the main piece of the tenon but simply mark out the area of the haunch and the stepped shoulder. These can be cut to depth after the full shape of the tenon has been cut.

The single haunched tenons can now be cut. The shape and dimensions of these are shown in Fig. 4. Then finish cutting the mortises, trying the tenon in the joint from time to time until the stepped shoulder and haunch are cut to the right depth.

The next step is to cut the mortise and tenon joints in the bottom rail and bottom of the stiles. Fig. 3 shows the shape and dimensions of these double haunched mortise joints. The tenon has two tongues, a stepped shoulder and a haunch.

First mark out the position of the mortises. Cut only that part of the mortise that will receive the two tongues of the tenon that go right through the stile.

Then cut the tenon. Score a line with a handyman's knife and straight edge along the squared lines you have previously marked on the ends of the bottom rails to indicate the width of the stiles. Set a marking gauge to one third the thickness of the rail. On the narrow edge of the rail score a line from the squared line to the edge of the rail, with the marking gauge. Continue this line on to the end grain and along the bottom edge to the squared line. Repeat this process with the block of the gauge flat against the opposite face of the rail.

A single tenon can now be cut to the overall size shown in Fig. 3. On the side that is not to have the rabbet, saw down the scored line through the end grain. Stop the cut when you reach the squared line. Then saw down the squared line through the face of the rail until you reach the original cut at right angles. Remove the waste. On the opposite face—the one at right angles to the edge to be rabbeted—mark inward a distance that equals the depth of the rabbet. This will indicate the size of the stepped shoulder. Square a line through this point and on the two edges. Saw away the waste as before.

Above. *For a children's room Dutch doors give necessary contact with the rest of the house at night but also prevent children wandering.*

Next mark out the positions of the two tongues on the existing single tenon. From the top of the tenon mark it into quarters along the end grain. Square lines through these points onto the face of the tenon. Mark the distance of the haunch, which is one third the length of the tenon, from the original squared line toward the end of the tenon. Cut out the waste down to this line to the shape shown in Fig. 3.

Now complete the mortise that will receive the double haunched tenon.

Cutting the rabbets

The inside edges of the rails and stiles are rabbeted to receive the tongue-and-groove boards. The rabbets are cut to a depth that equals one third the thickness of the lumber—½ in. if 1½ in. lumber is used. Their width along the wide surface of the pieces should be about ½ in.—giving you plenty of room to nail ½ in. thick boards in place.

The rabbets can be cut the whole length of the inside edges of the rails with an ordinary rabbet or plow plane. But along the edges of the stiles the rabbets run to ½ in. inside the squared lines that were marked to show the area of the joints. This will form a corner between the rabbeted stiles and rails. On the ends of the stiles mark a distance of ½ in. from the original squared lines toward the ends. From the squared lines that denote the position of the middle rails mark inward ½ in. toward the central waste area. Square lines through these points part of the way across the top surface and on the edge to be rabbeted.

The rabbets can now be cut in the stiles between these points. Stopped rabbets are difficult to make with either an ordinary rabbet plane or a plow plane. This is because the body of the plane projects about 3 in. in front of the blade and about 5 in. behind it. This means that you cannot stop the rabbet accurately at the point required. A special type of rabbet plane, called a bullnose rabbet plane, is specially made for stopped rabbets. Or you can do most of the job with an ordinary rabbet plane, and finish off with a chisel.

Assembling the door

The skeleton framework of the door can now be fitted together. Assemble the door on a perfectly flat surface, and check that the two parts of the door are the correct size. (Remember the ½ in. rabbet still to come in the middle!) When satisfied, glue the parts together with a waterproof glue such as resorcinol. Do not use white glue as this is soluble in water and is, therefore, not weatherproof. Check that the door is square with either a large square or by measuring the diagonals. These will be exactly equal if the door is square.

The next step is to drive thin, glue-coated wedges from the outsides of the stiles into the mortise and tenon joints with a mallet. Do this before the adhesive dries. The wedges will make a tight fit between the mortise and the tenon. If the door is not square the fault can be rectified by driving the wedges home with extra force at suitable corners.

Clamp the assembly using bar clamps with waste wood blocks between the shoes of the clamp and the edges of the stiles. Wipe excess glue from the surfaces of the doors.

Cutting the braces

The length of the braces will depend on the size of the door. Their width should equal that of the stiles, and their thickness should fill the space between the T & G and the back of the frame, so that stiles, rails and braces are all

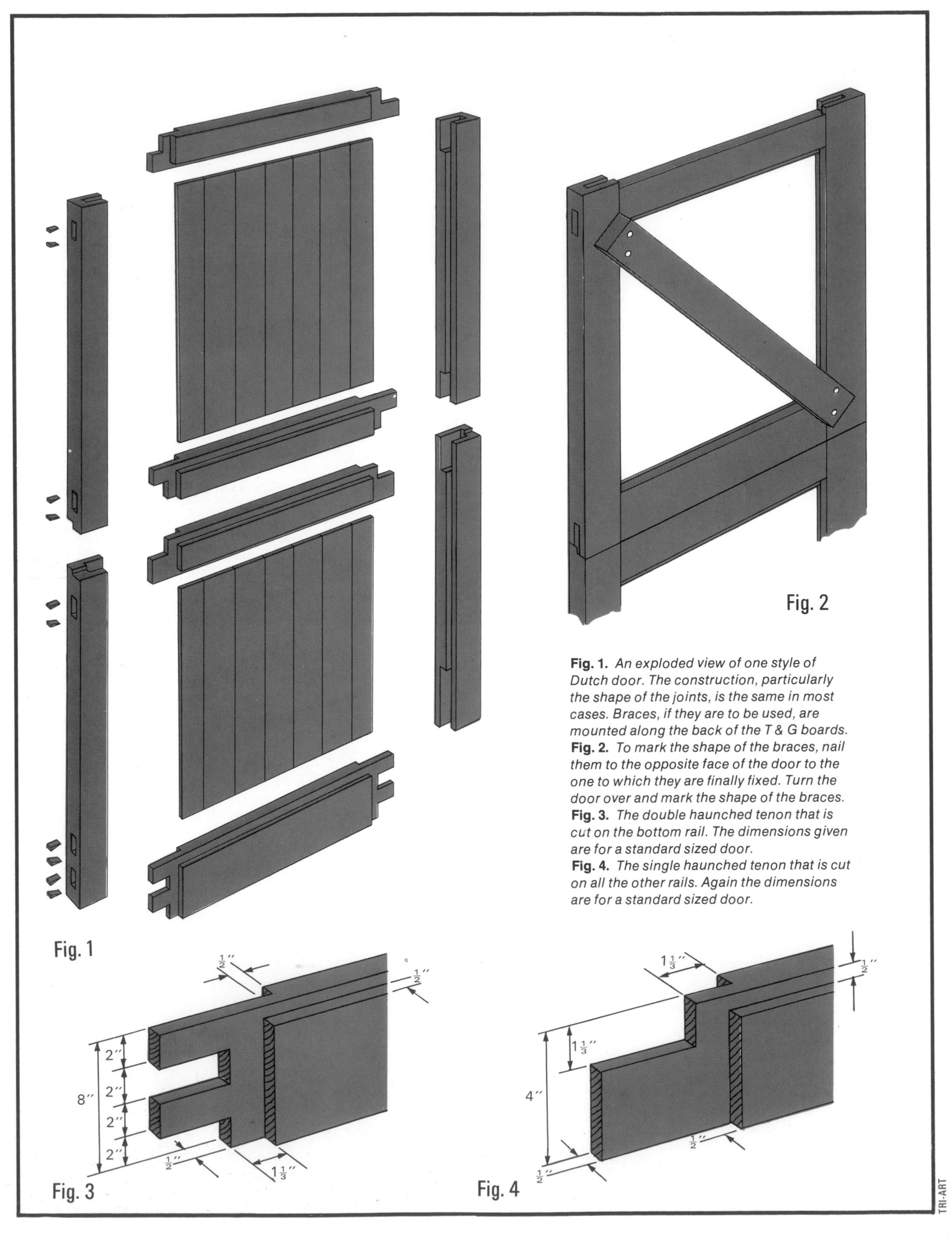

Fig. 1. *An exploded view of one style of Dutch door. The construction, particularly the shape of the joints, is the same in most cases. Braces, if they are to be used, are mounted along the back of the T & G boards.*
Fig. 2. *To mark the shape of the braces, nail them to the opposite face of the door to the one to which they are finally fixed. Turn the door over and mark the shape of the braces.*
Fig. 3. *The double haunched tenon that is cut on the bottom rail. The dimensions given are for a standard sized door.*
Fig. 4. *The single haunched tenon that is cut on all the other rails. Again the dimensions are for a standard sized door.*

flush. Cut the braces several inches over-size. Lay the braces across the rails so that they cover the corners, and lightly nail them in position. Turn the door over, and direct-mark the braces by running a marking knife around the corner. With a saw and smoothing plane cut down to these lines. Fig. 2 shows the braces nailed across the rails. Now remove the braces until the T & G has been nailed in place.

Now sand the back surfaces of the T & G with either an orbital (not rotary) sanding device fitted to an electric drill or sandpaper wrapped around a wood block.

Fitting the boards

The tongue-and-groove boards are fitted as follows:

Cut the T & G to the required length. Prime the ends (which fit into the rabbets) and the tongues and grooves with a waterproof sealant. Remove the groove edge from the first piece and place this piece into the rabbet, butting the inside of the stile, with the tongue side outward. Nail through the surface of the board into the stiles and rails with galvanized nails, clamping the boards.

Now mount the braces. They can be glued into place first. Then nail them to the T & G, nailing through its surface. Nail the braces to the stiles and rails by angling the nails.

Separating the two parts

The assembly can now be cut in two. Separate the two components of the door by sawing through the stiles into the middle waste area. Saw away any waste that protrudes from the ends of the stiles and remove any parts of the wedges and tenons that stick out at the joints.

The rabbets in the meeting rails

Cutting the rabbets in the meeting rails is the final step in the construction of a Dutch door. Set the rabbet plane to half the thickness of the meeting rails. Take the top half of the door. Clamp a waste block to the end to which you will be planing. This will avoid damaging the end grain of the stile. Lay the body of the plane along the face of the middle rail that will be the back of the door. Cut a rabbet along the rail and two stiles to a depth of ½ in.

Then cut the rabbet in the lower part of the door. The process is the same, but the body of the plane must lie on the surface of the rail that will be in the front of the finished door. Cut this rabbet to a depth of ½ in. also.

Finishing the door

Sand the stiles, rails and braces perfectly smooth. The two parts of the door can either be painted or varnished. If they are to be painted, apply a coat of sealer before applying the undercoat. The finish must be waterproof.

Hanging the door

Each half of the door is hung with two 3 in. butt hinges. Cut the positions of the hinges with care, first on the door, then on the frame. Hang the bottom door first. The top door can then be hung, but take care to ensure an accurate fit. Fit each part of the door with a bolt.

SYNDICATION INTERNATIONAL

Above. *This Dutch door's finish attractively contrasts with that of the table to add an extra touch of old-fashioned charm to this room.*

Below. *Here the Dutch doors, together with the table, chairs and tiled floor, give this otherwise modern kitchen a farmhouse look.*

VIVIEN HISLOP & JOHN PRIZEMAN/WESTINGHOUSE KITCHEN, ADVANCE DOMESTIC APPLIANCES LTD

NIGEL MESSETT

CHAPTER 24

Make a military chest

If you need drawer space and are contemplating building a chest of drawers, then you should seriously consider making a military chest. It takes a little longer to construct, but a military chest is more versatile, attractive, and just that bit different.

The military chest was really years ahead of its time—and is still an up-to-the-minute piece of furniture. It is probably the earliest example of unit furniture, because two chests can be built to stand together, or apart, either as a four-tier chest of drawers or as a pair of chests that might double as seating space. Not only that, in the past it also doubled as luggage! This versatility means a military chest could be an attractive addition to a modern home.

Origins of the military chest

Military chests first became popular as a convenient method of storage for soldiers and gentlemen making a sea voyage, from Britain to India for example, for long terms of office. It is designed in two sections, which makes it simpler to handle when moving. (This is still an advantage today if you are contemplating a move to a new home.) The top section rests on the lower section when the chest is in use. The chests were enclosed in a heavy canvas bag while on voyage to protect the surface from scratches. On arrival they would promptly be installed as units of furniture.

The handles are always made of brass, with brass corner brackets, and all the attachments are inset for protection in transit. There is provision underneath for detachable feet.

Often the top unit was a military *desk,* or contained a dressing table. You will find some examples in books about antiques and you may see specimens in antique shops. These will give you some alternative design ideas.

The originals were invariably made from teak, mahogany, or similar exotic woods; but the one shown here used more readily available—and more economical—pine.

Construction

This chest is 36 in. in length, 18 in. deep and 16 in. high, but the dimensions can easily be adjusted to suit your particular requirements. You can make two at once to provide the two sections of a four-drawer chest.

The woodworking joints on an original chest would almost certainly be dovetailed, but in this project the construction has been simplified and the joints are rabbeted, glued and screwed. For appearance the screw heads are sunk and the recesses covered with plugs of doweling or plugs of matching wood extracted with a power drill plug-cutter attachment available from cabinetmakers' suppliers.

The rabbeted joints are cut with a router such as a Stanley router as shown in Fig. 4, but the same result can be achieved using either a bench saw or portable electric circular saw.

The chest is 18 in. deep. It is difficult, to say the least, to buy wood of this width, so for the top, bottom and side panels two planks, each 9 in. wide, are butted and glued along their long sides and clamped tightly together while the glue sets.

To do this, lay a pair of planks on a flat surface with their long edges touching. Arrange this so that the faces with the most attractive grain are on the same side—this will be the outside of the chest—then check that the butting edges fit together with no unsightly gaps. If there are any high spots that prevent the edges butting evenly, plane or sand them down.

The edges must now be glued with a woodworking adhesive such as white glue. For this you will need at least two furniture clamps, preferably three. If you have no clamps, you can improvise by using strings and wedges.

Set out two clamps and lay two planks over them as shown in Fig. 3. Apply adhesive along both meeting long edges, place waste blocks or strips of wood between the outer edges of the planks and the jaws of the clamps to prevent the jaws marking the wood, then tighten the clamps. If you have a third clamp, place it in between the other two, but running *over* the planks. This will prevent the planks raising or "jack-knifing" when the lower clamps are tightened.

The planks must fit together perfectly with their surfaces completely level. If there is some uneven alignment, loosen the clamps a little, place a waste piece of wood on the raised plank and tap it lightly with a hammer. Then re-tighten the clamps.

Wipe any excess adhesive from the joint, using a soft damp cloth. Otherwise any glue left on the surface of the wood will show as a white patch through the polyurethane finish that is applied later. Leave for the adhesive to set.

Making the case

First cut and fit the case, the top, bottom and end pieces. As you can see in Fig. 5, the ends fit in between the top and bottom panels. Although no joints have been cut yet, briefly dry-assemble the case to check that the frame is square.

Next cut the rabbet recesses in the top and bottom panels to take the end panels. These are cut to the same depth as the thickness of the butting end panel, leaving a ¼ in. lip. This is shown quite clearly in Fig. 14. In this case the rabbets were cut with a router, which in fact cuts a groove. To make a rabbet, a groove of the required depth is cut along the end of an edge as shown in Fig. 4, then another groove is cut out at right angles to the first, creating the "L" shaped rabbet recess for the end panel to fit into.

Briefly dry-assemble the case and with a pencil lightly mark the four edges that will be at the back of the chest. Dismantle the case and cut a ¼ in. deep rabbet, ¾ in. wide around the inside edges of the rear. This creates a recess into which the plywood back fits, as shown in Fig. 13A.

Spread adhesive in the rabbet recesses of the ends of the top and bottom panels, fit the end panels in place and stand the frame on one end, with the opposite end clamped as shown in Fig. 6. With the clamps in position, drill the countersunk screw holes at approximately 3½ in. intervals, then drive the screws firmly in position and fill the screw recess with wood plugs as shown in Fig. 6. Repeat with the opposite end, square the case and leave for the glue to set.

The runners and cross member

The cross member—the center front in the cutting list—must be dadoed into each of the end panels at the front. This in effect makes the case stronger and keeps the drawers apart. It is located exactly halfway up each end panel.

Place the case on its back on a flat surface and, using the cross member for direct marking, measure halfway up each end panel. The cross member must be dadoed down into the depth of the panel so that it is flush with the front, and the ends must be dadoed into the thickness of the panels leaving ¼ in. on the outside, the same as with the rabbet recesses.

When you have marked the joints, cut them out with a sharp wood chisel. Then glue the ends into the recesses, securing the member with a countersunk screw into each end through the panel. There is no need to plug this screw recess because the head will be covered by an angle bracket.

Next fit the middle drawer runners. These are fixed level with the cross member, and the front of each runner is cut away so that it fits *around* the cross member as shown in Fig. 13. The cut away part is marked by direct marking—butt it against the cross member—and make two cuts down its length with a back saw, then chisel out.

Line the runners up with the aid of a square and fasten in position as shown in Fig. 7.

Now fit the bottom drawer runners in position. These are only glued and nailed in position in the bottom corners. Fig. 7 shows one being nailed.

The drawers

These are made in exactly the same way as the case except that the front panel, which is larger than the sides and back, has a rabbet on each side which is stopped at each end. Fig. 8 shows this quite clearly. To do this, lay one of the side panels against the end grain of a front panel so that it lies centrally, that is with the same amount of end grain showing at each end, and ¼ in. from the front of the panel, and

Left. *This impressive military chest is an example of all that is best in traditional design and will be the envy of your friends.*

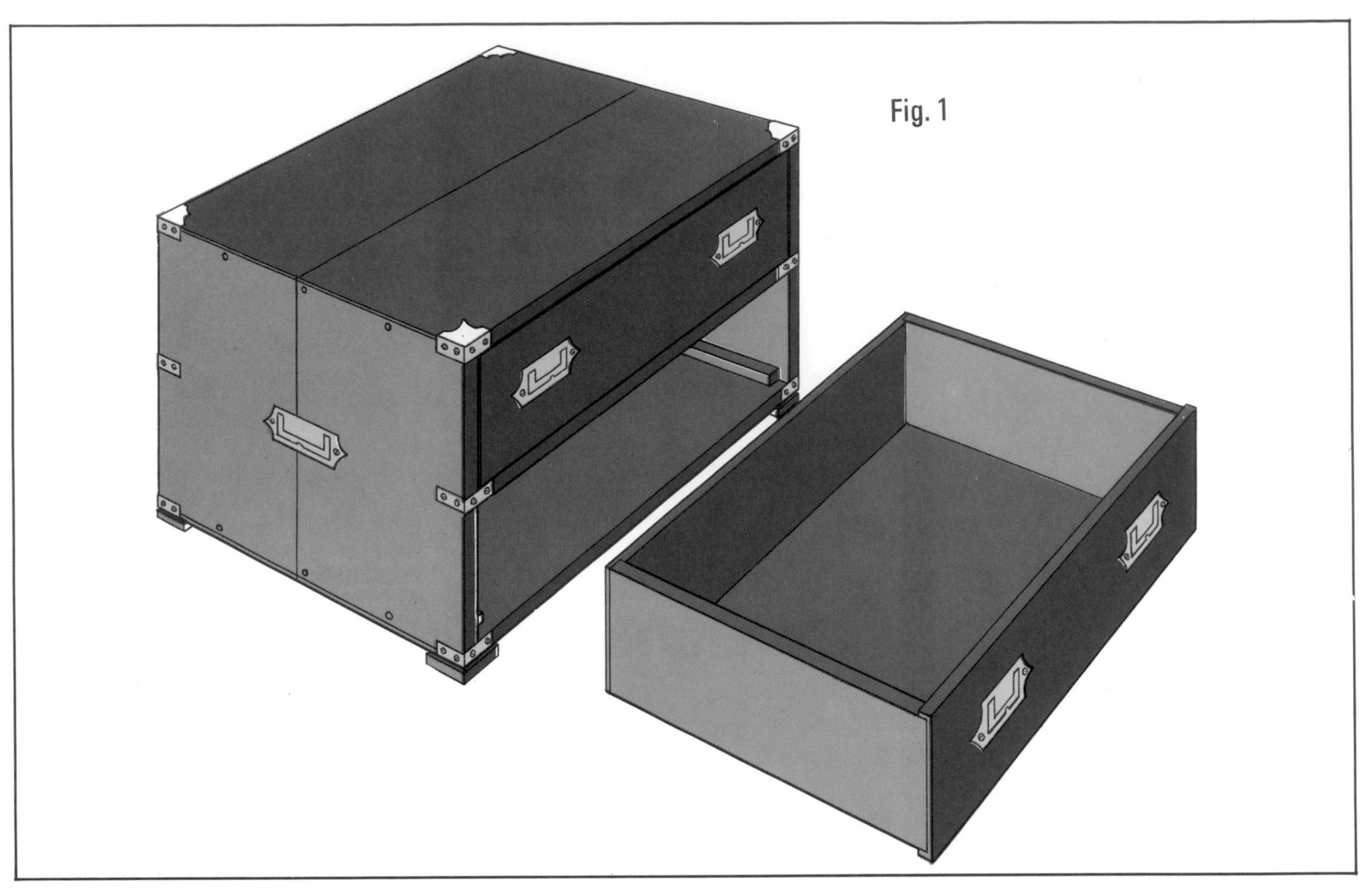
Fig. 1

3

4

5

6

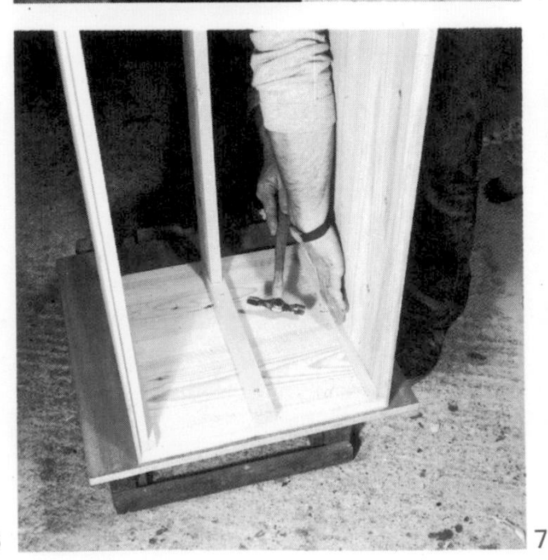
7

8

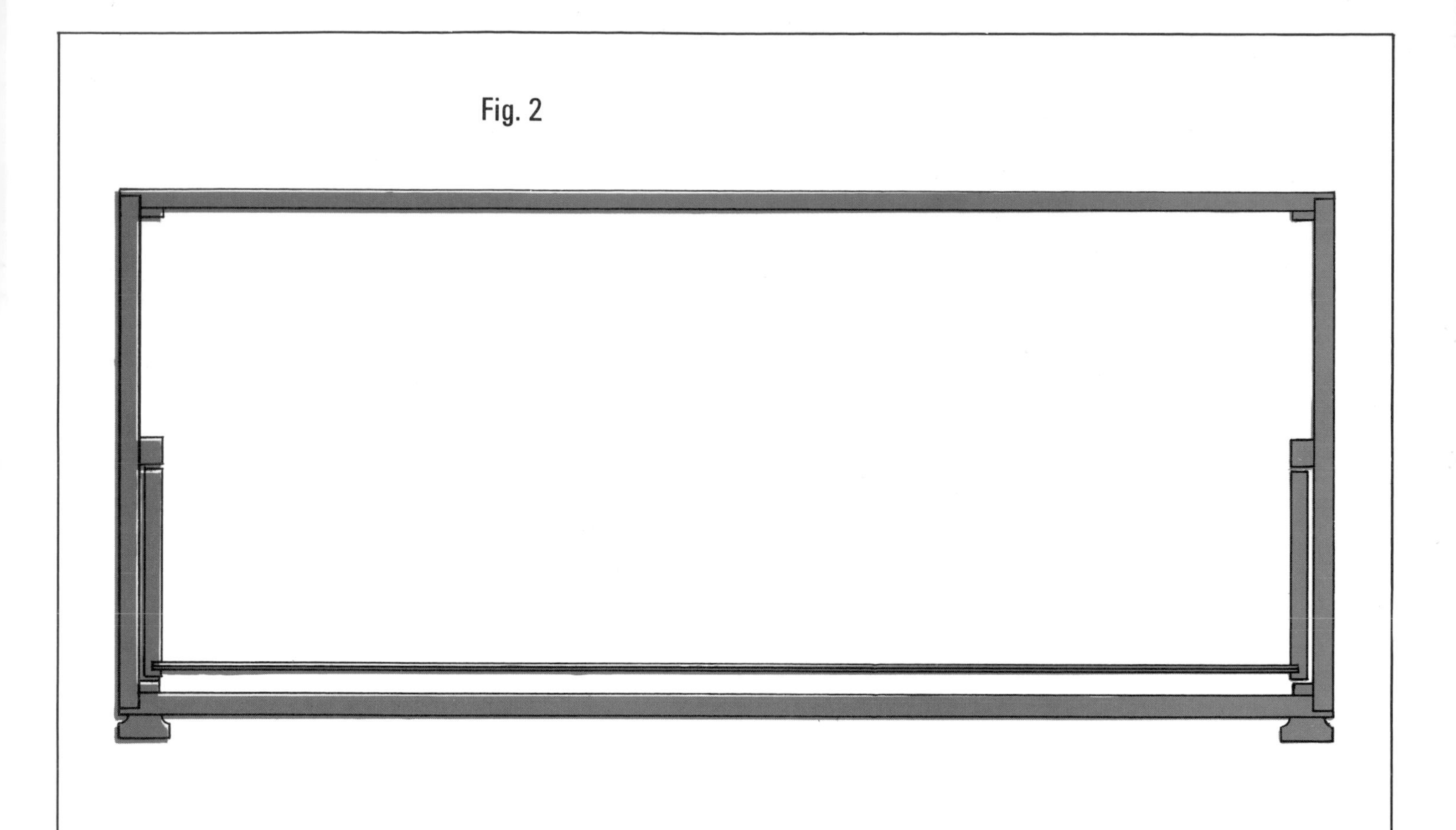

Fig. 1. *The complete unit. All brass fittings, including handles, are recessed flush with the surface.*
Fig. 2. *Front view of the chest case with runners and legs. The sides and base of the lower drawer are also shown.*
Fig. 3. *To join two planks of pine, spread adhesive along butting edges and clamp together with furniture clamps until the glue sets.*
Fig. 4. *One way of cutting a rabbet is to cut a groove with a router, then cut another groove at right angles.*
Fig. 5. *This shows how the top and bottom panels are rabbeted to take the side panels, which just butt into place.*
Fig. 6. *Three aspects of construction. Sash clamps are in place; screws driven in; and wood plugs have been fitted.*
Fig. 7. *The drawer runners and center front member in position. One end of the middle runners is cut away as shown in Fig. 13B.*
Fig. 8. *Side front of one of the drawers. Note how the side panel is recessed into the front panel, which is cut away to provide a rabbet which is stopped at each end.*
Fig. 9. *Locate the drawers so that they are flush at the front, then nail battening around the inside back to provide stops.*
Fig. 10. *Use each corner bracket as a template to mark its own outline, then chisel out to the depth of the brass.*
Fig. 11. *There is no need to use corner brackets at the bottom corners. Angle brackets are easier and cheaper to fit.*
Fig. 12. *Placing the back in position. This is simply glued and bradded.*

9

10

11

12

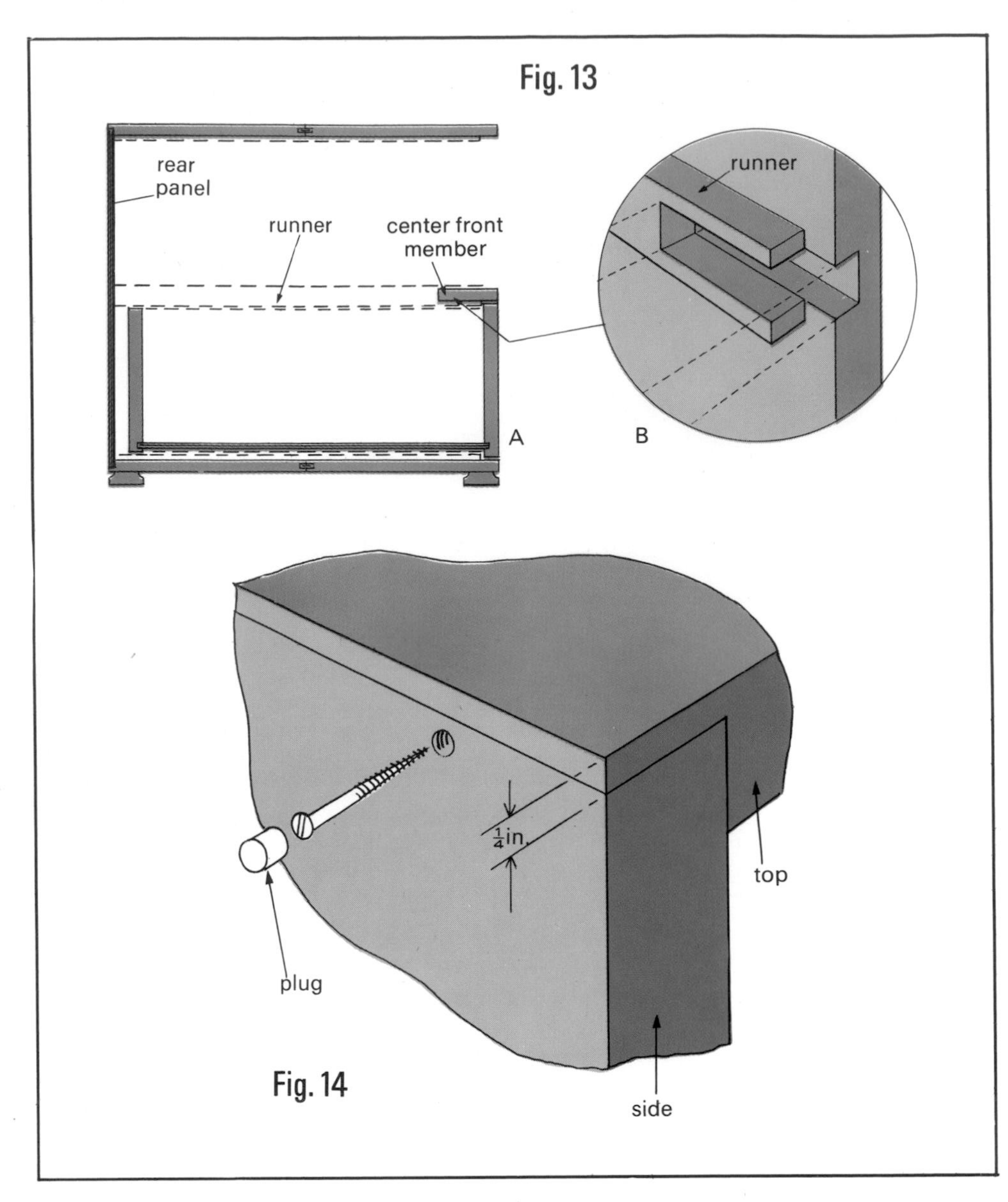

mark around the side panel on to the end grain of the front. With the router, cut two right angled grooves, stopping just short of each end, then chisel the rest out.

Mark and cut the rabbets to take the side panels on the rear panel, then dry-assemble the drawer to check for squareness. This is easier if you lay the front panel on its face, stand each side panel into the stopped rabbets, then place the rear panel on top.

If everything fits, mark a line around the inside of the drawer, 3/8 in. inward from the bottom. Dismantle the frame and with the router cut a 1/4 in. wide groove, about 1/4 in. deep, around the marked line. This will house the edges of the drawer bottom panel.

Now check that the drawer fits properly before you fasten the joints. Place the front panel in position in the front of the chest and, from the back, insert the two side panels in the stopped rabbets, slide the bottom panel through the dado, then place the back panel in place. If there are any badly fitting parts, these can be planed or sanded down.

Dismantle again, place the front panel face down on a flat surface, spread glue over the joints, set the bottom panel in place, assemble the frame, drill screw holes and drive the

Fig. 13A. *Side view of the chest. The rear panel has been glued and bradded in place and the center front member and runners fastened in position.* **B** *shows how the front end of each middle runner is cut away to allow the center front member to fit through it.*
Fig. 14. *Front corner section of the case, showing the joint used.*

screws home as shown in Fig. 8. The bottom panel (if plywood) may be glued. Repeat for the other drawer.

When the adhesive in both drawers has set, slide the drawers into the chest so that the front panels are flush with the front of the chest.

Finally, glue and nail some "drawer stops" in place at the back of the chest as shown in Fig. 9, to ensure that the front stops flush. Then glue and brad the rear plywood panel in place, driving a 1 in. brad in every 2 in.

Handles and brackets

These are all inset, and to ensure a perfect fit each one is direct marked and cut individually.

Lay a corner or angle bracket in position and mark the outline on the chest lightly with a pencil. The area inside the mark is then cut away with a chisel to a depth equal to the thickness of the brass. Figs. 10 and 11 show this being done.

Traditional military chests had a corner bracket at each corner, top and bottom. However, as modern chests rarely have to suffer the buffeting of an India passage, corner brackets are normally attached only around the top, where they can be seen, and angle brackets are attached around the bottom. It's really a matter of personal choice—but the latter method is cheaper and entails less work. These items are available from cabinetmakers' suppliers.

The handles are marked and cut in exactly the same way.

The position of the handles is also a matter of personal preference. Some people like them placed more toward the outer edges of the drawers, and some slightly toward the middle.

Finishing

Coat the chest with clear polyurethane varnish. Follow the varnish manufacturer's instructions regarding application and number of coats.

You are now the proud owner of a modern military chest that will be the envy of your neighbors. Of course you must now make a second one—unless you only need two drawers. In any case, before you make another one study alternative designs used in antique varieties. Your next effort might incorporate some of these ideas in modern form. The occasional writing desk would certainly be a good idea for many small homes. Or how about a small cocktail cabinet for the "top deck"?

CUTTING LIST

Pine:	Inches:
4 top and bottom	36 x 9 x 1
4 ends	15½ x 9 x 1
1 center front	35 x 4 x 1
4 drawer sides	15¾ x 6⅛ x 1
2 drawer backs	34⅜ x 6⅛ x 1
2 drawer fronts	34⅜ x 6⅝ x 1
4 top and bottom drawer runners	16 x 1 x ½
2 center drawer runners	16 x 1 x 1¼
4 feet	3 x 3 x 1
2 drawer bottoms	33¾ x 15¾ x ¼ ply.
1 back	35½ x 15½ x ¼ ply.

These sizes are for finished dimensions and do not include any allowance for waste. All pine is "nominal" thickness.
Also required for each chest: 4 inset drawer handles, 2 inset side handles (same, only larger), 8 corner brackets and 2 angle backets—all in brass.

This freestanding bar and wall of sliding cupboards faced with mirrors have been designed to give a modern look to this recreation area. The layout of the back of the bar can be seen in one of the mirrors.

HEIDEDE CARSTENSEN/STUDIO DIE WOHNFORM

CHAPTER 25

Easy-to-build cocktail bar

If you do a lot of entertaining at home, you may find yourself coping with make-shift catering arrangements—tables that aren't large enough, nowhere to keep the glasses, and so on. Build your own bar, however, and you can keep all your drinks and glasses together and pour a drink without having to move from one spot. The days of making-do will be over.

You can design your bar to suit your requirements exactly, so that it can take mainly bottles of wine and liquor or cases of beer. And you can make it just the right size, too, so that it fits your living room and stores away unobtrusively after use.

Bars tend to be very personal things, and a design may well please one person but not another. So a simple and readily-adaptable design for the basic frame of a freestanding bar is described here; it is shown in Fig. 2. There are also a number of ideas for features that will give your bar a character of its own, and these are illustrated in Fig. 1.

Design

The basic structure of this bar is rectangular and made from sheets of plywood so that it is surprisingly easy to construct and can be made in any size to fit your particular room. It can be freestanding in the middle of the room, joined to the wall at one end or, if you fit it with casters, can be pushed out for use and stored unobtrusively.

The unit is given a more open and less box-like appearance by cutting away the bottom of the front panel to leave a recess. This gives plenty of foot room if you want to stand up against the bar—or for an authentic touch, a tubular footrest can be fitted as shown in Fig. 2.

The actual size of the bar you build will depend on the size of your room, the number of

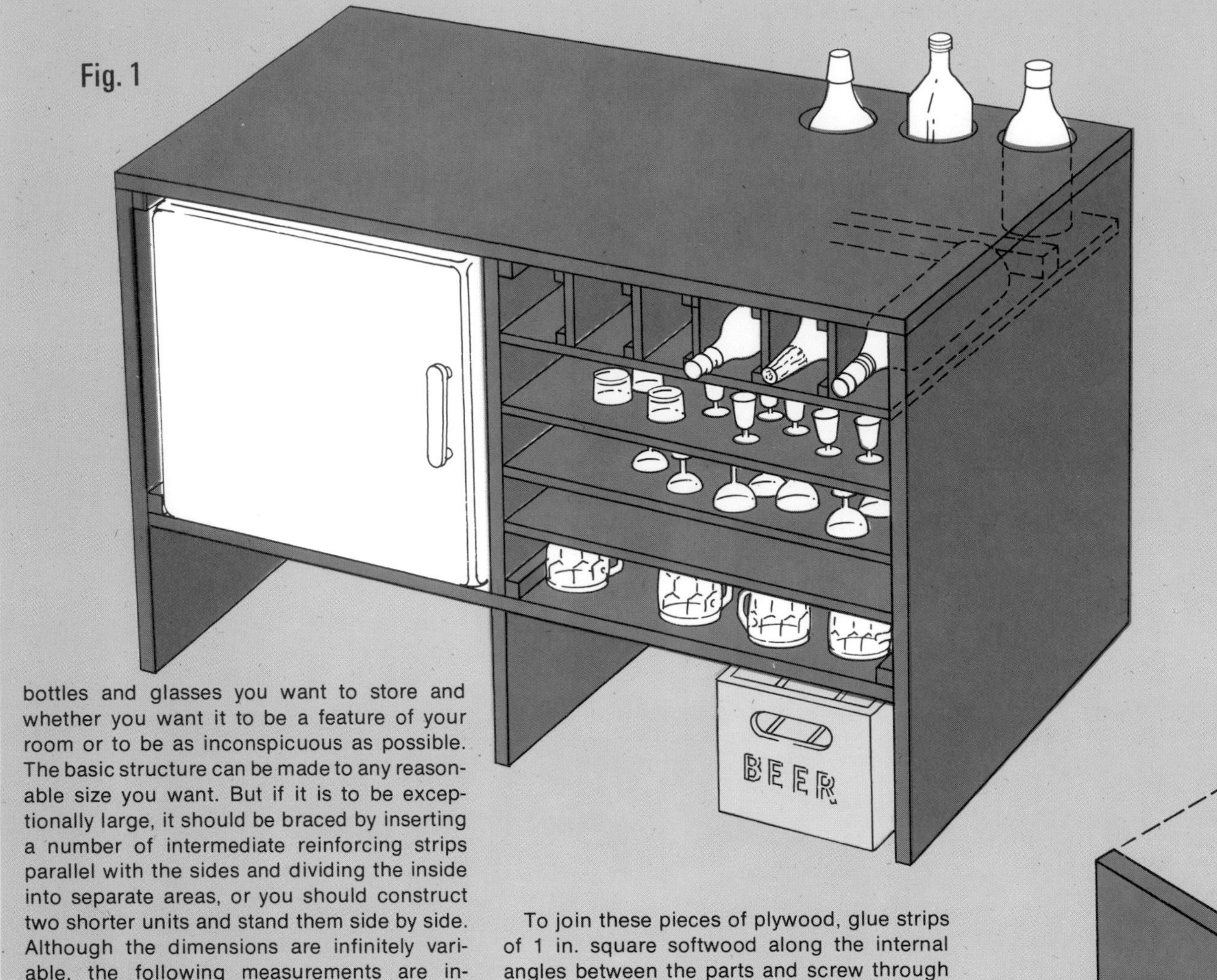

bottles and glasses you want to store and whether you want it to be a feature of your room or to be as inconspicuous as possible. The basic structure can be made to any reasonable size you want. But if it is to be exceptionally large, it should be braced by inserting a number of intermediate reinforcing strips parallel with the sides and dividing the inside into separate areas, or you should construct two shorter units and stand them side by side. Although the dimensions are infinitely variable, the following measurements are intended as a guide:

The height of the bar top depends on a number of factors. If you want to sit behind it, the bar top can be lower than if you want to be able to stand up at it. Or you may want it to match the height of the existing furniture and so continue the visual line around the room. In most cases, you will probably find that somewhere between 3 ft. and 3 ft. 6 in. is a convenient level.

The depth of the bar, that is, the distance from back to front, can vary too, but it must be deep enough for the unit to be stable and yet shallow enough so that when serving behind it you can easily reach to the front edge. In practice, something around 2 ft. will probably suit you.

The height of the recess at the bottom of the panel is entirely at your discretion, but note that in this design the bottom shelf (marked C in Fig. 2) also forms the top of the recess. You may want the shelf to be a specific height so that, for example, a case of beer can be stored underneath, and in this case the recess will be the same height as the top of the crate. You will need at least 1 ft. clearance to fit a footrest, and 18 in. should give sufficient space to store cases of beer under the shelf.

Case—general construction

The case is made from six pieces of plywood arranged as shown in Fig. 2.

To join these pieces of plywood, glue strips of 1 in. square softwood along the internal angles between the parts and screw through these into both pieces from the inside. (See "Assembly," below, for details.) Finishing nails driven in from the outside will provide additional strength, without marking the surface noticeably, but you are unlikely to need these except in a very large structure.

The front piece (A) provides much of the rigidity of the frame, as it overlaps the edges of the sides, top and bottom shelf (pieces B, D and C). The front piece is therefore the full length of the bar and its width is the full height of the bar less the height of the recess at the bottom.

The side pieces (B) are as wide as the depth of the bar from front to back, less the thickness of the front, and as long as the height of the bar, less the thickness of the top, which overlaps them. A long strip is cut out from the top front corner of each side as wide as the thickness of the front panel and reaching down to the top of the recess.

The shelf (C) runs between the side pieces and from the back edge of the bar to where it meets the front piece. It is as long as the bar, less twice the thickness of the plywood used for the sides. It is usually as wide as the depth of the bar, less the thickness of the front piece—but you can make it narrower and stop it short of the back edge so that you don't knock your shins on it when serving from behind the bar.

The top piece (D) is the full length of the unit and as wide as the depth of the bar, less the

Fig. 1. *This easy-to-build bar has been fitted out with a refrigerator, space for beer cases and shelves for a variety of glasses. The horizontal racks are ideal for bottles of wine and the holes in the bar top for bottles of liquor. These features could be altered to suit your needs: a garbage can may be more useful than the refrigerator and the wine rack could be extended downward as in Fig. 3.*

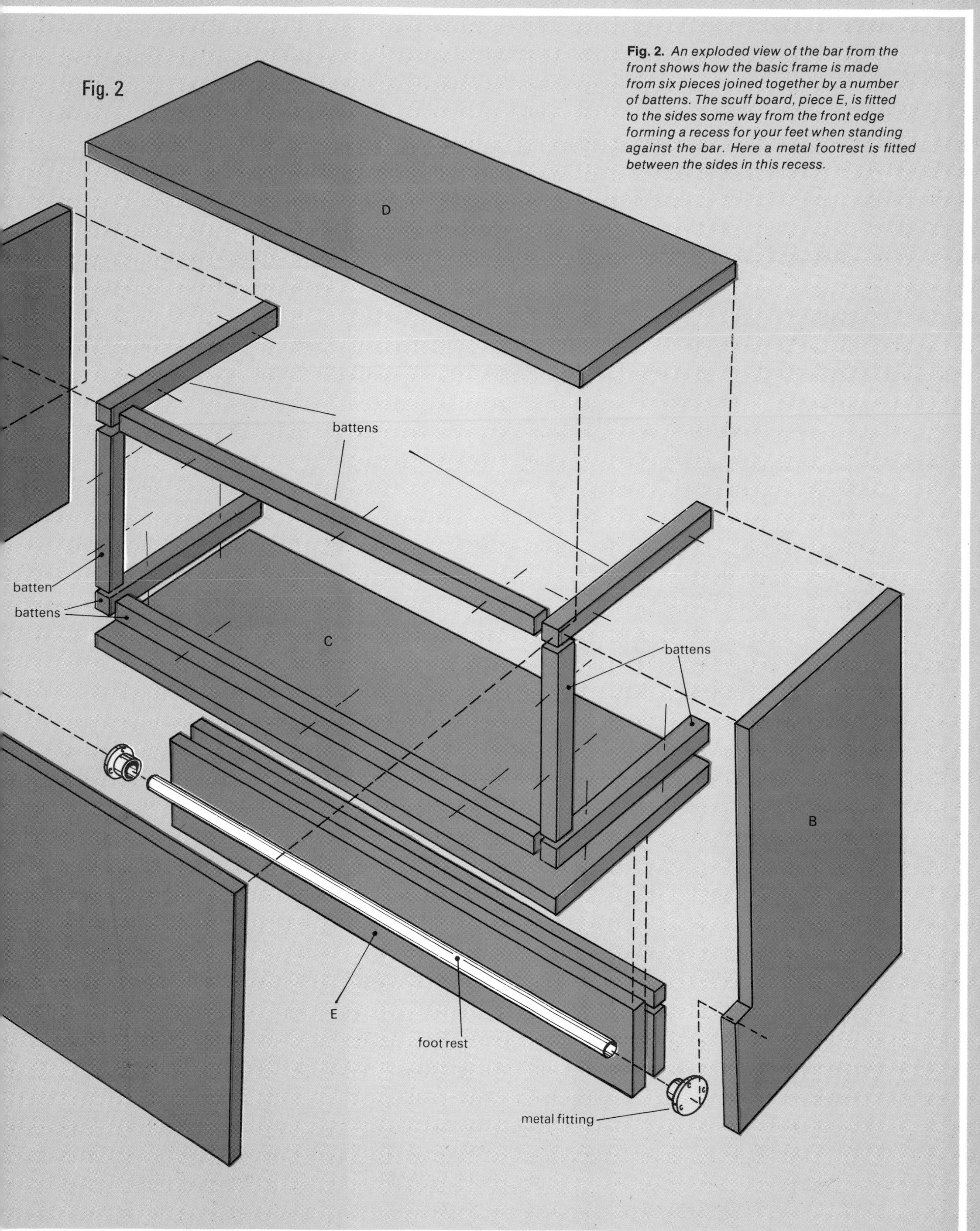

Fig. 2. *An exploded view of the bar from the front shows how the basic frame is made from six pieces joined together by a number of battens. The scuff board, piece E, is fitted to the sides some way from the front edge forming a recess for your feet when standing against the bar. Here a metal footrest is fitted between the sides in this recess.*

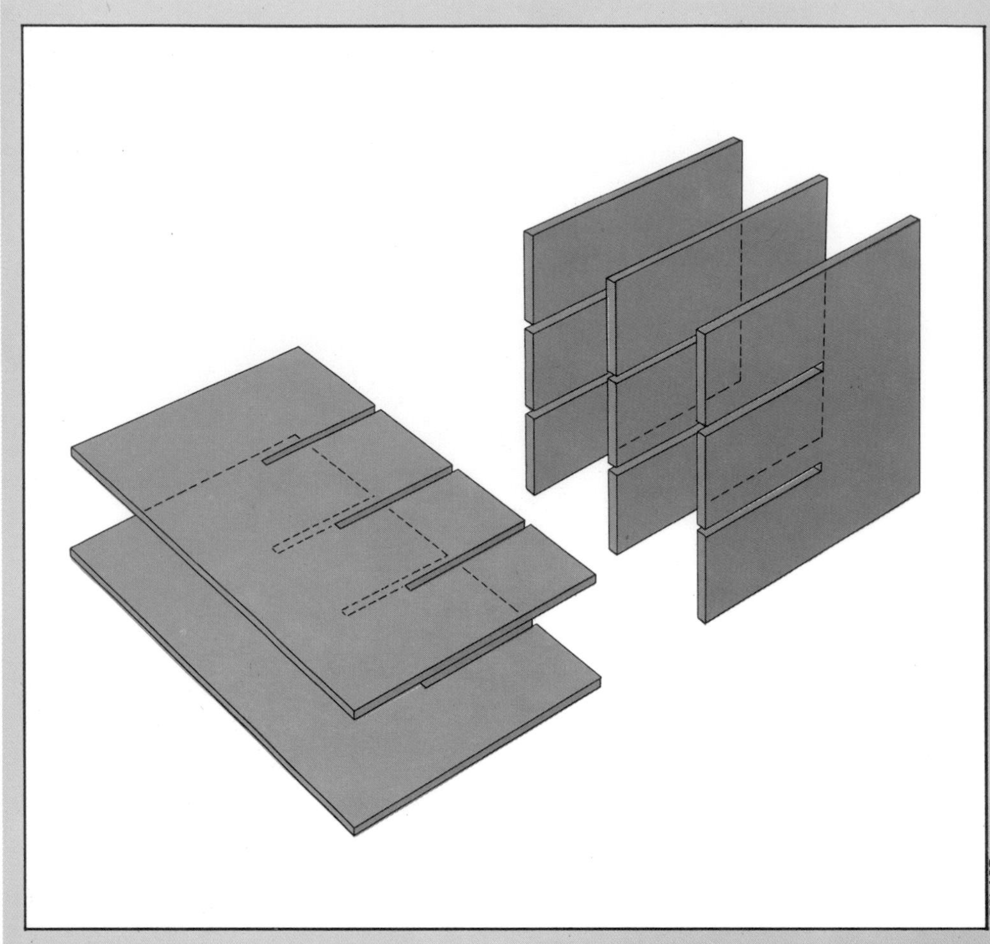

Fig. 3. *Exploded view of a wine rack to be fitted between shelves. Parallel slots are cut and chiseled halfway across the boards as shown and the boards slotted together.*

thickness of the front piece, and the *scuff board* (E) is as long as the shelf and as wide as the height of the recess.

Plan the pieces for your bar for cutting from standard sizes of plywood. If you plan the bar carefully you may be able to make it largely from standard-sized pieces (like half panels and quarter panels) and save yourself cutting a lot of straight lines. If you do have to cut out many pieces, use a power bench saw if possible, for the greatest accuracy.

At the same time, remember that it may be cheaper to cut all the pieces out of one large 4 x 8 ft. panel. When calculating the amount of board you need, include enough to make any of the extra shelves described below under "Additional features." Also buy sufficient ¾ x ¾ in. pine to brace all the joints.

The edges of the pieces in this design are clearly visible so you should cover the whole bar when it is finished with a covering such as one of the self-adhesive edgings available. (CHAPTER 36 gives full details of veneering and instructions on how to apply it.)

Case assembly

Mark and cut out the pieces shown in Fig. 2 to the dimensions you have chosen, as well as the pieces for any additional shelving. Check that the corners are all square and that the two side pieces (B) are identical—otherwise the whole bar will be out of line.

Check by direct measurement that the front panel fits into the cut-outs in the side pieces and projects beyond their tops by the thickness of the top piece. The bottom shelf and the scuff board must both be the same length and shorter than the length of the top piece by the thickness of the two side pieces.

Measure and mark the positions of the battens on the pieces as shown in Fig. 2—the battens are placed on the inside faces of the pieces and nearly all of them level with the edge except where shown. Cut the battens to length and drill holes for the screws both ways through the battens alternately at roughly 3 in. intervals. Hold the batten against the board where it is to fit and mark through the screw holes on to the board. Drill small pilot holes at this point with a small drill or awl—or in the case of particleboard, drill holes for, and insert, the fiber plugs to take the screws. Apply white glue to one side of each batten and stick it to the board. Then screw through the predrilled holes and allow the adhesive to set.

Now assemble the pieces in the following order. Place the front panel downward on the floor with one short end against the wall, then place a side panel on it leaning against the wall and glue and screw it in place with two or three screws only. Next, glue and screw the bottom shelf in place—this fits against the front and side panels, so that from this stage the case can support itself. Then fix on the second side, the top piece and the scuff board, again with only a few screws.

So far you have used only a few screws to each piece. Now, before the glue dries, stand the unit on its feet and check that everything is square—if necessary, use sash clamps to hold it in position. Then drive in the rest of the screws from the inside through the battens.

When the adhesive is dry, remove any clamps and smooth the joints with sandpaper. The basic case is now nearly ready for finishing and covering—but first add any extra shelves or other features that you need (see below).

Additional features

You can make this bar hold most of the things you need without resorting to fixtures on the wall behind. The space below the bottom shelf can accommodate a case or two of beer if you have made it high enough. In addition you need room to store glasses, to hold wine and liquor, a garbage can, a space for the ice bucket, and perhaps for a dishpan full of hot water. If the bar is really elaborate and permanently installed in a suitable place, a fully-plumbed sink set into a hole in a shelf below the top is extremely useful, but for most people this will be too ambitious.

A handsome footrest can be made from hard copper water tube mounted between adapters to threaded pipe flanges, and fastened to the side pieces as shown in Fig. 2.

A vertical divider of the type shown in Fig. 1, parallel with the end pieces reaching from the bottom shelf to the bar top adds greater strength to the unit and can support the ends of half-width shelves. It can be set between battens as in Fig. 1, or, if you don't want the battens to show from the back, screw down through the bar top and up through the bottom shelf into this divider.

If one side of the divider is fitted out with shelves for glasses, as shown, the space on the other side can be fitted with a garbage can, wine bottle rack, or anything else you need.

There are two ways of adapting a space to take wine or liquor bottles. Wine bottles are best stored horizontally so that the cork remains moist. This allows you to store the bottles on a shelf, or a number of shelves set close together, supported just below the bar top on battens, as shown in Fig. 1. Vertical dividers made of thin plywood stop the bottles from rolling about. If you want to store a lot of bottles, you can make a set of interlocking plywood shelves as shown. Remember that bottles vary in length and diameter, so check the size of the bottles you are most likely to store. Stick a batten at the back of the shelf at a suitable distance from the edge—this stops the bottle going back too far so you can't reach it.

Liquor bottles are best stored upright. One way, as shown, is to cut circular holes in the top of the bar so that the liquor bottles can be placed in them and supported by the shelf beneath. Position the top shelf so that a sufficient part of the bottle extends above the top of the bar to allow you to get hold of it easily (and also to see which bottle it is). If you cut these holes near the front edge of the bar it prevents people from leaning on the bar along its whole length, and leaves the back free for serving.

CHAPTER 26

How to update old doors

You can convert an old flush door into a modern paneled door, or re-style an antiquated paneled door to bring it in keeping with new decor. Simple tools and standard materials from your local lumberyard are all you need, and the job is easy and inexpensive.

The door at the left is an example of what you can do with an old flush door to add a decorative touch when you re-do a room. The panels shown are ready-made plastic ones mounted with an adhesive made for the purpose. If ready-made panels are not available in your area, however, or if the design you want has been discontinued, you can make your own by using stock moldings, mitered at the corners and mounted on a plywood or hardboard backing, as shown in the drawing at the bottom of page 126. The center square of that example was cut on a power saw but comparable forms can be made with hand tools by simply planing a bevel on the edges of a wood or plywood square, and mounting a layer of ¼ in. plywood or hardboard on top of the beveled wood square. The important point is that of improvisation. Look at the wide variety of moldings at your lumberyard, and plan your panel squares or rectangles. You can also use parquet flooring squares, cork flooring squares, or mirror tiles for exotic effects. Sand to the bare wood to provide a good gluing surface on old flush doors. Simply glue the panels to unfinished flush doors. You can also use bradded-on rim moldings to hold materials where glue alone might not be adequate.

To re-style antiquated paneled doors, fill the recesses with fiber wallboard, or wood, and cover the entire door with ¼ in. plywood or hardboard, nailed and glued. Then apply the new panels on its surface. Use white glue or aliphatic glue (like Titebond) and brads to hold the panels in place.

The shape and spacing of the decorative panels should be keyed to the effect you want.

NIGEL MESSETT

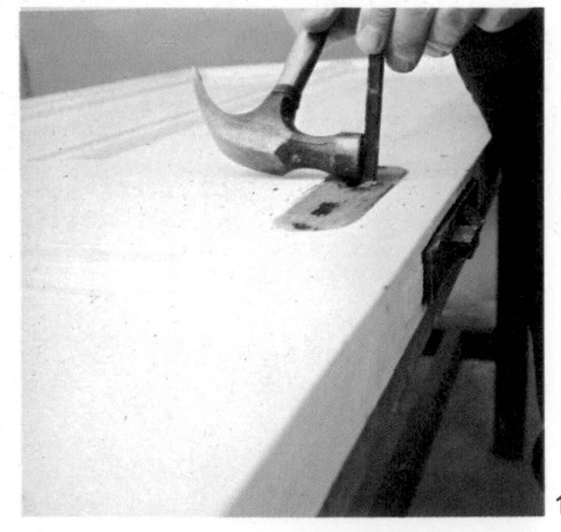

To make the door look higher, use vertical rectangles instead of squares. Hints on spacing the panels are given in the drawing.

Before you start your door revisions, inquire as to the ready-made panels that may be available at your nearby lumberyards. If your leisure time is limited, it pays to save working time whenever you can.

Fig. 1. *Start re-styling of old paneled door by removing surface plate and trim of lock.*
Fig. 2. *Use a straight edge and ruler like this to gauge thickness of filler pieces to use in old panels to provide firm backing for plywood.*
Fig. 3. *Evenly spaced strips may be used as filler pieces to back up plywood or hardboard.*
Fig. 4. *Use special finishing nails to fasten plywood or hardboard to door, avoid filler.*
Fig. 5. *Make trial spacing of decorative panels before gluing. Then measure and mark locations.*
Fig. 6. *After panels are glued and bradded, do the paint job. Use filler in any cracks or gaps.*

If A spaces are equal and B spaces are equal, door will be symmetrical in appearance. The C space at bottom may be greater without detracting. Drawings at lower left show typical dimensions of made-up panel. Adjust size as needed.

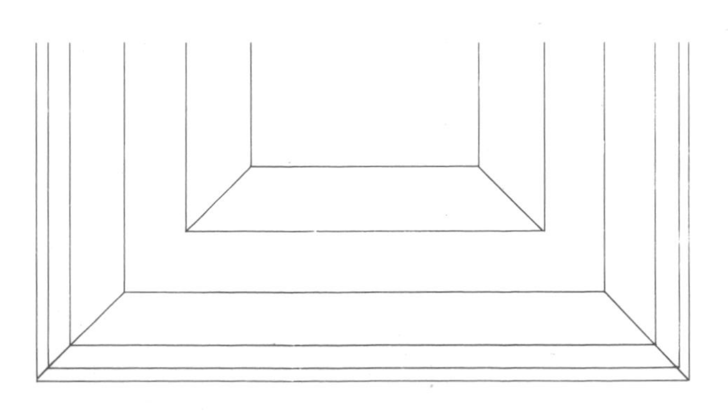

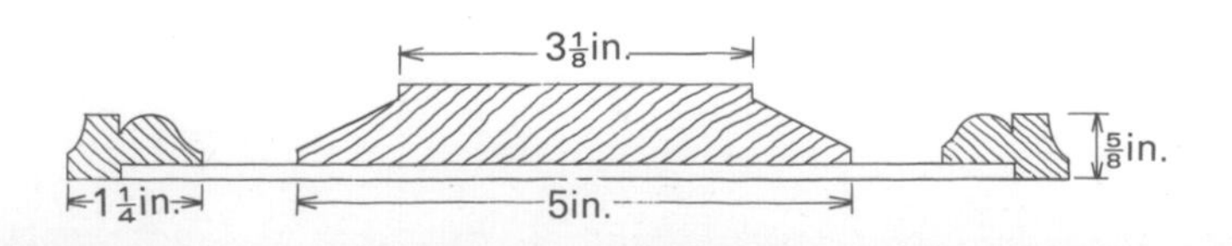

TRI-ART

CHAPTER 27

Tailor-made kitchen units:1

NIGEL MESSETT

Fitted kitchen units are one of the most popular home projects. For simplicity of construction, most commercial units are made up from prefabricated parts, but if you do the job yourself you will not only be able to introduce a distinctive design of your own, you will also be able to incorporate high grade materials and still save about one third of the price.

Manufactured units, however well made, can be produced only in a limited number of sizes. Making your own units means that you can tailor the units to fit exactly into the space available in your kitchen.

General construction details

Most home-built fitted units consist of a framework of lumber over which cladding of some sort is fixed. Although this is a satisfactory method of construction, the measuring, cutting and fitting of the frame does require a degree of technical competence.

This method uses, where possible, man-made boards such as particleboard or plywood. The thickness and rigidity of these materials eliminates the necessity for framing many pieces of lumber and reduces the cutting to the squaring up of several panels.

The boards are available in a range of sizes, and if the unit is designed around the standard sizes that are available in your area, you can eliminate a lot of sawing.

The basic construction is a box, as shown in Fig. 1, with butted joints that are either nailed and glued, or joined with screws and angle brackets. There are several types of hardware for making square joints, but you will have to decide whether one of these is suitable because some could in certain circumstances interfere with the fitting of drawers.

There is a wide variety of hinges available for fitting the doors, and of course the type of hinges would be dictated to a certain extent by the doors you use, which could be inset, flush, or sliding. A good hinge to fit is plastic or brass piano hinge. This hinge runs along the whole length of the door, distributing the weight and stresses evenly. It is sold in long strips and is easily cut to length—and you do not have to make dado joints to fit it.

Although any sort of handle can be fitted to the doors, this unit has inset handles made from "L" or "J" section aluminum strip as shown in Fig. 7. This is cut with a hacksaw, and can be drilled quite easily with a hand-drill. If possible, use aluminum screws so that the screw heads will match the strip.

Although screws are used for certain joints, care must be taken not to screw down too hard in particleboard (if you are using it), which has limited holding strength. A good method of screwing into particleboard is by drilling holes large enough to take fiber plugs such as Rawl plugs. The plugs are then glued into the holes with a woodworking adhesive, and provide a holding area almost as strong as solid wood.

Whenever nails are used in particleboard, they should be slanted for maximum holding power.

Planning and design

When you build your own units, you have the opportunity of making furniture to your own design and achieving something that is a little different and unobtainable in any shop. So give the design some careful consideration. Obtain information on the types of plywood available in your area so that you can select the veneer or laminate—or a combination of both—to suit your kitchen. Also list the dimensions of available panels, so that you can plan the sizes of your units.

List your requirements first in terms of measurements. Decide on an approximate height for the working surfaces of the units and the horizontal length into which they must be fitted, along with the necessary depth. You can then study the panel dimensions and plan the unit.

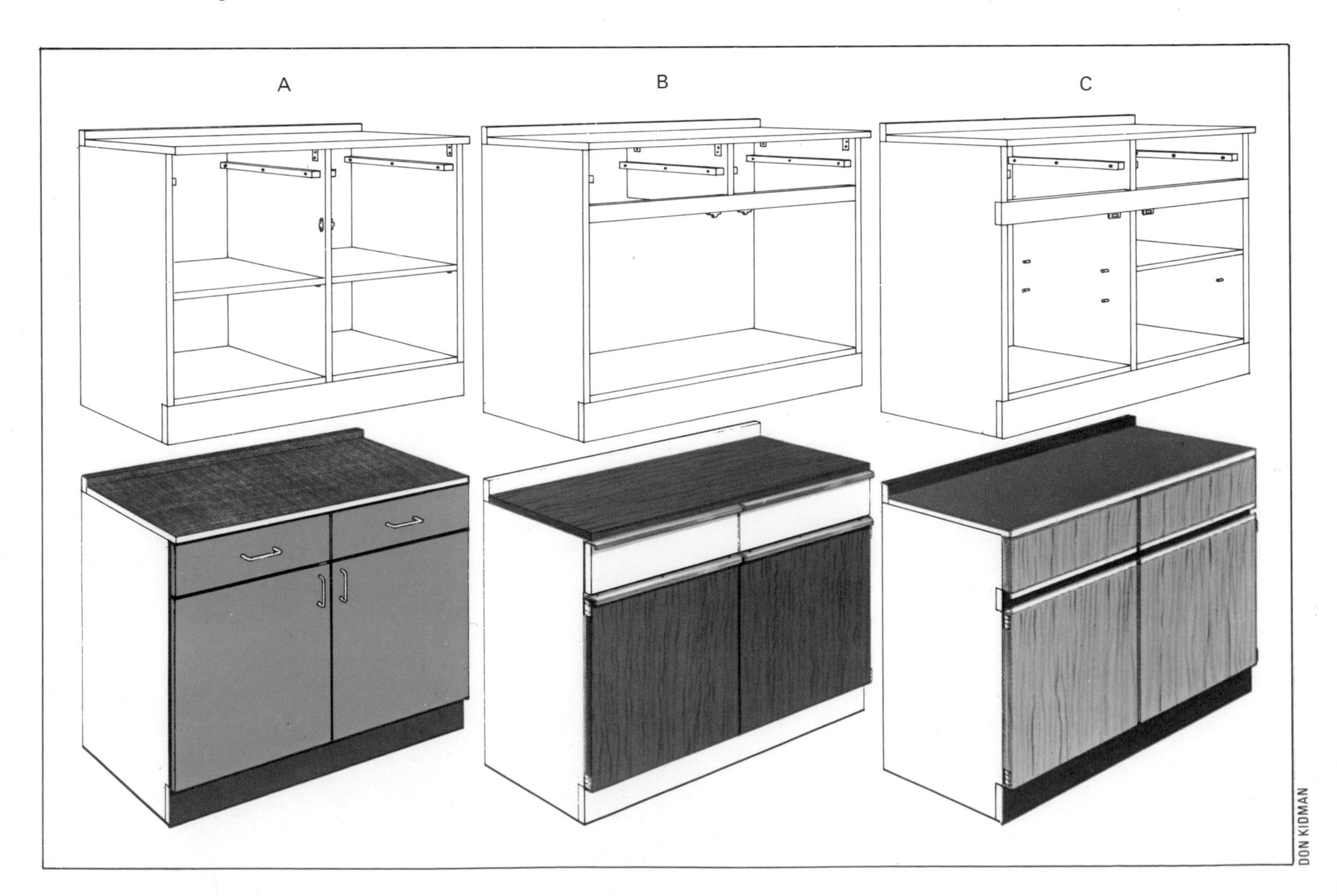

Fig. 1. *Although these three units are different in appearance, they are all constructed the same way. The shell in each case is identical, but the interiors and external fittings vary slightly.*

If you want to incorporate drawers in the unit(s), then this might alter the design aspect. This unit uses one of the many prefabricated drawer kits on the market, which saves a lot of time and is only fractionally more expensive than making your own. Using this method, you will have to obtain the drawer sizes first so that the unit is built with sufficient clearance for the drawers to be fitted. But if you are making the drawers yourself this point is unimportant as they can, if necessary, be fitted after the rest of the unit has been made.

Fig. 2. *Prefabricated drawer kits only require gluing and fitting together.*

Fig. 3. *When tapping joints together, a block of wood placed across the working surface will prevent unsightly hammer marks.*

Fig. 4. *The drawers are faced with a decorative panel glued and screwed to the front.*

Fig. 5. *One of the side panels. The battens for the plinth, bottom shelf and drawers have been screwed in place, and the recesses cut for the plinth and both cross members.*

Fig. 6. *The unit top is held in place with angle brackets or similar fittings after the cross members and plinth have been fitted.*

Fig. 7. *Fitting the drawers. Aluminum angle strip can be used in place of handles.*

Cost will obviously be a consideration. As mentioned, a decorative wood veneer on plywood costs considerably more than ordinary fir, depending on the species you choose. Also, if you want to paint the finished units, standard fir would be much cheaper.

If you are clever with your calculations, you will be able to design a unit that requires very little cutting. Where it is necessary to cut a panel, try to ensure that the cut edge is hidden from sight. For instance, in this unit a recess is cut out of the bottom front edges of the side panels, but the exposed plywood edge is covered by the "kick panel" or plinth which fits into it at each end, and the side edges of most other panels, if exposed, may be covered with veneer strips made for the purpose.

Making the drawers

The use of prefabricated drawer kits, while simpler from the aspect of construction, has one disadvantage in that you are limited to the sizes in the available kits. However this is not quite as serious as it appears. As shown in Fig. 4, the actual drawer is faced with a much larger panel. This means that you can get away with a drawer that is slightly smaller than the drawer opening by increasing the horizontal thickness of the batten used as a drawer runner on either side.

The alternative is to go all the way and build the drawers yourself. The method of doing this is described in CHAPTER 14.

Marking and fitting the shell

The shell is the unit minus doors, drawers and shelves.

The only involved piece of cutting in it—and even this is relatively easy—is in the sections marked A in Figs. 8 and 9.

Mark, and cut if necessary, the three A sections. Ideally these should be cut from standard plywood panels, veneered on the visible surface and veneer-stripped on exposed edges. Make sure that each section or panel has absolutely square edges by using a try square at the corners and measuring the diagonals. This applies throughout the construction of the unit.

There are three recesses to cut in each A section, to dado joints X, Y and Z. These are marked and cut, as shown in Fig. 8, to the thickness and depth of each piece that is to be inserted. The height of the top edge of joint Y is dictated by the depth of the drawers you are fitting. The height of the bottom edge of joint Z depends on the amount you want the splashboard to project.

When the cutting has been done, trial-assemble the unit by inserting sections C, D and B (Fig. 14) into their recesses, to enable you to check that the unit is square.

With a try square and straight edge or rule, mark the positions of all shelf and drawer supports, remembering to mark both sides of the center A partition.

Drill holes for the shelf support fittings.

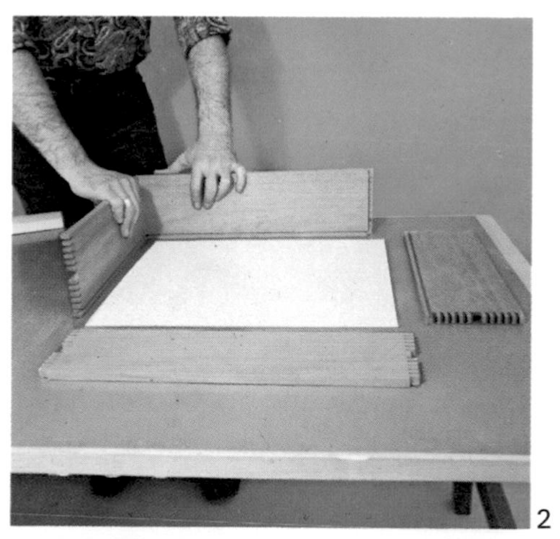

2

3

4

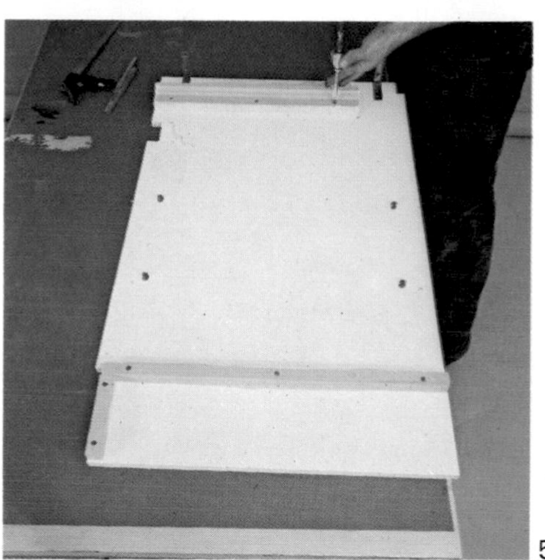

5

6

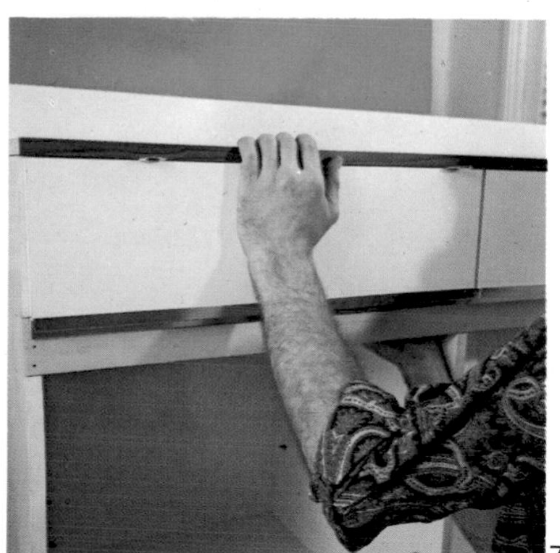

7

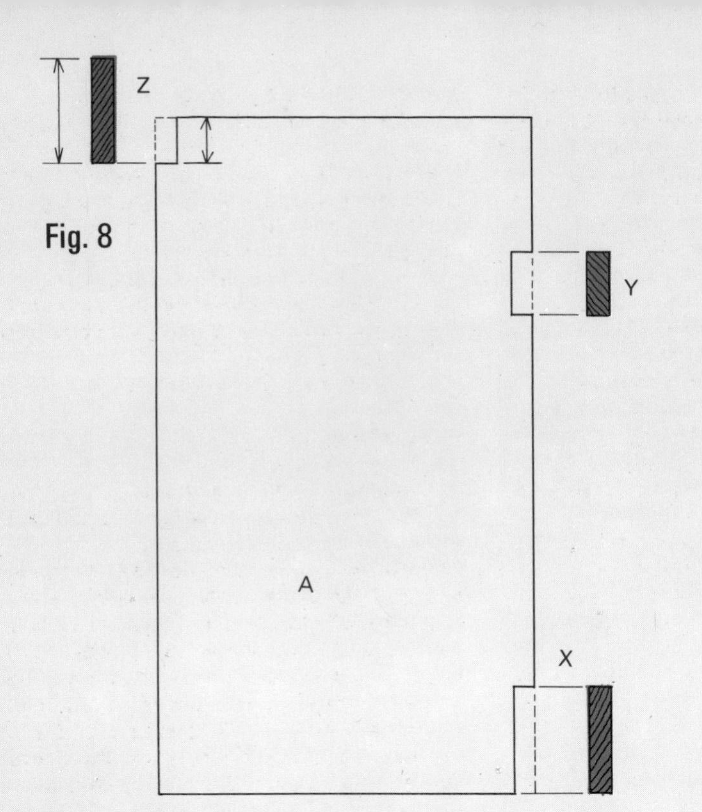

Fig. 8

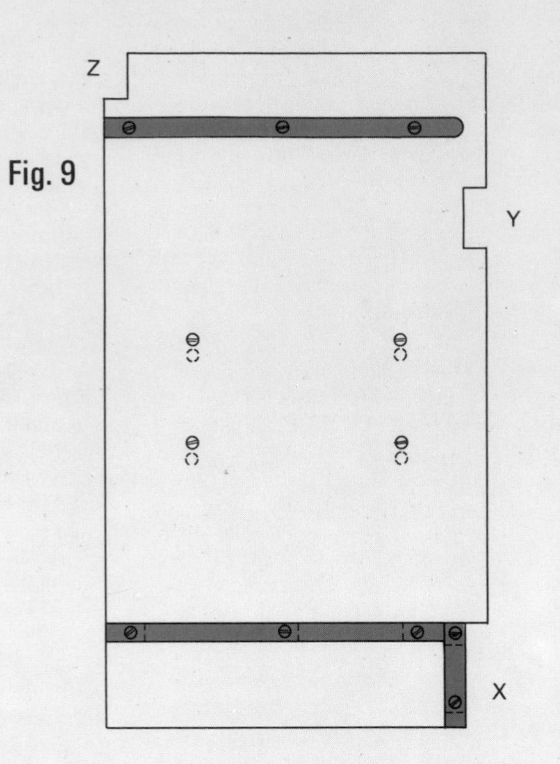

Fig. 9

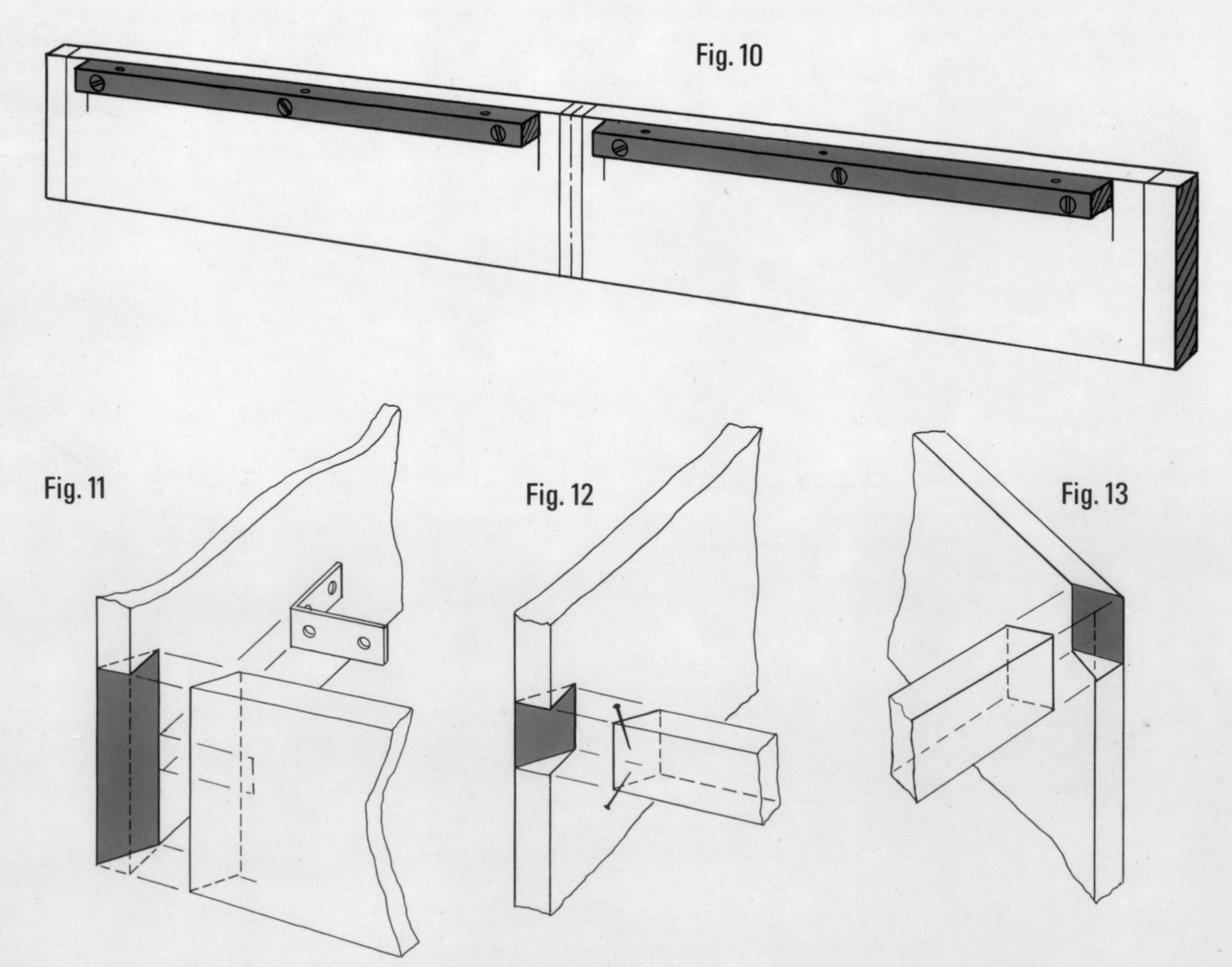

Fig. 10

Fig. 11

Fig. 12

Fig. 13

DON KIDMAN

Fig. 8. *The side panels—these are identical to the center panel if one is fitted—must be recessed to take joints Z, Y and X.*

Fig. 9. *Hardwood battens are screwed to panel A to provide support for the plinth and bottom shelf, and runners for the drawers.*

Fig. 10. *The rear of the plinth or kickboard has hardwood battening screwed on to provide a surface for attaching the bottom shelf.*

Figs. 11–13. *Cross-members need not be dadoed straight through; they can be mitered and fitted as shown if you do not want to cut away any of the surface veneer or laminate.*

Fig. 14. *Front view of the complete unit. This version has a center partition, but as you can see in Fig. 1, this is not essential.*

Fig. 15. *Three different types of drawer runner. The top version requires grooved drawer sides, but the other two do not.*

Cut the drawer runners to length, and drill and countersink them to take No. 10 screws. Smear glue on the backs of the runners, and secure with screws of a length ½ in. longer than the thickness of the runners, to take the weight of the drawer.

The bottom shelf supports are attached next. These can be of 1 in. x 1 in., as shown in Fig. 10, or you can fit metal angle brackets of the type used to fix the top panel F in place. In any case, screw on short vertical 1 in. x 1 in. battens flush with the front surface of the recess for joint X to take the kick panel, as shown in Figs. 5 and 9.

Trial-assemble the unit again, and check that panel F fits properly. Lay the angle brackets that will hold the underside of F to the sides of the A sections temporarily in place, and with an awl mark out the positions of the screws.

The inside face of each end A panel, and both sides of the internal A panel, should now appear as shown in Fig. 5.

Assembling the shell

Although you can do this by yourself, it is much easier if you have some assistance.

Lay section B down on a flat surface and glue and screw the A partitions to it as shown in Fig. 1.

Gently stand the unit upright and glue and nail sections C and D in position. Use 2½ in. finishing nails for this and pre-drill holes for them in C and D with a drill bit slightly smaller than the nail diameter. Check the unit for squareness and leave it while the adhesive sets.

Fit the base shelves E in position, holding them in position by screwing into the underside through the battening. Or, if you have fitted angle brackets, screw these to the shelves.

The unit top F, is now placed in position and screwed in place through its angle brackets.

The back of the unit is covered with a sheet of exterior grade hardboard, glued and nailed at 2 in. intervals. The proper fastening of this section is very important because it not only covers the back but also acts as a bracing for the whole unit.

Fasten the shelf supports in position.

Doors and drawers

All that is left is to fit the fascia panels to the fronts of the drawers, hinge the doors, and attach handles and door catches.

The drawer fascias are attached quite easily if you spread some contact cement over the front of each drawer panel and the back of each fascia. Fit the drawers in position, pushing them back as far as they will go. Place one fascia accurately over a drawer space, hold the fascia in position with one hand and, with the other hand, reach underneath the drawer immediately behind and slide it forward until it touches the fascia. The contact cement will hold the fascia in position until you can screw it in place from the inside.

The doors are cut from standard plywood panels.

The placing of the door hinges is particularly important. Whenever door hinges have to be screwed into plywood, it is essential to drill the screw holes to hold the screws.

Finish off by cutting the aluminum angle strip for the handles, drilling and countersinking holes in it, and fastening it to the edge of the door panels and drawer fascias.

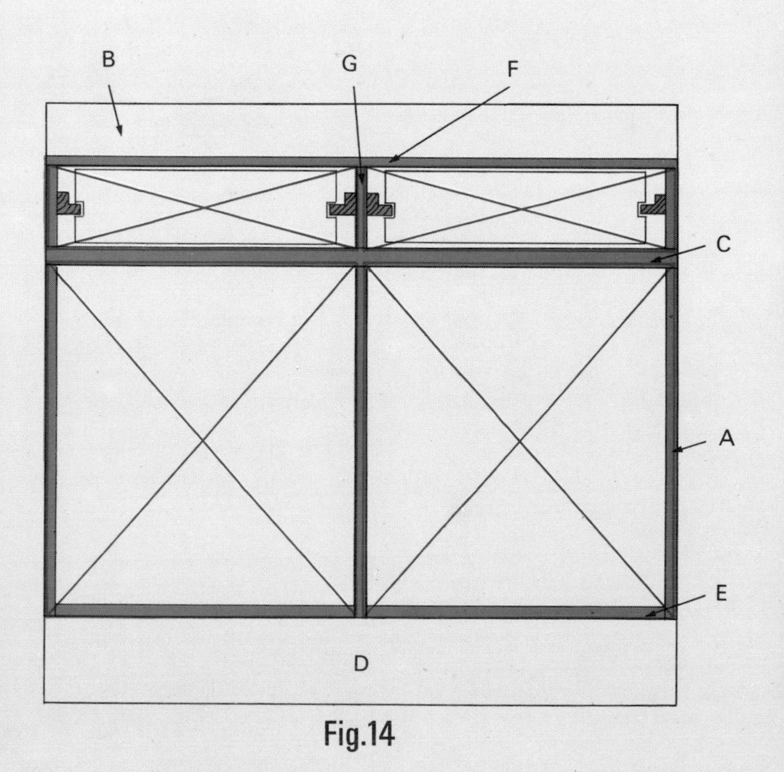

Fig.14

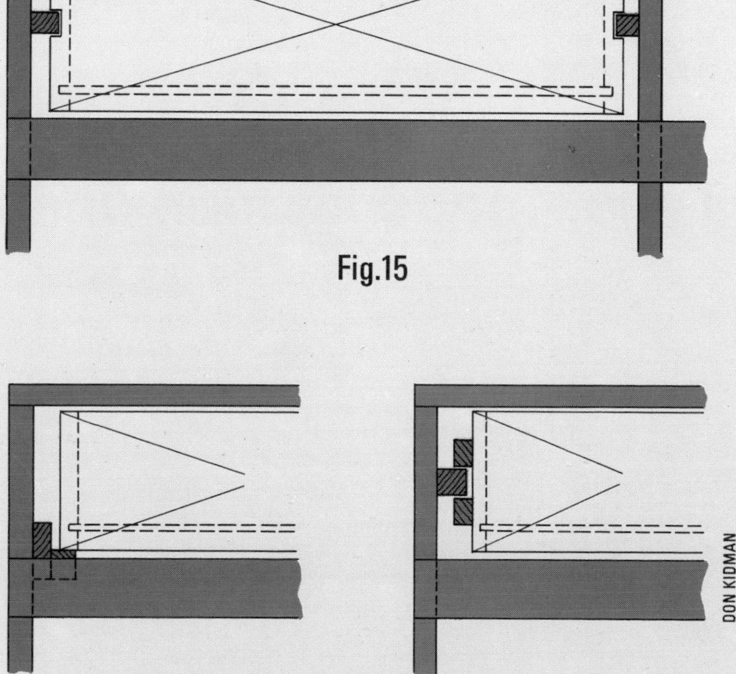

Fig.15

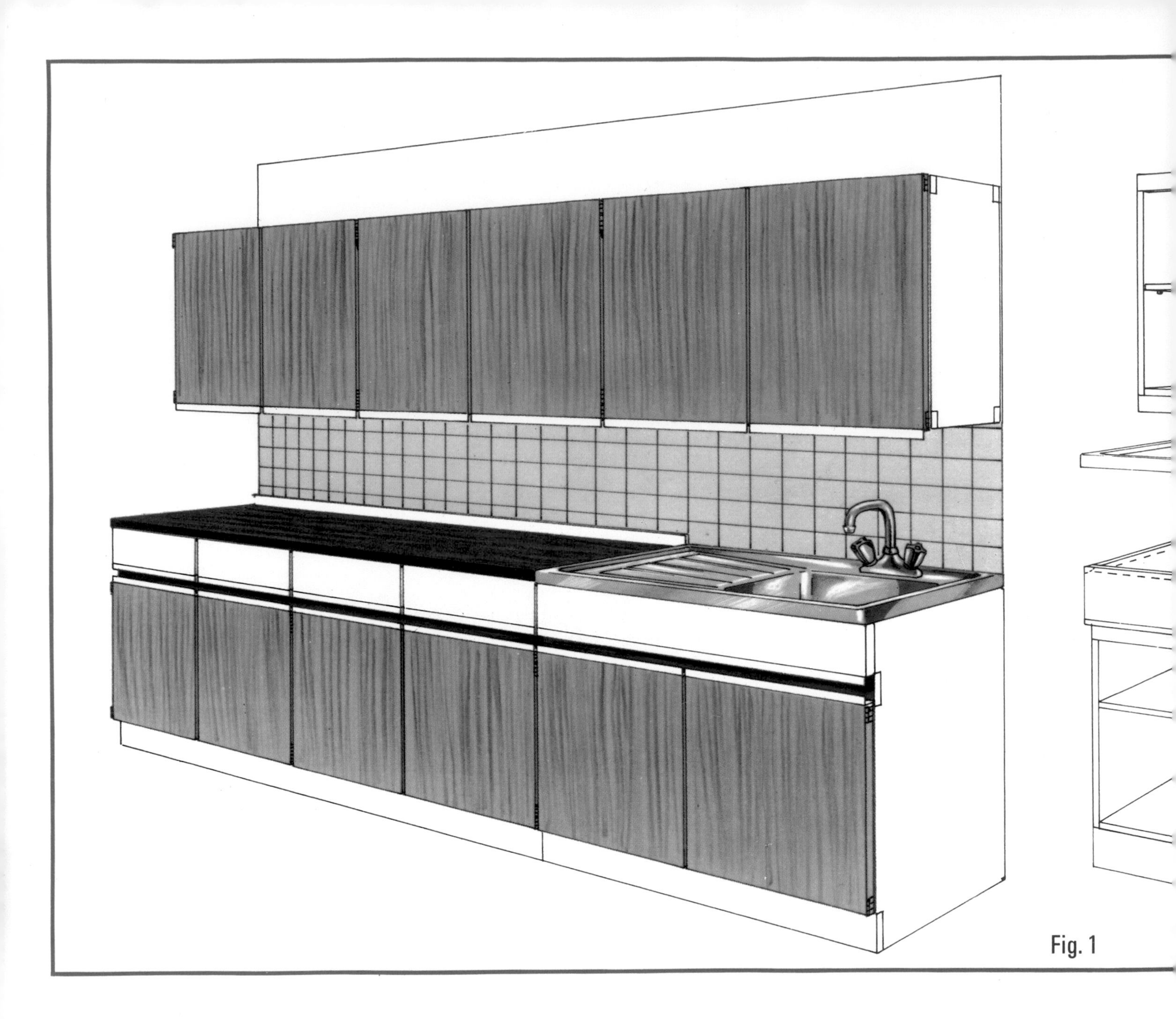

Fig. 1

CHAPTER 28

Tailor-made kitchen units: 2

When you have built the kitchen unit, the same techniques can be applied to make hanging wall units to match and a base to house the kitchen sink.

Building and fitting the sink unit will of course involve some basic plumbing, but this is not at all complicated—even if you want to change the location of your present taps. If not, the work only involves removing and replacing a few nuts and washers.

General construction details

The woodworking techniques involved here are virtually the same as those used in the first stages of this project, but some parts of the units have to be strengthened because of the different stresses involved.

For instance, the hanging wall cupboards are attached to the wall by two cross-members at the back of the unit, and the weight of the unit and its contents creates severe strain on the side panels where the cross-members join them. So these joints have to be stronger than in an ordinary freestanding unit. If the sink unit has a single bowl sink set to left or right, it may usually have a centrally located partition, as shown in Fig. 1.

The sink itself can be fitted as a complete top, as shown in Fig. 1, or inset as in Fig. 2. With the second type, you do not have a drain board, although a portable one could be devised. In some cases, for example in a very small kitchen, this disadvantage could be more than offset by the increased working area.

The hanging wall unit

Resist the temptation to make these units too deep. As already mentioned, there is considerable strain on the points where the unit is attached to the wall, and it is surprising how heavy it can become when filled with dishes and the like.

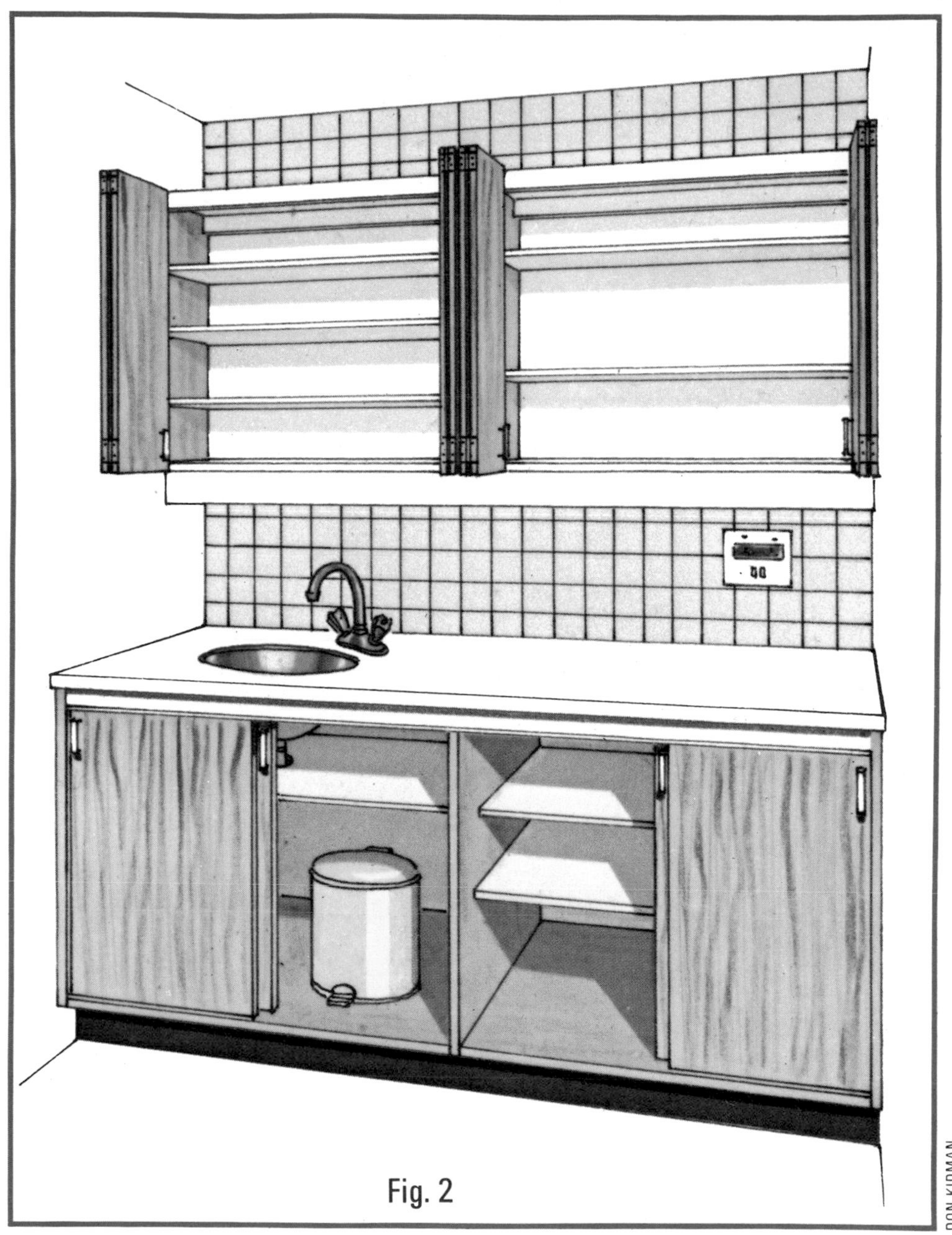

Fig. 1. *This shows a run of kitchen units, one of them fitted with a sink and drain board, and hanging wall cupboards above. A construction outline is on the right. It is important that the sink unit has a central partition to add strength to the frame when the sink is holding the considerable weight of several gallons of water.*
Fig. 2. *The sink can be fitted flush with the work surface to save space.*

If you plan to tile the wall underneath the hanging units, then try to arrange the distance between the top and bottom units to equal the number of whole tiles that come nearest to 20 in. as shown in Fig. 1. This will save you the bother of having to cut tiles to fit.

It is not necessary to use expensive paneling for the top of the unit. Hardboard or thin plywood is sufficient. But the top must be screwed—not nailed—in position to add strength to the case. This applies throughout; all joints should be glued and screwed.

Apart from the middle shelf, which is optional, and the doors, the hanging unit consists of the following: two sides and a base of ¾ in. plywood; four lengths of nominal 1 x 2 pine for the cross-members; three lengths of ¾ x ¾ pine for the supports underneath the base panel; and hardboard sheets as needed to cover the top, back and bottom.

To construct the unit, first cut out the recesses for the ends of the cross-members. As shown in Fig. 4, these are all 1½ in. high and ¾ in. deep and located at each corner of each panel. Cut the recesses carefully, using the end of a cross-member for direct fitting.

Cut the cross-members to length, then glue and screw the ¾ x ¾ battening supports for the bottom shelf along the inside base of the side panels and along the inside edge of the front bottom cross-member. This side battening is fixed in position so that the top edge is level with a line drawn between the tops of the recesses shown in Fig. 4.

Trial assemble the unit, check for squareness, then cut the base panel or shelf to length and glue and screw it to the battens.

Drill screw holes through the ends of the cross-members, and also into the edge of the side panel to take 1½ in. No. 8 flathead screws. Note that the screw holes for the rear members may be dovetailed for extra strength, as shown in Fig. 5. (Dotted lines show the paths of the screws.)

When all the screw holes are drilled, apply glue and assemble the sides, base and cross-members. (A multi-bore drill bit, like the Stanley Screw Mate, matched to the screw size simplifies the drilling.) Check for squareness and leave for the adhesive to set.

Screw the hardboard rear covering in place. Don't be tempted to leave this part out, because it braces the whole case. If you don't want to see a hardboard panel when you open

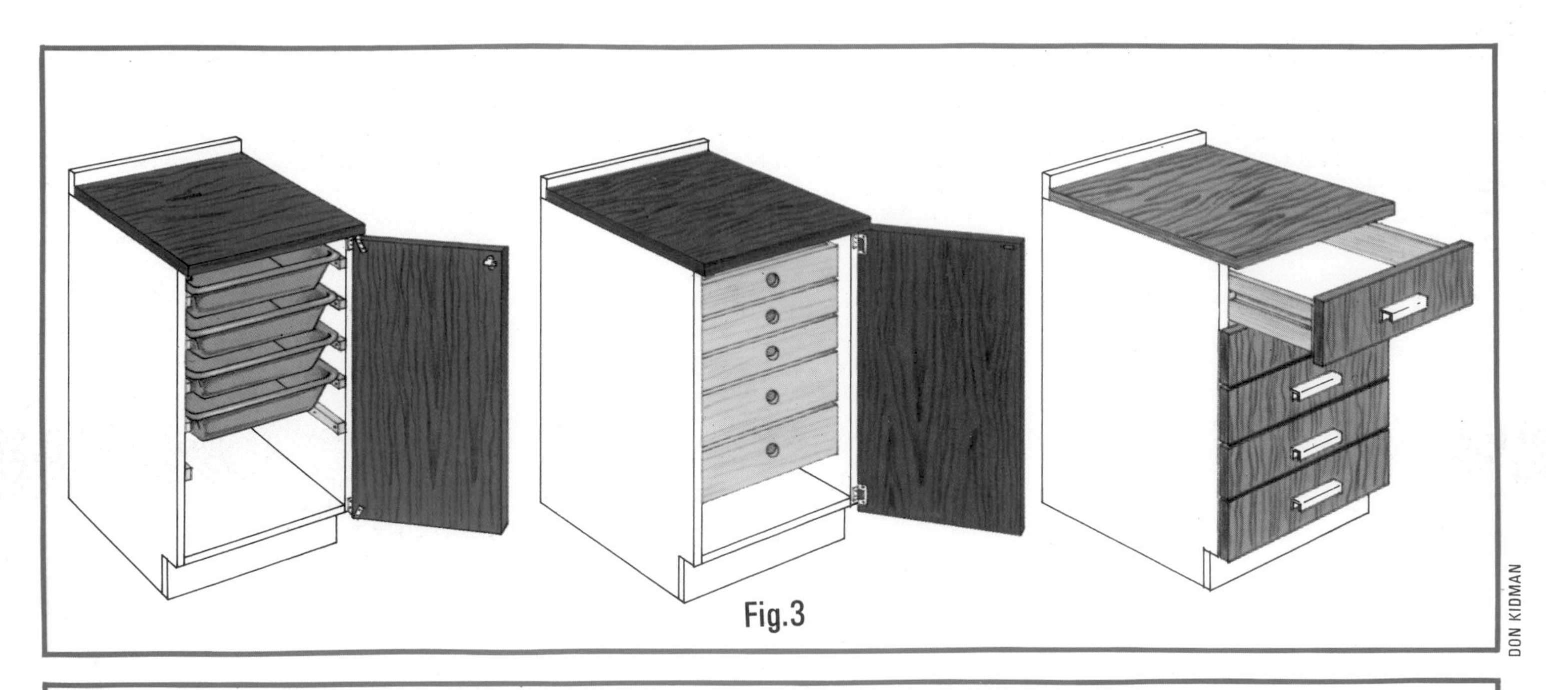

Fig.3

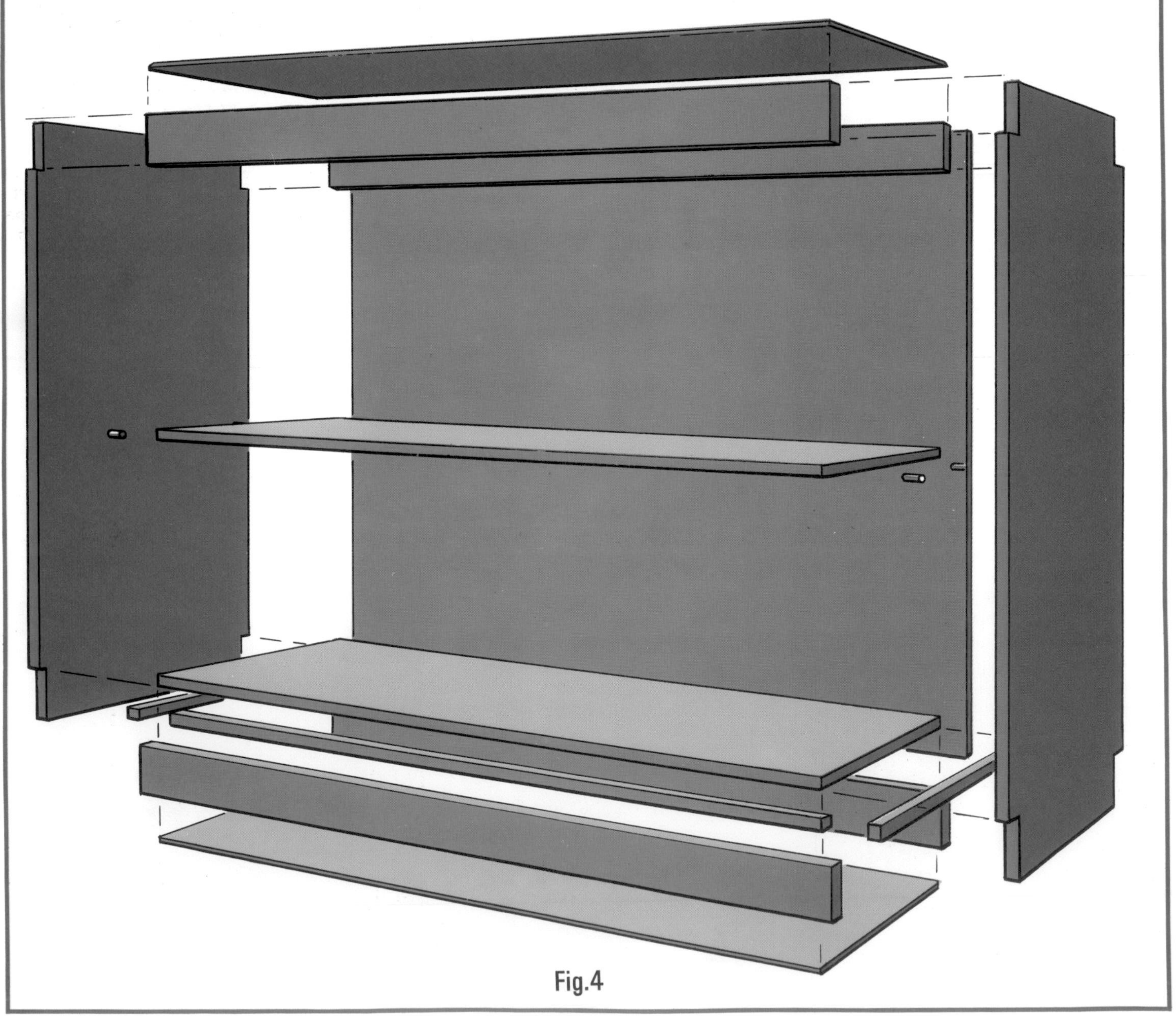
Fig.4

the doors, paint the hardboard or use plywood instead.

The hardboard top can be fitted at this stage but this will make it more difficult to screw the rear cross-members to the wall. If you handle the unit gently, it is generally more convenient to drill holes for the top panel, then leave the fitting until the unit has been mounted on the wall.

The inside joints at each end of the top rear member must now be reinforced with two angle brackets, as shown in Fig. 5.

Cut and fit the doors. The method for this has been described, but note that for these units the aluminum angle strip that forms the handle is placed underneath the bottom edge of the doors, in the same way as the handle fittings for the drawer fronts.

The unit is secured to the wall by 3 in. No. 10 screws passed through the rear cross-members so as to go into the studs (posts) inside the wall. Use a stud-locator from the hardware store, or locate the studs by tapping lightly with a hammer.

Fig. 3. *There are many variations you can make to the basic unit. Some of these are shown in the preceding chapter. Here, the storage receptacles are simple slide-in trays or drawers. The handles shown are made from short lengths of aluminum U-channel.*

Fig. 4. *Exploded view of the hanging wall cupboard. Because of the considerable weight the unit will hold when laden, all joints and the top, bottom and rear panels, must be securely screwed in place and glued.*

Fig. 5. *The two rear cross-members will take the most strain, so the securing screws may be dovetailed and the joints reinforced with brass or steel angle brackets.*

Figs. 6 and 7. *Edge detail of sink unit types.*

The sink top

Broadly, there are two types of sinks: the integral bowl and drain board, shown in Fig. 1, and the bowl that has to be inset into the surface, as in Fig. 2.

When you are purchasing your sink top, bear in mind the possibilities of fitting a waste disposal which will be a great asset in the kitchen. These units fit most sink plumbing, but check this when you buy your sink.

If you want to fit a sink that has an integral drain board, you must make sure that the external dimensions will fit the top of the unit. The unit shown in Fig. 1, for instance, is 21 in. deep from front to back. For wider counters use a sink that fits into a matching opening cut in the counter top. Some sink tops rest on the top of the counter, and are sometimes secured with screws or bolts passed through holes in a turned-in lip on the underside. Others are rimmed with a separate trim piece. Your best bet: examine the various types and select accordingly. Two types are shown in Figs. 6 and 7.

To make a plastic-surfaced plywood counter top with a suitable hole to take a bowl, buy the laminate and plywood separately. Cut to identical sizes. Glue the laminate to the plywood using contact cement, then cut the hole with a fine-toothed sabre saw.

The area of wall behind the sink will have to be covered with a splash-proof surface of some kind. This often takes the form of ceramic tiles, as shown in Fig. 1, though you could use a readymade splashback often available to match readymade countertops.

The sink base

This is constructed in the same way as the unit described in the first part of this project, except that this unit has no drawers, and the top front part of the unit, where the drawer fascias would be, has a 6 in. wide length of plywood running across it instead, as shown in Fig. 1. If sink and plumbing permit, you may fit a central partition to support the sink, as already described.

Plumbing

Unless you wish to change the location of your faucets, this will only entail turning off the water supply at the main valve. The next step is the disconnecting of the faucet and drainage pipe fittings, replacing or re-fitting the sink, then re-connecting the faucets and sink trap.

If the faucets are to be relocated, then you will have to extend or shorten the existing pipework. This involves cutting and fitting pipes. Your plumbing supplier can give you information on these jobs.

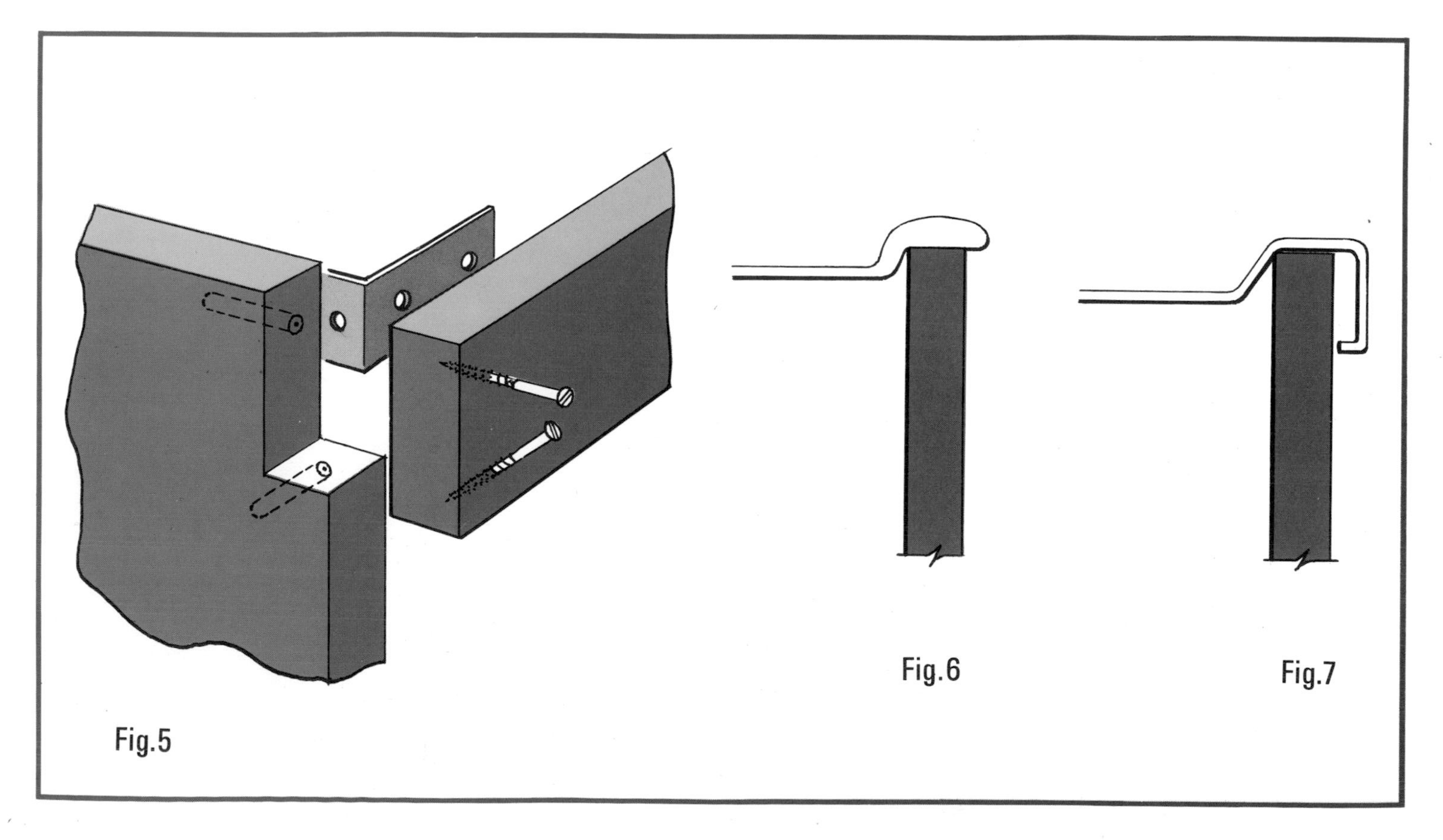

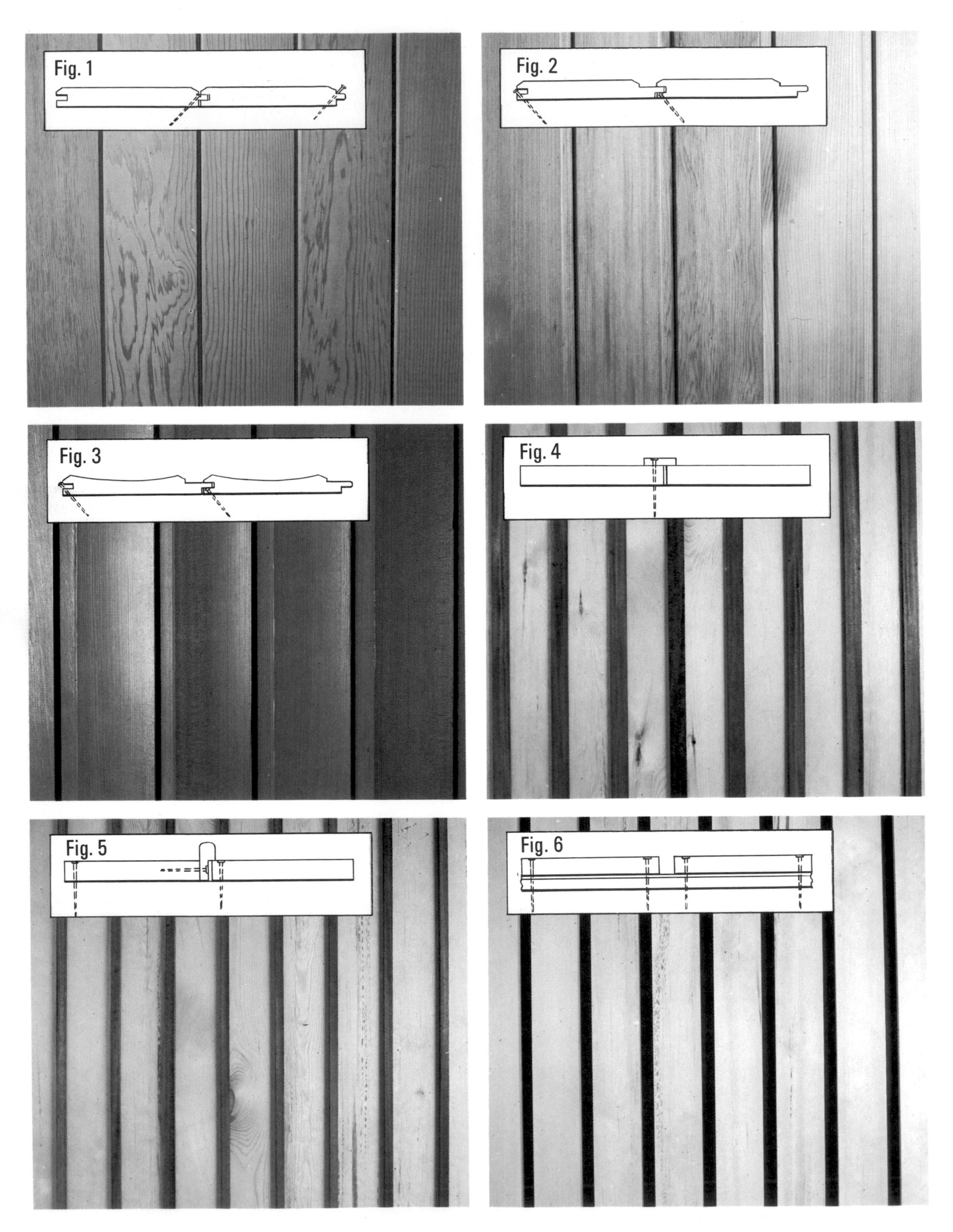
Fig. 1
Fig. 2
Fig. 3
Fig. 4
Fig. 5
Fig. 6

CHAPTER 29

Wood paneling with a difference

The rich appearance of solid wood paneling can be spoiled by careless installation. Always check as you go along to see that boards are straight and joints are even.

Horizontal positioning

Horizontal T and G boards are mounted in the same general manner as vertical ones. It is best to position the first board an inch or two above floor level, with the tongue upward. This board should be skew-nailed into the underlying battens through the surface of the paneling board about ½ in. up from the bottom edge. It should then be skew-nailed through the battens along the edge of the tongue so that the groove of the next board will cover the nails. Use a spirit level over this first board to see that it is level. The battens, or "Furring strips" should be of nominal 1 x 2 lumber nailed through the wall into the studs inside. They should run vertically, as in Fig. 8, for horizontal or diagonal paneling; horizontally for vertical paneling.

Continue mounting the boards up to ceiling level, close-fitting each board and checking your boards every once in a while to see that they are level. If they are slightly out of line, you can make adjustments by fitting edges very closely where necessary. Just below the ceiling, leave a ¼ in. gap.

If you cannot buy boards which will fit the entire breadth of your wall, you will have to join two or more lengths. Use a *scarf joint* (Fig. 7), so that no gap shows at the joint.

Diagonal positioning

An unusual and often effective means of applying boarding is diagonally. Visually, the best angle for this is 45°.

Instead of beginning with short lengths of board in a corner, begin at a position along a wall where your first board will be a full length one, as in Fig. 8 (you can save short offcuts to fill in the corners later).

Opposite page. *Solid wood paneling comes in a variety of styles, using either T & G boards or cheaper square-edged planks.*
Fig. 1. *Conventional, v-jointed T & G.*
Fig. 2. *Flat surfaced, extended tongue T & G.*
Fig. 3. *Concave surfaced, extended tongue T & G.*
Fig. 4. *"Board-and-batten"—hardwood batten over square-edged planks.*
Fig. 5. *Square-edged planks with molding fixed to one board at joints.*
Fig. 6. *Square-edge planks mounted over hardboard which has been pre-painted in a contrasting color.*

To establish the length of the full-length boards required, cut 45° angles in two boards with a miter block (Fig. 9). Be sure to cut these angles in the direction in which you want the boards to slope. Position these boards one beside the other and hold them against the wall, adjusting them until they are straight and you get the correct length. Hold them together by nailing two short battens across them (Fig. 10). These boards will give you a gauge from which you can measure the first full-length board. The others can be marked from this—but check as you go that the lengths do not need to vary slightly because the floor and ceiling are "out." If your walls are very high, it is not advisable to attempt to cover them in diagonal boarding, since you may not be able to buy long enough lumber to avoid a jumble of joints.

For boards which are to slope upward from left to right, work from left to right across a wall, with the tongue on each board facing toward the right. Hold your first board in place with nails partly driven through the surface of the timber on the right-hand side only. This first board (A in Fig. 8) is mounted loosely so that later boards can be "sprung" into position, as previously explained. Blind nail the remaining full-length boards into position by skew-nailing them through the tongues and punching the nail heads below the surface of the timber. Clamp each board as you go along and check, occasionally, to see that all the boards are at the correct angle of 45°.

To fill the top left-hand corner (above board A in Fig. 8) start by temporarily stacking a row of offcuts against the wall along the left-hand edge (you will need to hammer in a few nails to keep these boards up) until you can accurately judge the size of the board that will fit into the extreme corner (C in Fig. 8). Cut this board, which will be a right triangle, about ¼ in. oversize along the two sides forming the right angle, and fix it in position. (Cutting the board oversize will ensure a tight fit later.)

Once you have mounted board C into the corner, you can remove the offcuts from the wall and with boards which have been cut to the correct size, work progressively back toward board A, clamping each board along the tongue-side as you go. However, the last three or four boards should not be nailed into place at first, but sprung into position. (See page 35.) If the last board does not quite fit, the tongue-edge may be smoothed with a plane slightly so that it can be eased in. After these boards are in place, nail them through the surface and punch in the nailheads. Fill the holes with a wood filler.

To fill the bottom right-hand corner, follow roughly the same process. When you have six or seven boards to go, stack offcuts as you did before until you can ascertain the size of the corner piece (see Fig. 8). However, do not mount the corner piece first. Rather, work from left to right, springing the last three or four boards into place and surface nailing them.

Doors and windows

When mounting panel boarding around doors or windows, one of the most important considerations is to see that the unsightly rough edges of the planks do not protrude beyond architraves or flush edges. Usually, the thickness of the battens plus the thickness of the planks will be greater than the thickness of your architraves. If you have doors or windows which are set into walls without any architraves, you will encounter the same problem. And anyway, most elaborate molded architraves will look awkward set against this sort of boarding.

Such problems can be solved by first removing all existing architraves. Try to pry them off from their external edges so as not to damage adjoining woodwork. Windows may present many more problems than doors, since you will have to remove not only the architraves but also the window sills (or trim them flush with the wall surface).

On *masonry* walls, replace the old molding (or, in the case of doors and windows without any architraves, make a new molding) with square-section lumber (1 in. x 1 in. or whatever size necessary) which will be slightly thicker than the *combined* thickness of your furring strips and your planks (see Fig. 11). The same procedure will work on wood framed walls, but only if "nailing" is available around the door or window—either to existing studs or to battens which you provide.

When putting up vertical planks around a door, work progressively across one side of the wall from a corner to the door, then across the wall space above the architrave, and then across the wall on the opposite side of the door to the end of the wall. If you are planking around a window, follow the same plan—side, top, side—and then fill in the area across the bottom of the window. To fit boards around the corners of architraves, scribe and cut boards as in Fig. 12.

When putting up horizontal boarding around doors, begin on one side of the door and work up the wall to the top of the architrave, butting the boards against the architrave as you go.

Then, work up the wall on the other side of the door in the same way. Finally, mount boards across the wall area above the architrave, staggering any scarf joints which you may need to make. For boarding around windows, follow essentially the same procedure, beginning with full-length boards at the bottom of the wall and working up and around the walls on either side of the window and then across the top of the window. It is important that you stagger any scarf joints across the wall so that they do not form a conspicuous line.

Fixing the board diagonally around doors and windows often is not visually pleasing, and problems can arise in matching the slope of the boards and the angles. Consider carefully before you attempt to apply diagonal boarding on a wall area broken by doors and windows.

If you do use diagonal boarding, be absolutely sure that you measure and cut your boards so that they can be butted smoothly against any architraves and are all positioned at a 45° angle.

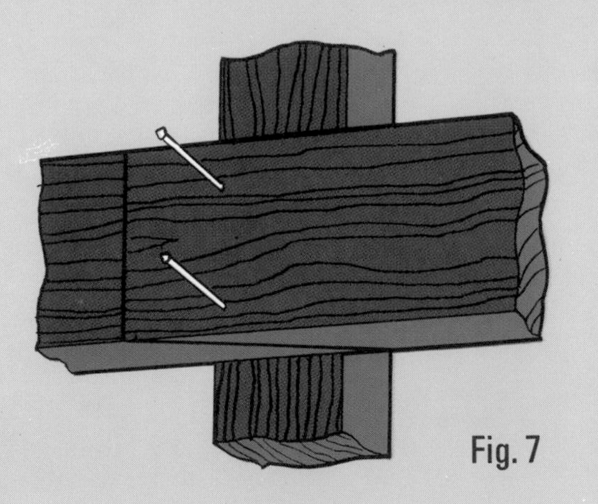

Fig. 7

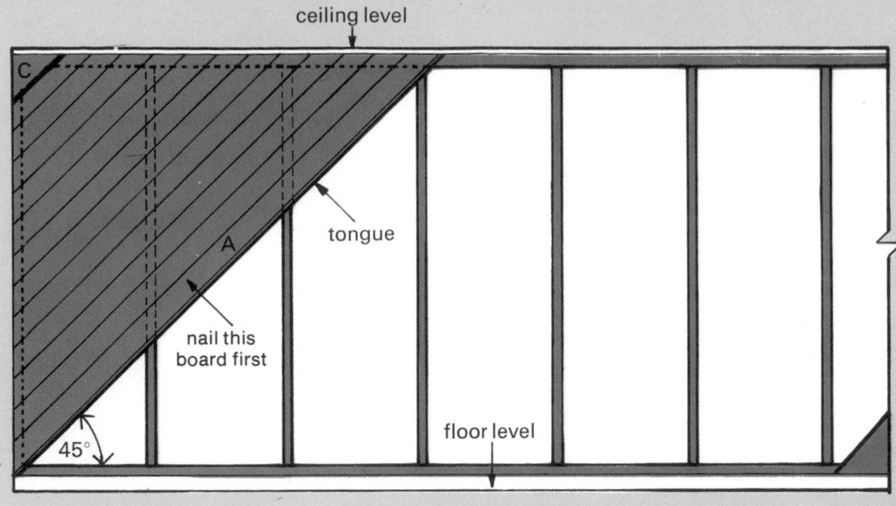

Fig. 8

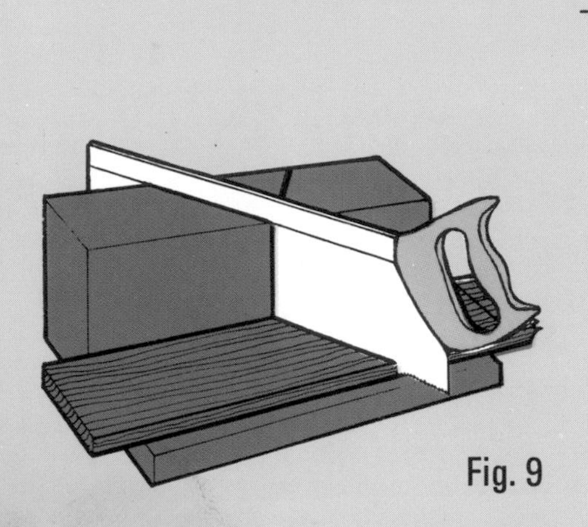

Fig. 9

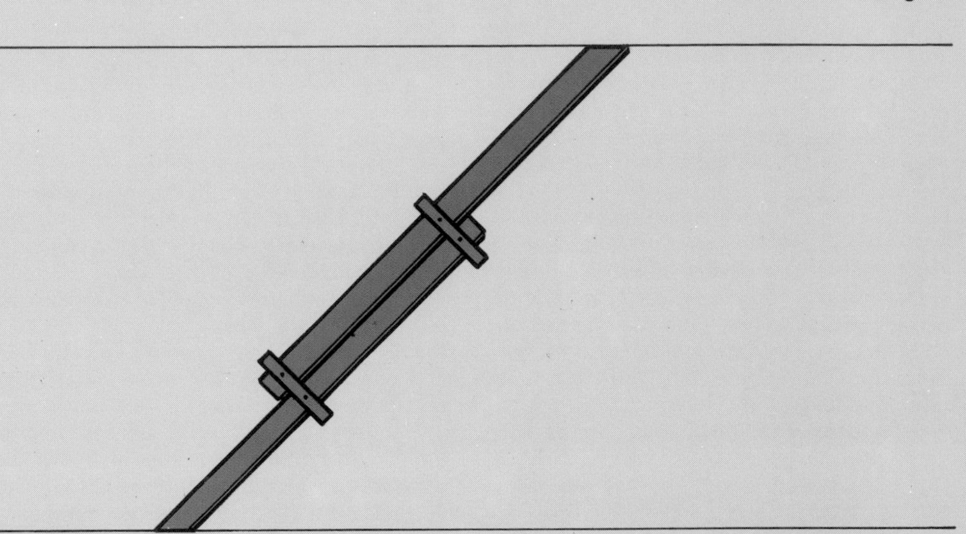

Fig. 10

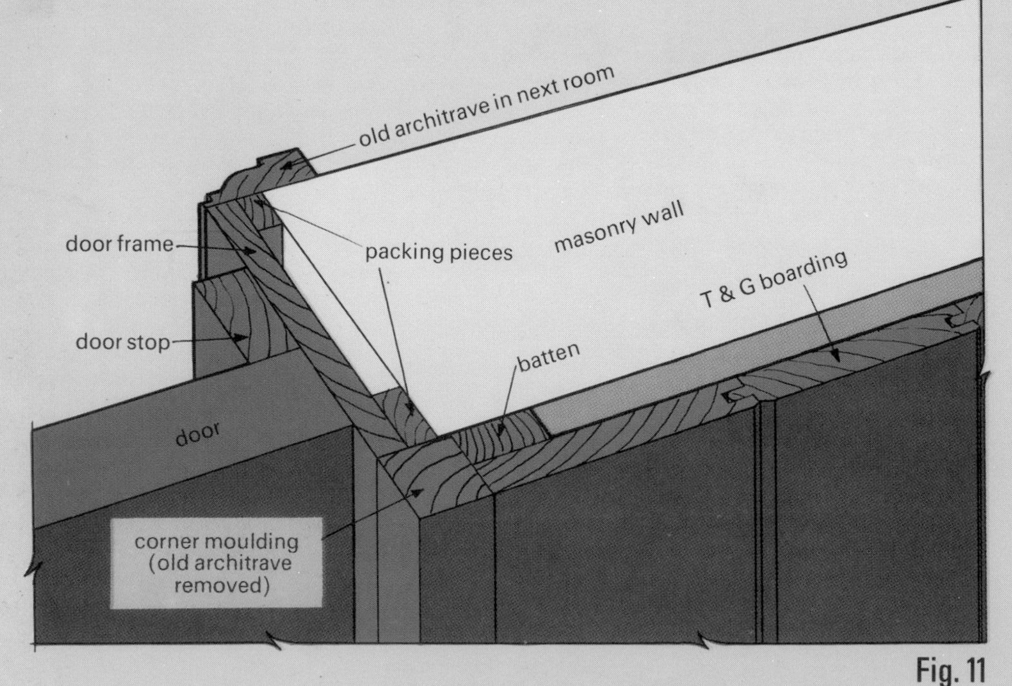

Fig. 11

Fig. 12

Fig. 7 *(top left). A scarf joint leaves no gap between boards joined lengthwise. It is made by clamping two boards together and sawing through both before trimming them to the final length.*

Fig. 8 *(top right). For diagonally positioned boarding, nail full-length boards first and use offcuts to find the size of corner pieces.*

Fig. 9 *(center left). Use a miter box to hold lumber that is to be cut at a 45° angle.*

Fig. 10 *(center right). To find the size of a full-length, diagonally installed board, cut 45° angles in two boards, hold them against the wall and when the correct length is found, fasten them together with short battens. Use these boards as a gauge to cut your full-length one.*

Fig. 11 *(left). A cross-section through a wall and door frame, showing how the old architrave has been replaced by a square-section corner molding.*

Fig. 12 *(bottom left). Boards must be scribed to fit around architraves. Mount "A" loosely where the last full-width board will fall, and using "B," scribe the portion of "A" to be cut away. Then spring the boards into place and surface nail.*

CHAPTER 30

Good looking bed platforms

Bed platforms can be an attractive addition to any room. A platform covered in carpet can integrate easily into an old-fashioned setting while a naturally finished wood platform complements the clean lines of a modern home setting. Whatever shape or finish you choose for your platform, it's easy. And all you need to complete it is an innerspring mattress.

The basic construction of all bed platforms consists of the sides, the top and internal support framework. These are essential to prevent the top sagging. The method of framing is the same whatever type of platform is built—the components are evenly spaced, run the length and width of the box construction, and meet the box sides and each other at right angles. They are as wide as the distance between the floor and the underside of the top of the platform.

HEIDEDE CARSTENSEN/STUDIO HUELSTA

CAMERA PRESS/IMS

Design considerations

Before constructing a platform think hard about the shape of the room and the position the platform will occupy. Remember that the platform will be an almost permanent feature—even if it is not nailed to the wall and is free standing it will be too heavy to move with any ease.

It is a good idea to make a scale drawing of the room and position the platform on it. This will avoid mistakes in positioning the platform—the worst being not allowing sufficient space for the room door to open fully when the platform is in place.

The nature of the room in which it is to be used will influence the shape of the platform you choose. A rectangular bed platform will fit into any setting if it is finished suitably. The height of the finished platform is a matter of taste but again the nature of the room should influence your decision. A very low platform will look out of place in a room with a high ceiling whereas a high platform will tend to overpower a low room.

Bed platforms can be either free standing or attached to the wall. The advantage of free standing platforms is that they can be moved more easily than fixed platforms. Free standing platforms do, of course, involve using more materials than fixed platforms.

If possible construct the platform in the room in which it is to be used. It will be fairly heavy and will be awkward to move from place to place.

Materials

A variety of materials and combinations of materials can be used to build a bed platform.

—Stock lumber is easy to work with, but it may prove expensive to make the whole platform with this.

—Lumber and plywood is again expensive, but the pieces can be easily fastened together and will be very strong.

—Lumber and particleboard is cheaper and will make a platform of adequate strength.

—Particleboard alone is not really practical. It is cheap and light, but fastening the parts together strongly enough will prove difficult.

—Lumber with a ¼ in. hardboard top is workable, but the top may need extra support to prevent it sagging.

Finishing the platform

The way in which you intend to finish the platform will affect the type of material you use and how you fasten the pieces together.

The finished platform can be carpeted with either the same type of carpet used on the floor or a contrasting one. This method of finishing thoroughly integrates the bed platform with the rest of the room. If this finish is used, the sides of the cabinet do not have to be particularly well finished—they could, for example, be made of particleboard with the constituent parts screwed in place through the outer face.

The sides can be covered in plastic laminate. Again fastening can be done through the outer surface of the sides before the laminate is applied. Apply the plastic laminate in the usual way, using a contact cement.

A natural solid lumber finish will make a feature of your bed platform. The constituent parts should be screwed together into the inside face of the sides or fastened through the sides with brass screws. The sides can be sanded smooth and treated with several coats of clear polyurethane varnish.

Shape of the platform

A rectangular platform is the easiest to construct and its shape is conventional enough to fit all types of rooms. It can be made slightly wider than the mattress to allow space at each side for bedclothes.

Platforms that are constructed exactly to the mattress width can have a rim or lip incorporated in them. Here the top of the platform is recessed below the level of the top of the sides. The disadvantage of this design is that if the rim is more than a few inches deep, making the bed will prove difficult.

Freestanding bed platforms

With rectangular bed platforms the sides are constructed and fastened together in a box shape first. If you use stock lumber and the bed platform is to be over 12 in. high you may have trouble getting boards of this width. You can overcome this by matching the long edges

Above left. *In this bed platform, the top extends beyond the sides. It has an attractive clean finish, though the butt joints at the corners must be filled.*
Above right. *Bed platforms add an extra touch of distinction to the modern bedroom. The colors of the mattress cover and bolster are attractively mirrored in the blinds.*

of two planks, applying glue and butting them together to give a board of the required height. If plywood is used there will be no problem cutting sides high enough for the job.

If you intend to cover the sides of the platform, they can be cut square and the ends simply fitted at right angles to each other. These joints will, however, be obvious if a natural wood finish is intended.

One good way of making neat corner joints is to miter the end edges of the sides. Cut the side pieces to the exact size required. Set a bevel gauge to 45° and, with a pencil, mark inward from the outside edge of the ends of all side pieces. It is a good idea to stand the pieces together in a rough box shape—this will ensure that the miters are marked in the correct way. Then saw the miters. Clamp a piece of waste wood to the end to which you will be planing (if trimming is required) to avoid knocking the corners off the ends, and plane down to the lines, holding the boards upright in a vise. Check that the miters are flat by drawing a straight edge along their surfaces.

Other neat corner finishes can be achieved with exposed dovetail joints (described in full in CHAPTER 11) or rabbeted and butted joints (see CHAPTER 9). This last-named is probably the easiest to make.

If the sides are to be covered in plastic laminate then the shape of the corner joints will not matter. The platform may, however, have a rim with the top panel inset to take the mattress. From above, a joint made by butting the end edge of one piece to the end of the wide surface of another piece will look unsightly. In this case a hardwood edging strip, mitered at the ends, can be glued to the tops of the side pieces after they have been screwed together in a box shape.

If you are using a natural finish on the top of your platform, two of the sides will have to be rabbeted to accept the ends of the top planking. This is to avoid having a lot of end grain showing. The depth of the rabbets should match the thickness of the planking. And you will have to cut the internal components that much narrower.

Assembling the frame

The method of fixing the sides together, and the internal frame members to the sides and each other, is to use wood blocks on the insides of the corners. Blocks of nominal 2 x 2 fir are glued and screwed to the two pieces which meet at right angles. If it does not matter whether screw heads appear on the outer surface of the sides, the pieces are fastened through the sides, screwing into the wood blocks. If it does matter, you screw through the blocks and into the inner surface of the sides.

Start by fastening together the four sides to form an open box. As well as four 2 x 2 wood blocks cut to size, you will need eight pieces of 2 in. x 1 in. lumber the same length as the wood blocks, and four G clamps. With a coarse-toothed saw, cut a groove lengthwise down each of these pieces of 2 in. x 1 in. Stand the sides of the box together, apply white glue to two surfaces of each of the glue blocks, and put the blocks into place. Now use a pair of the 2 in. x 1 in. pieces on each corner as packing for the G clamps—the notches will stop them sliding off the corners (Fig. 4). Make sure the box is in square, then leave the glue to dry.

Before the glue sets, fasten the blocks more securely. Depending on the considerations given above, screw the blocks to the sides through either the blocks or the sides.

Building the inner frame

The next step in making the bed platform is to construct the internal support components. The space between them will depend on the type of material to be used for the top—a particleboard or 1/4 in. hardboard top will require more support than solid lumber boards. If solid lumber is used the frame members can be placed about 2 ft. apart, but space them about 18 in. or less apart for particleboard or hardboard.

For a simple rectangular platform the struts

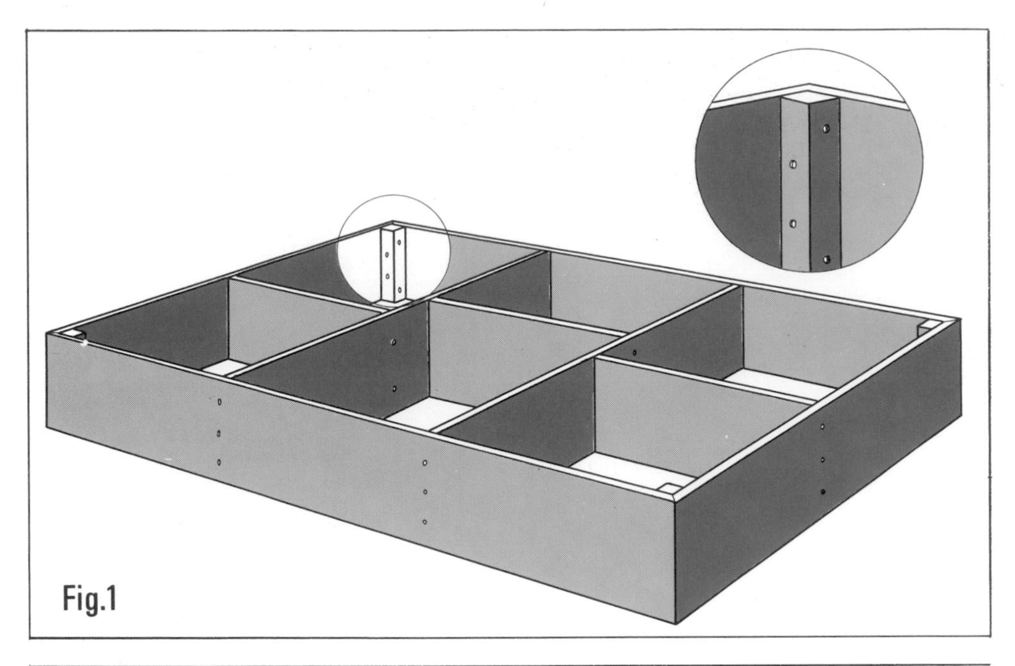

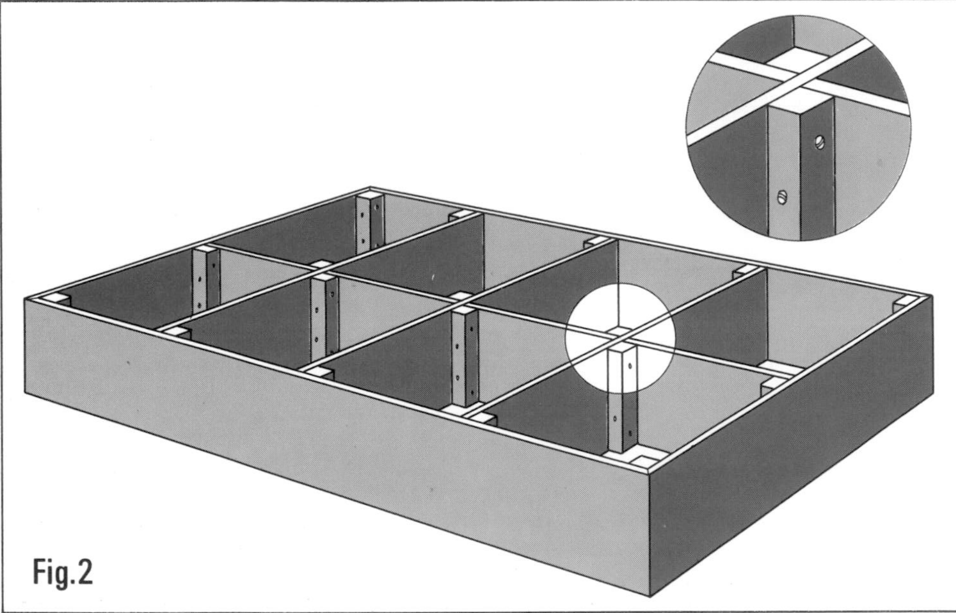

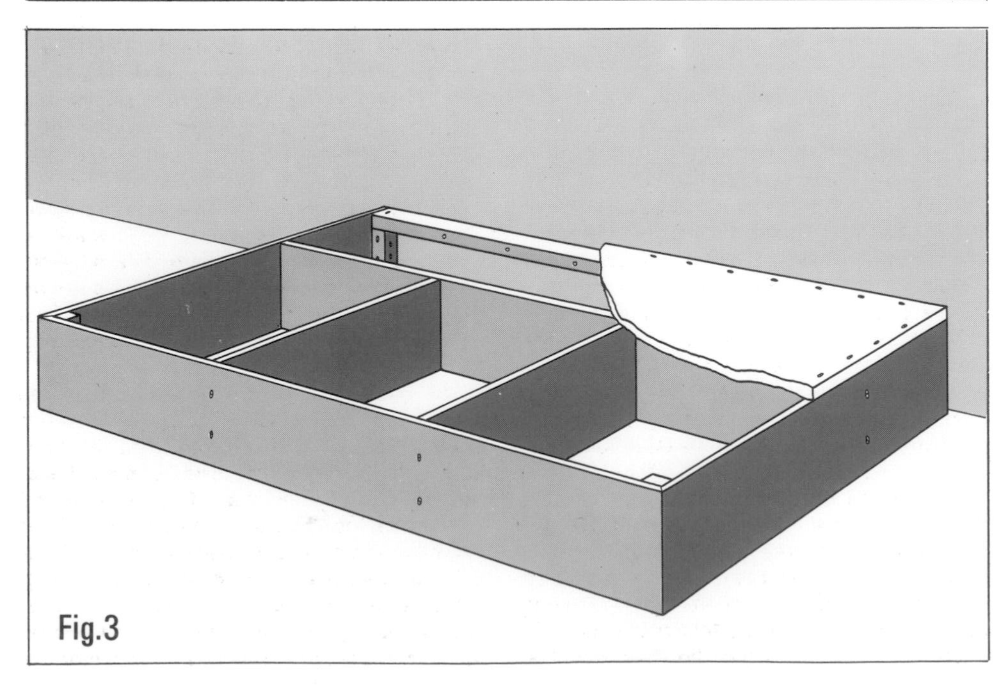

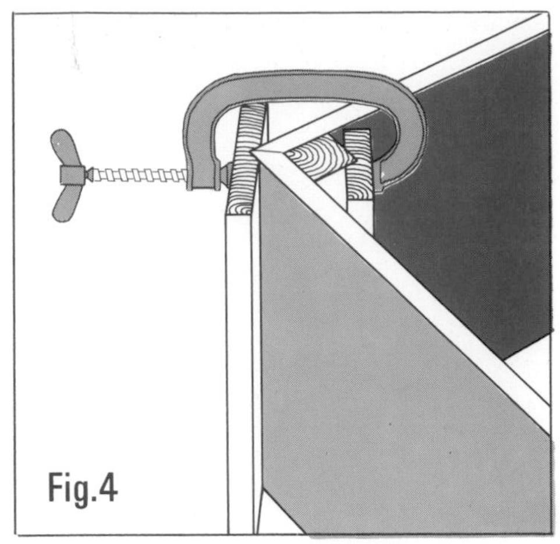

Fig. 1. *The inner parts in this platform are staggered so that they can be screwed directly to each other. The parts are fixed to the sides from the outside of the box.*
Fig. 2. *If particleboard parts are used it will be difficult to screw them to each other, so glue blocks are used at all the corners.*
Fig. 3. *In a fixed platform, pieces of lumber are attached to the wall to support part of the top. The long member, which in a free standing platform would be centrally placed, is offset so that the short ones can be attached to it and not to the wall.*
Fig. 4. *This method of gluing and clamping wood blocks at the corners ensures that they do not move out of place while the glue dries.*

should be as deep as the sides of the box. For all platforms the inner members should meet at right angles. Fig. 1 shows an ideal method of placing framing within a solid lumber box where screwing from the outer face of the sides is suitable. The members that run parallel to the long sides are staggered to allow them to be screwed to the ones that run parallel to the short sides.

Fig. 2 shows how the struts are placed in a particleboard box. Wood blocks are used at every internal corner as well as at the corners of the four sides.

The top board or boards can now be glued and screwed into place with countersunk screws. If one single sheet of particleboard or hardboard is used, cut it slightly oversize. Be sure two of the sides that meet at right angles are exactly square. Measure and mark, on the top of this board, the positions of the internal frame members—this will tell you where to drive screws later. Spread white glue on the top edges of the frame members and platform ends and sides. Lay the top panel in place, marked side upward, with the squared edges flush with two edges of the box. Screw the board to the sides and to the internal members, the positions of which you have marked. Remove any waste from the top panel with a smoothing plane so that its edges are flush with the sides of the box.

If the top is to be formed of solid lumber boards then they are attached so that their long edges butt together. Cut each plank slightly oversize. Square a line on the wide surface of each plank, close to one end. Saw and plane to these lines taking care not to damage

Above. *This brightly finished platform can be moved about on small casters.*

the corners of the boards when planing.

The boards can either run the long way or the short way across the box. Running them the long way will involve less sawing and planing. Lay the first board with its long edge and squared end flush with the sides of the platform and glue and screw it into place with countersunk screws. Drive the screws into the edges of the side panels and into the supporting members that run at right angles to the first board of the top. Lay and fasten each subsequent board in the same way as the first. When all the boards are in place, remove any waste wood from the unsquared ends with a plane.

For a recessed mattress

A variation on the simple rectangular design is to construct the platform with a lip or rim. The mattress fits snugly between the sides and on the recessed top. In this case the top edges of the internal supporting members are below the tops of the box sides—they are of a depth sufficient to give you the depth of recess you require. Fasten the sides together in a box shape in the manner described above and fix the internal members in place. Now place the platform top inside of the platform sides so that it rests on the inner frame members.

The top panel or boards for this type of platform will have to be cut exactly to size before being screwed and glued into place. When this is done fasten the top in the manner described above.

Fixed platforms

If your bed platform is to stand against a wall you may decide to fix it permanently in place. To do this the top board or boards will have to be screwed to a support fastened to the wall and the side pieces fastened to the floor.

First, however, fasten the sides together, using wood corner blocks. Then mark the position the bed platform will finally occupy on the wall. Make the marks where the insides of the side pieces meet the wall at right angles. Draw lines through these marks at right angles to the floor. Along these lines measure and mark a distance that equals the depth of the side pieces of the platform, measuring from the floor. Draw a line through these marks parallel with the floor. Screw a length of nominal 2 x 2 fir to the wall so that its top edge is exactly on these lines. Fasten a block of wood to the end of each bottom edge of this piece, at right angles to it and running to the floor. Now fasten the ends of the sides that run to the open end of the platform to these wood blocks.

The next step is to position the inner framing. The members that run to the wall from the side of the platform parallel to it can be fastened to wood blocks screwed to the wall. Or, if you prefer, one of the internal members that runs parallel to the open wall side can be moved further toward the wall than would be the case in a four-sided platform. This will reduce the strength of the platform a little but the primary purpose of inner framing is to support the top. This arrangement of components will do this as adequately as any other (see Fig. 3).

The platform can now be fastened to the floor. Again the method of doing this is to use wood blocks—this time screwed to the floor and to the bottom of the inside of the sides of the platform. The number of blocks required will depend on the size of the platform. If, for example, the platform measures 6 ft. 8 in. x 5 ft., six blocks of nominal 2 x 2, 6 in. long, will be sufficient. Place two on each long side, each about 18 in. from the ends, and one on each short side, positioned near the center.

PICTUREPOINT

CHAPTER 31

Fascias, soffits and bargeboards

Fascias, soffits and bargeboards sound like exotic pieces of equipment. Perhaps you think you haven't got any. But the chances are that if you live in a house you have them on your roof—and they may need looking at.

Fascias, soffits and bargeboards are the facings of the eaves and gables of a house. They cover up the ends of the rafters and the gap between the top of the walls and the edge of the roof. And their style varies in different parts of the world. Often, merely changing to a different style of one of these details can revitalize the appearance of an otherwise ordinary house.

The positions of the different boards are shown in Fig. 1. A *fascia* is set on edge under the lower edge of the roof covering and is nailed to the ends of the rafters. Buildings with hip roofs have fascias all around; those with gable roofs have fascias along the horizontal sides only.

A *soffit* is a horizontal piece reaching from the lower edge of a fascia to the wall behind it. Some houses with narrow eaves have no soffits; instead, the fascia is mounted directly against the wall.

A *bargeboard* is like a fascia, but runs up the slope of a gable instead of horizontally. For this reason, bargeboards are generally found in pairs. They frequently have soffits underneath them where the sloping roof edge projects beyond the flat gable front.

None of these pieces plays any part in supporting the roof. They are for weather protection and to improve the appearance of the outside of the house. When they rot or break, they can easily be removed or replaced—and should be before the damage goes too far. A leaky fascia can admit rainwater to a house. Gutters are generally mounted on fascias, too. If the wood becomes rotten, gutter fittings may fall off—a remote possibility, but an important one.

Opposite page. *The old houses in this photograph have a rich variety of bargeboards, fascias and soffits decorated with fretwork.*

Ladders and scaffolds

One of the main difficulties in repairing the upper parts of a house is reaching up there in safety. An ordinary ladder is sufficient for inspecting them, but not for repairs. This is because the pieces of wood you will be working with are large and heavy. You will almost certainly need a helper to hold them up while you remove or fasten them in place—and two people on one ladder won't go.

The simplest, but not the most satisfactory solution, is to use two ladders, one for each person. They should be securely fastened in place at the top and bottom—otherwise you could have a serious accident.

The trouble with this method is that you will have to keep moving the ladders along the house, which will involve a lot of climbing up and down, tying and untying of ropes and a lot of other work. What you really need is a platform near roof level, so that two people can move about with reasonable freedom.

The cheapest way to achieve this is to go to a tool rental firm and rent a scaffold. Get complete instructions in its assembly and use when you rent it. You'll also need a ladder to reach it.

A ladder scaffold offers a useful increase in mobility. But it has to be dismantled to move it to a new part of the roof.

A better idea, though it may be more expensive, is to rent a mobile scaffold. This is a structure made of prefabricated scaffolding-like sections that fit together to reach to any height within reason. At the top, there is a generous-sized platform. The whole structure is mounted on wheels, so that it is easy to move around on a flat surface. The wheels lock with screws to hold the tower still. It can also be set up on uneven ground by means of adjustable *jacking bases.* The top platform is reached with an ordinary ladder.

Fascia repairs

Fascias can be of various types, and before starting work you should find out what kind you have. Its details often vary with the type of roofing, and the age of the house. If yours is an uncommon type it may match one of the variations illustrated.

They may be fastened direct to the ends of the rafters, without a soffit—this type of fascia is generally found in houses over 60 years old.

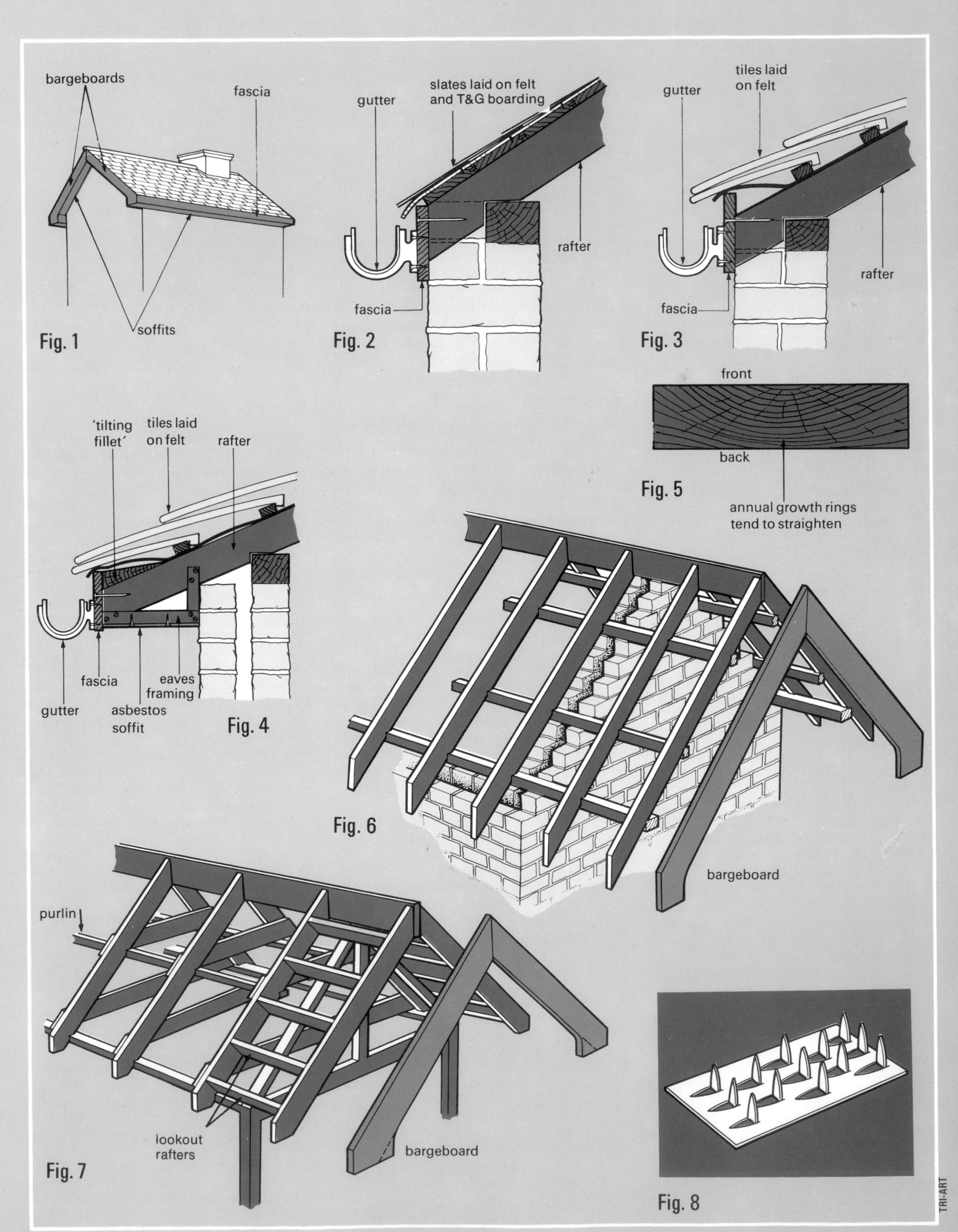
bargeboards
fascia
soffits
Fig. 1
gutter
slates laid on felt
and T&G boarding
rafter
fascia
Fig. 2
gutter
tiles laid
on felt
rafter
fascia
Fig. 3
'tilting
fillet'
tiles laid
on felt
rafter
gutter
fascia
asbestos
soffit
eaves
framing
Fig. 4
front
back
Fig. 5
annual growth rings
tend to straighten
Fig. 6
bargeboard
purlin
lookout
rafters
bargeboard
Fig. 7
Fig. 8
TRI-ART

The fascia may not quite touch the underside of the roof covering (Fig. 2), in which case all you need to do is pry it off. Or it may hold up the bottom row of tiles or slates (Fig. 3) in that type of roof. In this case, removing the fascia will cause the tiles that are resting on it to sag. But they are (or should be) quite adequately fastened in place at the top, and will not fall off. Check that this is so by lifting up one tile and pulling at it to see if it is firmly attached. If it comes away, you will have to remove the bottom row of tiles before you lever the fascia off, and replace it afterwards.

Fascias with soffits behind them rest against (but are generally not nailed to) the soffits (see Fig. 4). There are various other fascia-soffit constructions, too, depending on the type and age of the house. If you examine any of them carefully, however, you can usually determine the best way to remove a damaged or rotted part.

Fascias nearly always have gutters attached to them, and the brackets which hold them should be carefully removed from the fascia and fastened to the new fascias after these are installed.

The order of work for replacing a decayed fascia is as follows. First, inspect the fascia and determine whether it needs replacing or just repainting. Small rotted areas may be hardened with marine rot-hardening preparations sold by boatyards.

If it does need replacing, hire and erect a suitable scaffold. Unscrew the gutter brackets and carefully take down the gutter—you will probably need a friend to help you. While the gutter is down, you might as well repair and paint it.

Carefully pry off the old fascia, trying not to break it. There are several ways of doing this. You can use a hammer and nailpunch to drive all the nails right through, or drive wooden wedges up behind the fascia. Or use an old chisel to cut away enough wood around the nails to reach their heads with a crowbar or a claw hammer. Carry the fascia down, measure it, and order a suitable length of new lumber in the same size, or in a size that can be cut to fit.

Rake out all the old birds' and insects' nests and dirt between the rafters and inspect the adjacent roofing. Patch it if necessary. Paint the exposed ends of the rafters with primer to protect them from rot.

Fig. 1 *(opposite). A guide to where the various eaves facings of a roof are fitted.*
Fig. 2. *A fascia may be completely separate from the rest of the roof covering.*
Fig. 3. *Or it may support a row of tiles.*
Fig. 4. *A modern roof with wide eaves, a hardboard sheet soffit and a "tilting fillet."*
Fig. 5. *Mount the new fascia this way.*
Fig. 6. *The roof structure of a typical British house, showing the bargeboard mounting and how the end rafter is supported.*
Fig. 7. *The roof of one Australian type of frame house. The end rafter, on which the bargeboard is mounted, is supported by a ladder-like row of "lookout rafters."*
Fig. 8. *A gang nail or ply plate used in roof trusses.*

Examine the new fascia. The annual growth rings of the tree it was cut from will show at the ends. The fascia should be installed so that the *concave* side of the rings is at the front (see Fig. 5). When it warps (as it sometimes will), if it is this way around, the curvature it takes on will tend to straighten the rings and so press the top and bottom of the board against the wall, providing an efficient seal against moisture. Cut a small notch out of the back to mark it, then prime and paint the fascia on both sides, and, most important, both ends, since the end grain of the lumber is the most vulnerable to rot.

When the paint is dry, carry the fascia up. Hold it in position (pushing the tiles up to their original height if necessary, if you have a tiled roof) and nail it on with *one* large nail into each rafter, hammered through about 1 in. from the top edge. This will allow it to expand and contract with changes in humidity. Use galvanized nails for this and all other outside jobs.

Finally, screw the gutter back on, using a spirit level to ensure that it slopes down at least 1 in. in every 10 ft.

Soffit repairs

Soffits may be made out of solid wood, hardboard or plywood. The tendency in modern houses is to use hardboard or plywood if the soffit is a wide one. Asbestos-cement board is also used for some soffits.

Asbestos is rotproof but fragile, and needs replacement only if it is broken. It is tricky stuff to work with because if you press it too hard it cracks. The best saw to cut it with is a compass type hacksaw, which has a replaceable hacksaw blade mounted so that there is no frame to get in the way when cutting along wide sheets.

An alternative method of cutting asbestos-cement board is to treat it like glass. Scribe a deep scratch along a straightedge where you want to cut it; the tang of a file cuts as good a scratch as anything; or you can use an old, discarded chisel. Then lay the sheet down on the straightedge so that the scratched line is immediately over it and snap it along the line by pressing down hard on both sides of the straightedge.

Holes can be drilled with an ordinary twist bit. Use a cheap one, as you'll dull it. You have to drill holes for nails as well as for screws. With some types, however, you can nail straight through. But it is wise to drill holes at corners; otherwise they might crack off if you nail too near them.

To replace an asbestos soffit, first try to unscrew the old one, or pull out the nails if it is nailed on. Use the old soffit as a template to mark out the new one. If it is too badly broken to do this, measure the space where it fitted.

Cut the new soffit to shape and drill and countersink screw holes in it. If you are replacing several pieces, they should be butt-jointed, not overlapped, and both adjoining pieces screwed to the frame they meet under. It is essential to give asbestos-cement board adequate support so it will not break again.

Install the new soffit, being careful not to over-tighten the screws. Use galvanized screws to prevent rust marks. Leave the new sheet to weather for a while before painting it.

Wooden soffits are easier to work with. Often, however, tongue-and-groove boards are used. If you are replacing only one of these boards, you will have to cut off the tongue on one side of the old board to get it out. This is best done by inserting a knife in the seam and working along the length. Remove the tongue from the replacement so it can be fitted in place.

Sometimes, particularly in old houses, plain-edged boards are found. These are very simple to replace.

The eaves framing is sheltered from the weather, and is unlikely to rot until the rest of the roof does. If it is rotten or broken, simply replace the bad parts with new ones of the same size screwed in place.

Bargeboard repairs

Bargeboards are very similar to fascias except that they are sloping. They are fastened in various ways and may be flush with, or projecting beyond, the gable end.

In houses with brick gables a bargeboard is generally nailed to the end rafter of the roof, which is supported on timber strips running between the end rafter and the stepped top of the brickwork (see Fig. 6). In the equivalent arrangement for a timber gable, the overhanging roof is supported on small members called lookout rafters parallel to the purlin and holding up the main end rafter. The bargeboard is nailed to the main end rafter (see Fig. 7). Sometimes, in the case of a rigid roof such as a corrugated-iron one, the bargeboard replaces the main end rafter. In this case the last row of sheet roofing may have to be removed to gain access.

Whichever way the bargeboards are fixed, they are only nailed on and can easily be pried off. The angled joint at the top is often braced at the back with a *ply plate* or *gusset* (see Fig. 8) and cannot be dismantled until you have taken both boards down together. This is a two-man job, and best done by loosening the ends first and working toward the middle.

Some old houses have decorative bargeboards cut into intricate shapes. These can be copied with a sabre saw by tracing the shape direct from the old board to the new one. Or you can put up a plain replacement. To get the correct angles at the top and bottom, the most accurate way is to trace the angles from the originals.

Even the plainest bargeboard generally has a wider part at the bottom end to hide the soffit. The best way to make this is to glue a piece on with resorcinol resin glue and cut it to shape when the glue is dry, using a sabre saw.

New bargeboards are installed in exactly the same way as fascias. It is best to prime the endgrain, then nail both boards of each gable together with a plywood gusset at the correct angle and put them up in one piece. This gives extra strength, even if it does make them rather hard to lift.

Screws and bolts

Screws and bolts provide enormous holding power but are simple to use. The types shown here are the ones you are likely to come across when fastening both wood and metal, together with the most commonly used accessories.

Screws—common types

1. Flathead wood screw. For general use; head let in flush with wood surface.
2. Pozidriv head countersunk screw. Driven with special screwdriver. One of several cross-slot types.
3. Oval head countersunk screw. Decorative head designed to be seen.
4. Round head screw. For fastening without countersunk holes in wood.

Screws—special

5. Lag screw. Extra-large wood screw with square head; tightened with wrench.
6. Self-tapping screw. For sheet metal; cuts its own thread as it is screwed in; has slot, Phillips (cross), or Pozidriv head.
7. Dowel screw. For invisible fastening; two pieces of wood twisted together to tighten.
8. Handrail screw. For "pocket" screwing; head screwed on from side with screwdriver.
9. Cup hook and screw eye. Large number of shapes and sizes available.

Screws—accessories

10. Flat washer. For round head screws; spreads load to give a strong grip.
11. Screw cup—raised type. For flathead or oval head countersunk screws; spreads load, improves appearance.
12. Screw cup—socket type. For flathead screws; hammered into pre-drilled hole for a completely flush fastening.

Bolts

13. Machine screw. Not a screw, but a bolt. Available with (l to r) round, pan, binding, or flat heads, and others.
14. Hex head machine screw. Also available with square heads.
15. Carriage bolt. Large bolt with a square collar under the head that sinks in wood, stops it from turning when the nut is tightened.
16. Foundation bolt. For bolting wood or metal to concrete; jagged head is set in wet concrete and holds bolt firmly when concrete dries.

Nuts

17. Hexagonal nut. Commonest type, available in a wide range of sizes.
18. Square nut. Also available in many sizes.
19. Flat square nut. Small sizes only; thinner than 18 in proportion to width.
20. Handrail nut. Used on handrail screw (8) and in other places where nuts have to be tightened from the side in a small space.
21. Wing nut. Tightened by hand; for uses where nuts must be loosened quickly.
22. Domed nut. Decorative nut, generally chromium-plated.
23. Locking nut. For places where vibration might make nuts loosen; has fiber ring inside to make it hard to turn.

Bolts—accessories

24. Flat washer. Same as 10; used in same way; also makes nuts easier to turn.
25. Helical spring lock washer. For metal fastening only; spring shape prevents bolts from loosening.
26. Internal and external tooth washers. Gripping teeth keep bolts from loosening.
27. Timber connector. Used between pieces of wood bolted together; teeth prevent slippage.

Screws – common types

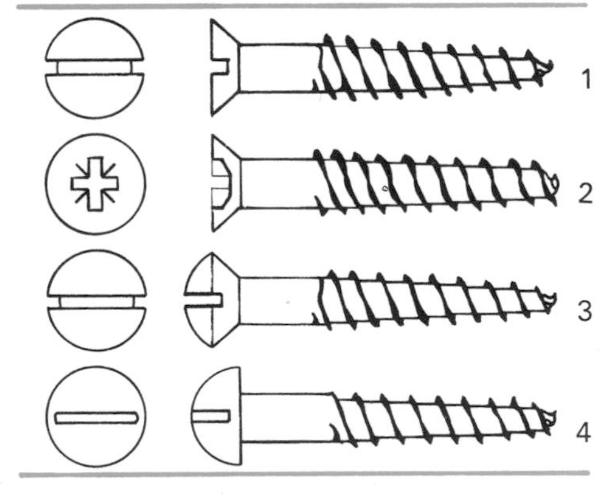

Screws – special

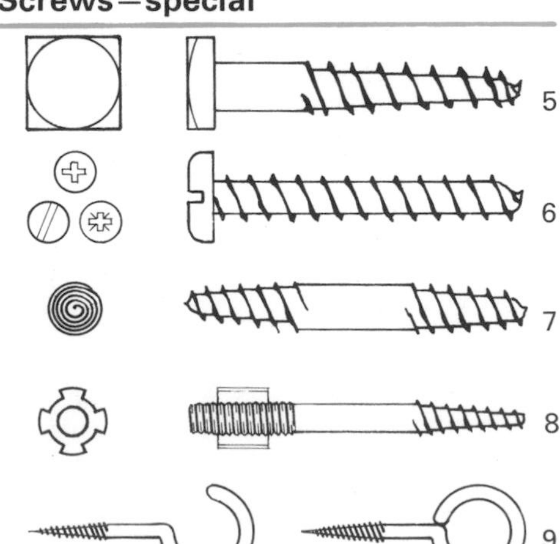

Screws – accessories

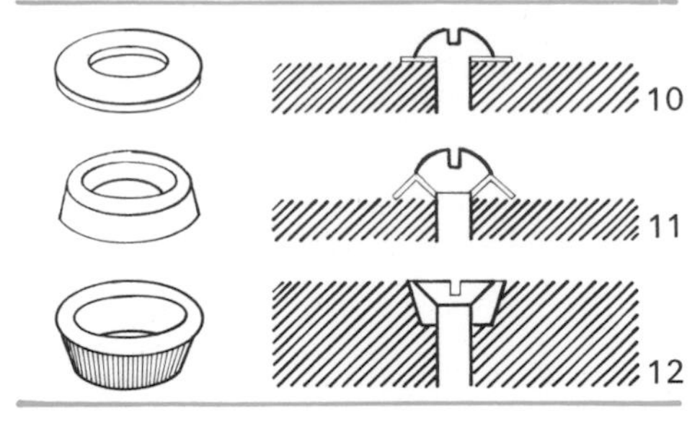

Bolts

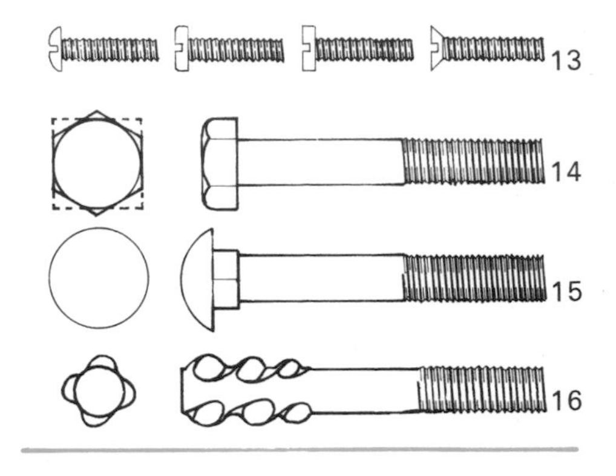

Nuts

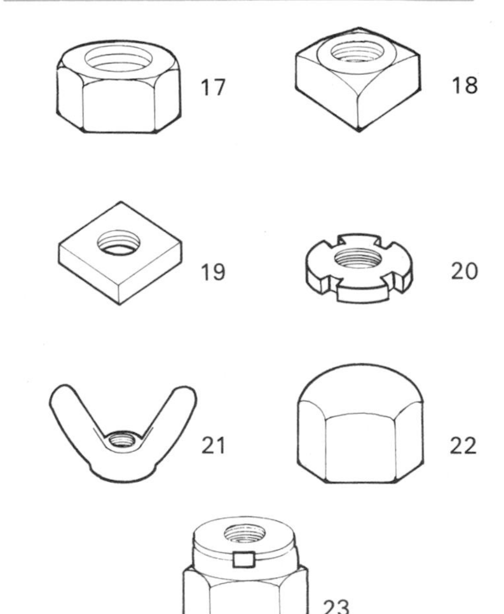

Bolts – accessories

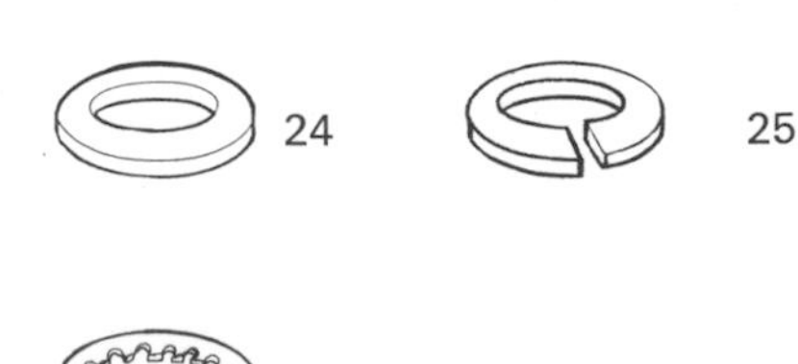

CHAPTER 32

Garden fences

Aside from providing an attractive addition to your home, garden fencing can lend an air of privacy to your property. In some cases it may even offer some protection from the elements.

To plan your fence, look at the photos and drawings on these pages and also at fences you find attractive in your neighborhood. Then you can either copy the style or modify it according to your taste and budget. The height of the fence is up to you. In general, it ranges from around 4 ft. for ranch types to 6 ft. or more for "stockade" fencing. Whatever the height, the in-ground portion of the fence posts should be about one third of the total post length. So plan accordingly. And soak the in-ground portion of the posts in a wood preservative like creosote.

The materials list that follows will give you an idea of what is needed for 8 ft. of the fence type shown at the right. Let the list also serve as a guide if you make up a list for another type or create a fence style of your own. By calculating the materials needed for a single post-to-post section, you make it easy to figure the materials required for the total fence length, and to know the cost in advance.

As you can see below and on the pages that follow, garden and lawn fencing can be made from an almost limitless variety of lumber. If you live in a wooded area you can even make a fence from tree branches woven on a framework of saplings. If you prefer a more formal style you can build your fence from stock sized lumber along the lines shown at the bottom of

HEDRICH-BLESSING

JOHN HOVELL

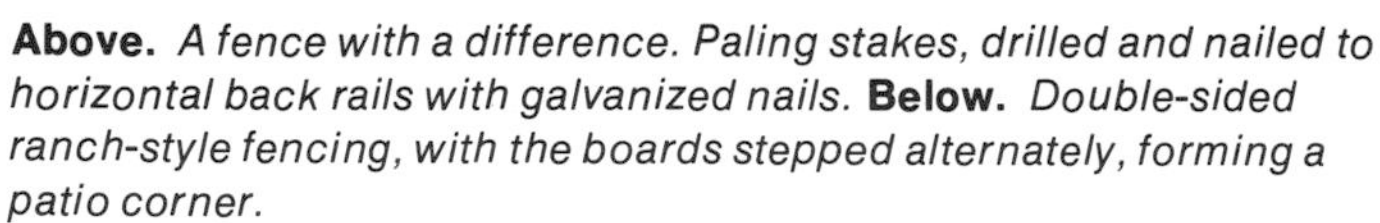

Above. *A fence with a difference. Paling stakes, drilled and nailed to horizontal back rails with galvanized nails.* **Below.** *Double-sided ranch-style fencing, with the boards stepped alternately, forming a patio corner.*

Above. *Rustic fencing simply trellised to form an attractive boundary. The hedge forms an integral part and acts as both screen and background.*
Below. *Plain ranch fencing painted in four harmonizing colors.*

HARRY SMITH

JOHN HOVELL

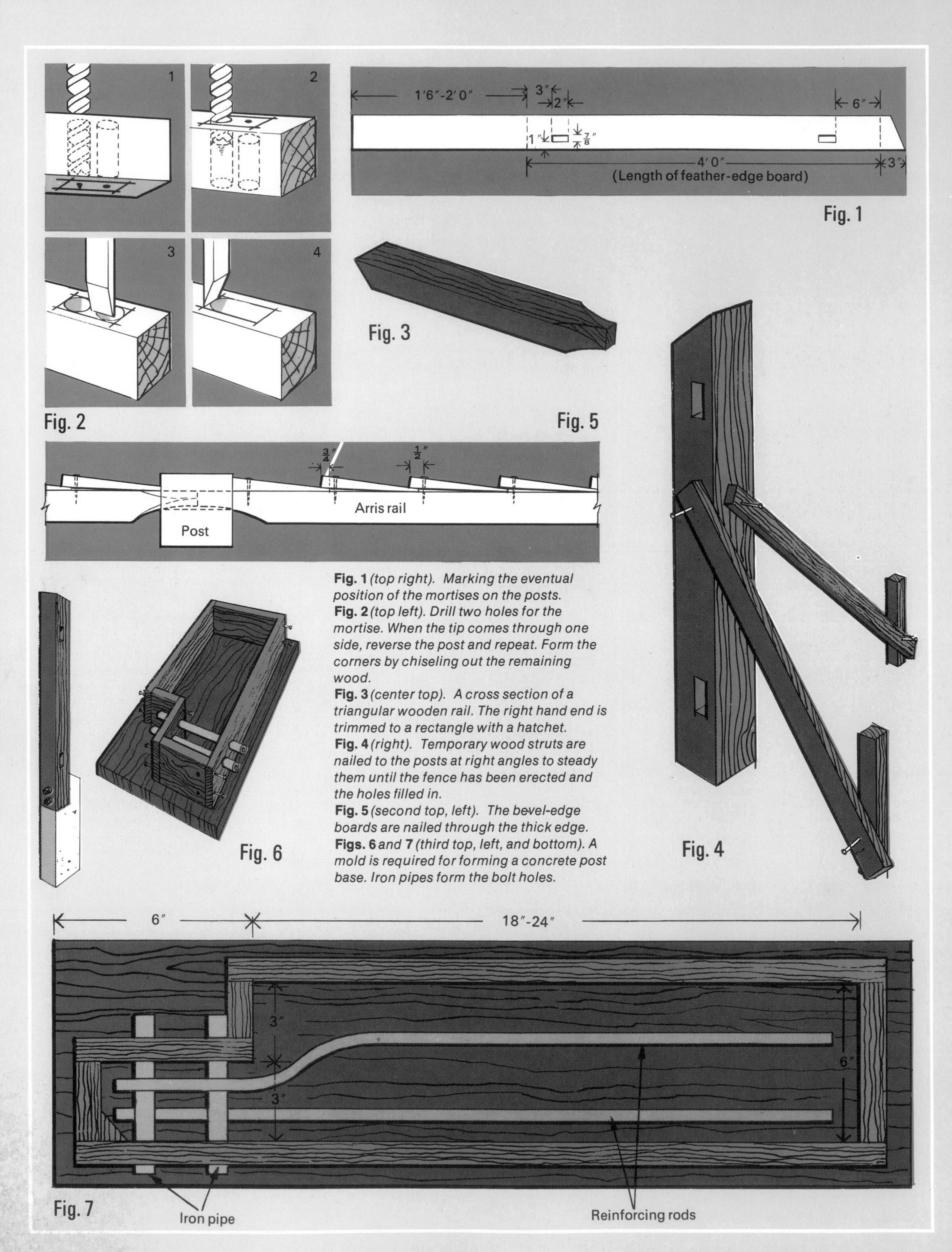

Fig. 1 *(top right). Marking the eventual position of the mortises on the posts.*
Fig. 2 *(top left). Drill two holes for the mortise. When the tip comes through one side, reverse the post and repeat. Form the corners by chiseling out the remaining wood.*
Fig. 3 *(center top). A cross section of a triangular wooden rail. The right hand end is trimmed to a rectangle with a hatchet.*
Fig. 4 *(right). Temporary wood struts are nailed to the posts at right angles to steady them until the fence has been erected and the holes filled in.*
Fig. 5 *(second top, left). The bevel-edge boards are nailed through the thick edge.*
Figs. 6 and **7** *(third top, left, and bottom). A mold is required for forming a concrete post base. Iron pipes form the bolt holes.*

this page. And if the overall length of your fence is enough to make materials costs a strain on your budget, look to the building wreckers for your supplies. You are likely to find used lumber of most common types selling at wreckers' yards for about half the price of new lumber.

For rustic fencing, a country sawmill is often a source of bargain materials. And if you make enough inquiries in outlying suburbs where woodland is not too distant, you can still find such mills.

If you are short of leisure time for projects like lawn fence construction, however, you can buy many popular types of fencing in precut form or even in readymade sections at your local lumberyard. This way, you'll spend a little more but you'll save time and effort.

Materials needed

For each 8 ft. panel:

Lumber for single upright post:
1 length of 6 ft. 6 in.–7 ft. nominal 3 x 3.

Hewn rails with a triangular cross-section:
2 lengths of 8 ft. (Rails and post are often available in sets from lumberyards, though rail length may be 10 ft.)

Lumber for the "gravel board" running along the bottom of the fence:
1 length of 8 ft. nominal 1 x 6.

Lumber for mounting the gravel board:
1 length of 1 ft. nominal 1 x 2.

Bevel siding:
24 lengths of 4 ft. 6 in. wide siding.

These boards are generally sold in random lengths, and may be bought used from house wreckers. Plan lengths for minimum waste.

Galvanized nails, as required, 2 in. length.

Rent a post-hole digger to dig the holes. It's an easy tool to use.

Creosote.

Hatchet.

Planning the fence

Before buying any wood, it is essential to determine exactly how long the fence will be, and whether there will be any special problems in constructing it. It is a sensible precaution to draw a scale plan of the site and mark the position of the posts on it. This will not only tell you how many posts you need but also ensure that none of the posts is to be placed in an impossible position. If you discover that one of the post sites is in solid rock, you can insert a shorter section in the fence so that the posts miss the obstacle. There will usually be a short section at one end of the fence, unless you care to space *all* the posts less than 8 ft. apart to even up the spacing.

If the ground under the fence is uneven, ensure that there are no sudden changes in height in the middle of a section. The sections can be at any angle to the horizontal—or, better still, stepped to follow the contours of the ground—but obviously they cannot be bent in the middle.

The posts are going to be sunk 18 in.–2 ft. into the ground, so check that there is enough soil or loose rock to allow this all along the run.

The bevel siding boards are attached after the gravel boards are attached to the mounting blocks. The gravel boards may be omitted.

Preparing the lumber

First mark out and cut the posts. For a 4 ft. 6 in. fence they can either be 6 ft. 6 in. or 7 ft. long, depending on whether you mean to sink them 18 in. or 2 ft. into the ground. The longer posts are recommended where the fence is to be set in soft soil.

To increase weather resistance, the tops of the posts should be cut at an angle. This will prevent water from collecting on the top. Mark the holes for the mortises (unless you buy pre-mortised posts) in the positions shown in Fig. 1. The size of the mortises depends on the size of your rails, but standard-sized rails will probably allow you a maximum size of 2 in. x ⅞ in. The closer to this size that you can make the mortises, the stronger the joints will be.

If there are any sharp drops in the ground under the fence, it may be necessary to cut the mortises lower on one side of a post than on the other to create a step in the fence. Slight irregularities can be taken care of by simply angling the rails, however. There is no need to angle the mortises—it is easier to cut the tenons at an angle.

The mortises should be cut by drilling two holes with a bit the same width as the mortise (see Fig. 2). The wood between the holes and in the corners can then be taken out with a chisel. The holes should go right through each post unless, of course, the mortises on each side are at different heights.

The tenons can be made more quickly. First cut the rails to length and set aside a short offcut to practice on.

Shape the wood with a small one-handed hatchet.

Cut the bevel siding boards to length. Also cut the 1 ft. length of nominal 1 x 2 in half at an angle so that it forms two 6 in. pieces with slanted tops. This will enable them to resist the weather, and also to fit snugly under the rail when they are nailed to the posts as gravel board mountings. Leave the gravel boards over-length for the time being.

Give all the wooden parts a generous coat of creosote or other preservative. The bottoms of the posts, which will be most exposed to decay, should be stood in a bucket of the liquid for a few hours, to make sure that it soaks into the wood.

Work on site

While the preservative is drying you can begin preparing the site. Hammer in stakes at both ends of the proposed length of the fence or, if the fence does not run in a straight line, at the beginning and end of each straight section. Tie some string firmly to the tops of the stakes. To prevent the string from getting entangled later with the uprights, the stakes should be about 2 ft. higher than the eventual height of the uprights. This will be your guide for keeping the fence straight.

Dig the first hole at one end of the site.

The simplest way to space the second and subsequent post-holes is to lay rails end to end along the fence site and dig at the ends. If the preservative is still wet, however, you can just as easily measure the distance with a tape.

Erecting the fence

Fit the first post into its hole and pack soil around its base to stop it slipping sideways. Then lock the post upright by nailing a pair of wood struts between the post and pegs set into the ground. The struts should be set at right angles to each other (see Fig. 4).

Position the second post loosely in its hole. Then take a pair of rails and knock them into the mortises in the first post with a mallet. The flat side of the rails should normally face outward from your property. Push the tenons at the other end into the mortises in the second post and drive them home by hitting the post. Check that the second post is upright (use a spirit level), ram in a little soil and prop the post upright with one temporary wood strut at right angles to the line of the rails.

Continue in this way until you have put up all the fence posts and rails. The last pair of rails should be trimmed to length on the spot. Apply preservative to the cut ends before inserting them in the mortises—there is no need to wait for it to dry.

Nail the gravel board mounting blocks to the fence in a vertical position one inch from the flat side of the fence with their slanted tops resting against the rail. This will provide a valuable guide to whether all the posts have been sunk the same distance into the ground. A gravel board is to be nailed between each pair of blocks, so the ground must be level between the posts or there will be a gap under the fence.

Check all the posts for straightness using a spirit level, and for alignment by looking at the fence from one end. When you are satisfied, fill the holes and tamp the earth firmly around them. An alternative method which prolongs the life of the posts is to fill the holes with a rather dry concrete mix. Slope the top of the concrete so that water will run away from the post. Do not remove the bracing struts till the concrete has set hard.

Cut each gravel board to the exact length of the space between the pairs of posts and treat the cut ends with preservative. Then nail the boards in place so that their lower edges rest

AUSTRALIAN HOME BEAUTIFUL

AUSTRALIAN HOME BEAUTIFUL

AUSTRALIAN WOMEN'S WEEKLY

Top left. *An unusual fence of branches. These are nailed to top and bottom rails which are supported by preservative-treated poles.*
Top right. *Closely spaced ranch fencing.*
Bottom. *White-painted palisade fencing looks simple and, at the same time, elegant.*

on the ground. These boards provide an easily replaceable base for the fence.

Finally, nail the feather edge boards to the rails. (This may be omitted if all you want is a rail fence.) Each board should be secured by only two nails, one each into the top and bottom rails. This allows the boards to expand and contract with changes in humidity—an important consideration in outdoor carpentry. The spacing of the boards and the position of the nails is shown in Fig. 5. The nails must go through the thick side of each board about ¾ in. from the edge. The boards should overlap by about ½ in.

The best way to space the boards equally is to cut a small wood block ½ in. narrower than the width of the boards. Use this block and a pencil to step out the distance along the rails, but remember that *all* the width of the last board will show.

Extra-strong posts

The fence posts already described are quite strong and should last for years if properly treated with preservative. But in areas where the soil is unusually damp, or attack by pests is a problem, you may prefer to make concrete bases for the posts and bolt the wooden part to them above ground level.

The formwork (mold) for the posts is shown in Fig. 7. It should be a wooden box lightly nailed together so that it can be taken apart to release the post. A couple of iron pipes can be passed through holes in the side of the mold to form bolt holes for the attachment of the post. Two short reinforcing rods should also be inserted. The whole inside of the mold, particularly the pipes, should be heavily greased to stop the concrete from sticking to it. Old oil from a car is good enough for the job.

The moldings should not be removed until the concrete is hard and should be treated with care for a further four days. Obviously, you will need several sets of formwork if post bases are to be produced at a reasonable rate.

An even stronger fence can be made with solid concrete posts. In this case the formwork should be in the shape of one complete post, with reinforcing rods the whole way up. Remember to insert a couple of *full* 1 in. x 2 in. wood blocks to mold the mortises.

Other types of fences

The bevel board fence described here is sturdy and long-lasting, and may be painted. The same building technique can be used to make a fence of flat boards if available at lower cost.

Ranch-style fencing is made by nailing nominal 1 x 4 or 1 x 6 boards to the sides of the fence posts. No mortises are needed. The boards can be fastened either on alternate sides or all on the same side with alternate boards stepped up on small wood blocks to give a varied effect. One panel of each type is shown in Figs. 8 and 9. The joints in the horizontal boards are sometimes staggered to make the fence stronger.

The horizontal boards can be fastened edge-to-edge as shown to make a solid fence, or spaced apart to make it possible to see through. Nails used in its construction should be galvanized. Of course, this type of construction is unsuitable for concrete posts.

This type of fencing looks best painted. But this means that you will not be able to creosote the posts except for their in-ground portion. To make the job weatherproof, the knots in the wood should be sealed and an undercoat applied to the separate pieces *before* erecting the posts or putting the fence together.

A fence can also be made with horizontal bevel siding nailed to nominal 1 x 2 battens running down the sides of the posts (the same way as the gravel boards were mounted on the original fence). The horizontal boards will need to be nailed to vertical stiffeners between the posts (see Fig. 10) to make the structure more rigid. These stiffeners are not posts and should not touch the ground.

This type of construction looks good with wavey-edge boarding, from a lumberyard or sawmill, which has the bark left on one edge to give a rustic effect.

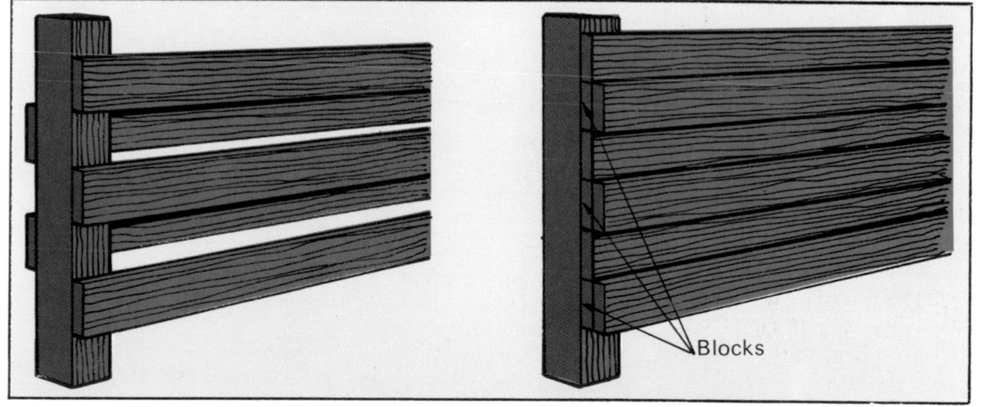

Figs. 8 and 9 *(top left and right). Mortises are not necessary for ranch-style fencing. The boards are simply nailed to the posts.*
Fig. 10 *(bottom). Bevel siding can be mounted horizontally by attaching it to battens on the sides of the posts.*

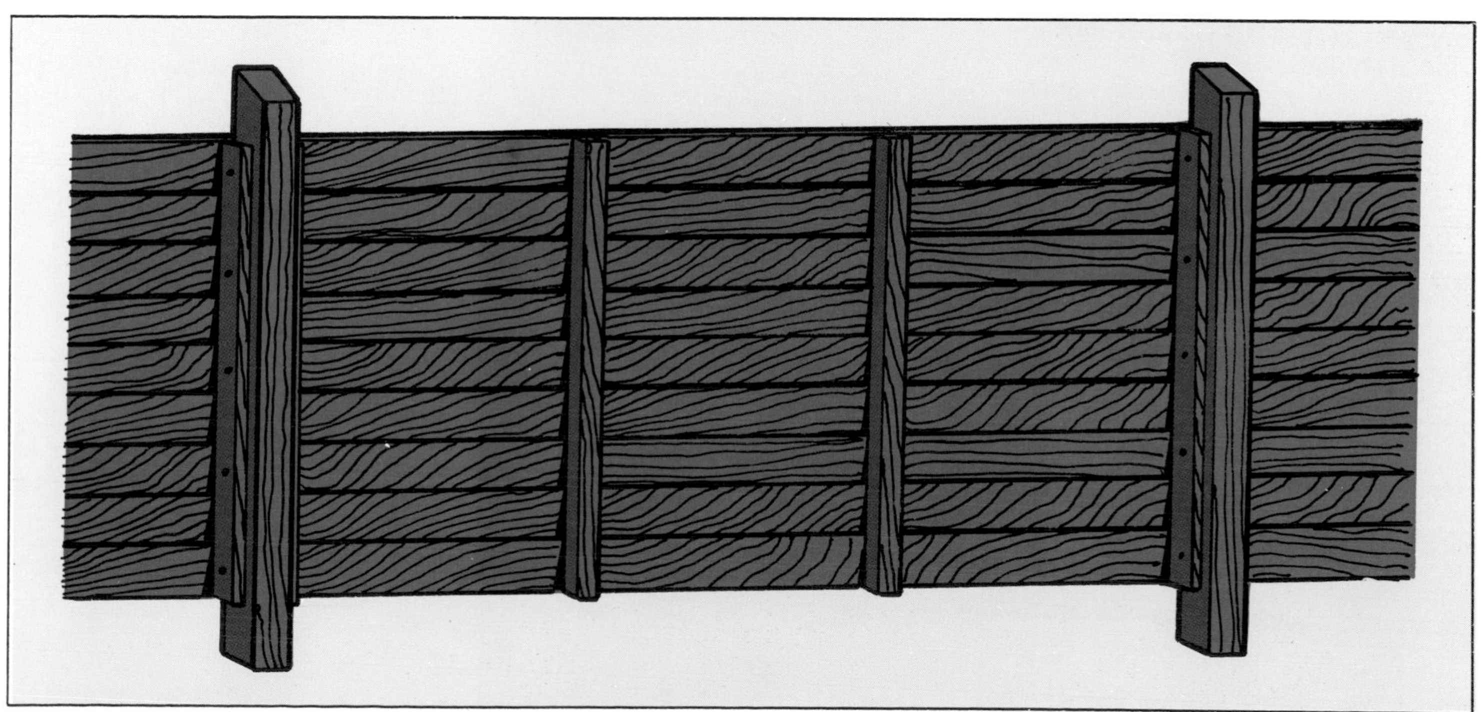

CHAPTER 33

Power tools: 1

One of the dullest and slowest jobs in carpentry is making a large number of identical pieces—sawing the same shape over and over again, hand-drilling row after row of holes, sanding everything to shape, and so on. If you have to do this, you may easily lose interest. So why not speed up the routine work with a power tool?

Power tools greatly speed up some of the most time-consuming work in carpentry, such as drilling, sawing and sanding. Anyone who intends to do a lot of carpentry would be well advised to invest in the power tool they need most. It would soon pay for itself in time saved.

Many types of individual power tools are made, such as power saws, orbital sanders, lathes and so on. But the small job woodworker can save some of the money required to buy this extensive equipment by buying a power drill and a range of attachments to fit it.

Power drills

Handy power drills do a lot more than just drill. A huge number of fittings can be attached to the basic drill unit, enabling it to do almost anything that can be done by a specialized power tool—perhaps not quite so fast but certainly well enough for general use.

Basically, a power drill is a compact electric motor geared to a projecting shaft at one end on which is mounted a *chuck*—a revolving clamp that grips and drives drill bits or other attachments. The motor unit is held in the hand by a pistol grip, and the motor is started by pressing a trigger mounted on the grip. For safety reasons, the motor stops if pressure is released on the trigger, but most drills have a locking pin that can be engaged to hold the trigger in the "on" position.

Electric power is supplied to the drill by a cable that enters the machine through the bottom of the handle. On double insulated drills, a system of insulation is built in to keep the user from getting an electric shock in case of internal drill damage.

The motor is cooled by a built-in fan that draws air through slots in the side of the drill. These slots must be kept uncovered and free of sawdust or the motor may overheat and burn out.

Many drills can be adjusted to run at different speeds. A few are two-speed types wired to run at such speeds as 1,000 and 2,500 rpm. These two speeds are suitable for most household jobs, and make the tool handier than a single-speed model.

Variable-speed drills, where the speed can be infinitely varied by an electrical device, are also made, and have even greater versatility. If you buy several single-speed power tools it may pay to buy a separate electronic speed control and plug the tools into it when they are used.

The most suitable speeds for various woodworking and other operations are shown jn Fig. 12.

Drills come in various sizes, which are graded by the capacity of their chuck—i.e. the largest drill bit that can be fitted into it. Common sizes are ¼ in., ⅜ in. and ½ in. The larger ones have more powerful motors. A medium-sized drill—say, ⅜ in.—should be adequate for all ordinary jobs, though the ¼ in. size can handle average household jobs.

An indispensable accessory that every drill user will need is an extension cable. This enables him to use the tool in places remote from a power outlet. Cables are available in standard lengths from 6 ft. to 100 ft., or you can make up your own. The longer the cord, the larger the wire it needs to prevent power loss. Heavier machines also need heavier cords. Recommended cable sizes are shown in Fig. 13.

Drill bits and fittings

Many types of drills are sold for making different sizes and shapes of holes in different materials (see Fig. 14). The everyday sort are *twist bits,* used for drilling a wide range of sizes of holes in metal, and in wood. The smallest common size of twist bit is 1/16 in., and sizes increase in steps of 1/64 in. up from this.

Larger holes in wood are drilled with *Power bore bits,* which make clean-cut holes. *Jennings bits* are often used in a drill press. They cut rapidly in soft wood and expel chips as they work. *Spur bits* are also widely used in drill presses and are fast and clean cutting in most woods.

Very large holes (up to 3 in.) are drilled with *hole saws.* The *flat bit* has a flat, spade-shaped cutter with a central spur to hold it in place. It is an inexpensive and fast-cutting type for holes to 1 in. diameter in wood.

Other types of bits include *countersink bits,* for countersinking screw holes, and *multi-bore bits* or *"screw sinks,"* which are specially shaped to drill and countersink (or counterbore) a hole for a particular size of screw.

A *plug cutter* is often used in conjunction with a screw sink to conceal screw heads in wood. The screw sink is used to counterbore a screw hole—i.e., to recess the screw head some way into the wood—and then a plug like

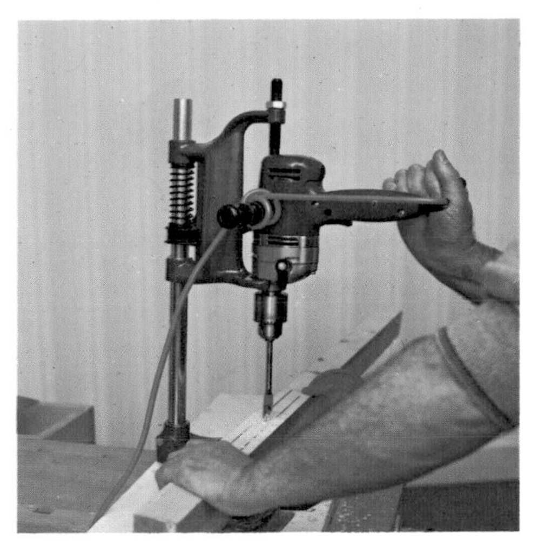

Fig. 1. *A drill stand is one of the most useful accessories for your drill. It allows you to make perfectly accurate vertical holes in wood, metal or plastic.*

Fig. 2. *A sabre saw attachment. Here, a cut is being started in the middle of this wooden door by tipping the saw on its nose and gradually lowering the blade into the wood.*

NELSON HARGREAVES

Fig. 3. *A more powerful integral sabre saw. Here, the saw is being slid along a wooden straight edge to make a straight cut in a sheet of plastic laminate.*

Fig. 4. *When sanding a flat surface, you should hold the rubber disc at an angle to the wood to minimize swirl marks.*

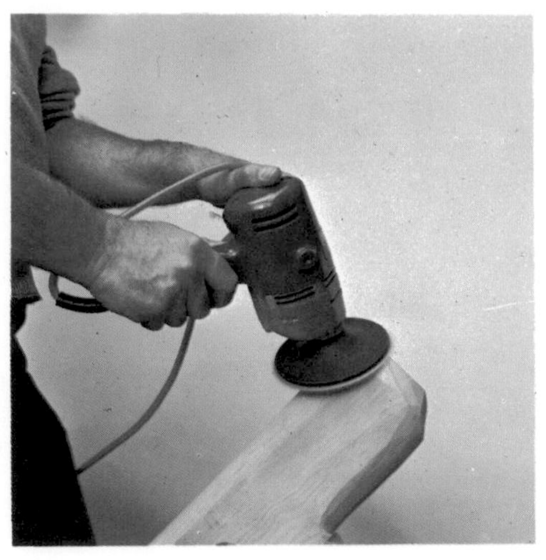

Fig. 5. *The same disc can also be used, with practice, to sand rounded objects. Note how the cable is being held away from the disc.*

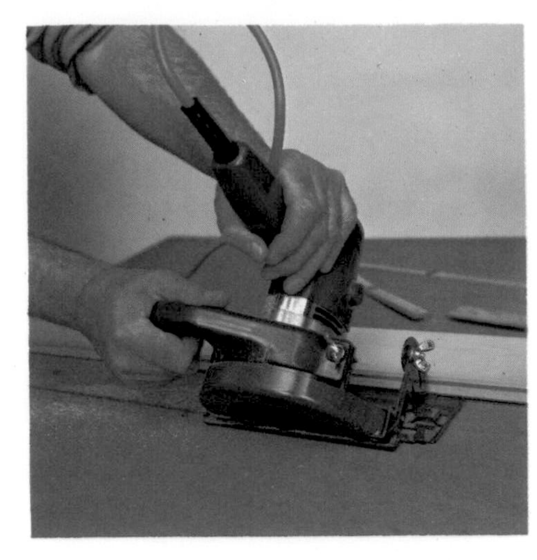

Fig. 6. *A drill-powered circular saw being used to cut a bevel, or miter, along the edge of a piece of lumber.*

NELSON HARGREAVES

Fig. 8. *An easy way of sharpening chisels without a grinder. First mark and cut slots in a fine-grade sanding disc . . .*

Fig. 9. *. . . then apply special disc cement to the revolving sanding plate and press the sanding disc on to its tacky surface.*

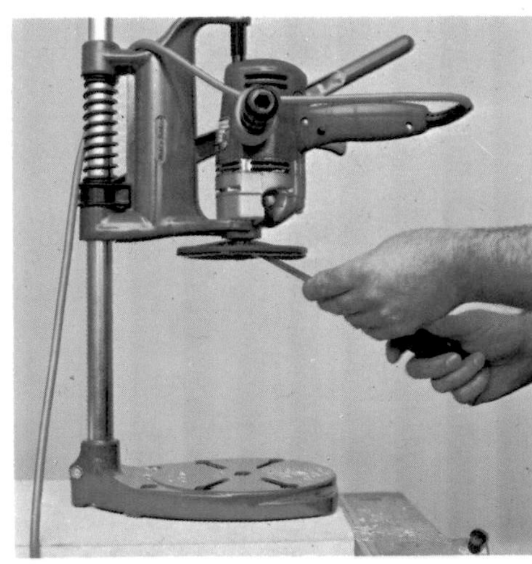

Fig. 10. *Start the motor and hold the chisel tip to the underside of the disc, angled so it will not catch in the slots.*

a short length of dowel is cut from a matching piece of wood, glued into the recess over the screw head, and planed flat to give an almost invisible result.

For drilling hard masonry, special *masonry bits* are made—they look like twist bits but have cutting tips made of a special hard alloy.

Glass and tiles are best drilled with a special glass drill, which also has a hardened tip.

Special *right-angle* and *flexible drive shafts* are made to allow drilling in awkward corners that could not normally be reached.

Drilling techniques

Whatever you are drilling, the position of the hole should be clearly marked before you start, and a small indentation made in the workpiece with a center punch (or, in a pinch, nailset or big nail), to stop the drill from wandering in the first few seconds of work. Clamp the workpiece down securely, or it may start to revolve.

It is essential that the drill should be at right angles to the surface to be drilled. You can line it up with a try square before you start—though of course the drill always tilts a bit once you start drilling.

Simple drill guides are made (the Stanley Drill Guide is an example) that hold the drill at right angles to any flat surface. Or you can buy a drill stand, which holds the drill vertical on a frame. The drill can be moved up and down by a lever. The workpiece is placed underneath on the base of the stand (see Fig. 1).

Drill bits should be prevented from overheating through friction. If they become too hot, the metal loses it temper and becomes soft. This is a particular problem with ordinary carbon steel twist bits. Special high-speed steel bits, made for drilling hard metals, are more resistant—but also more expensive. Masonry bits are very prone to overheating.

When drilling wood with a twist bit, remove it occasionally to check that the spirals are not clogged with wood dust, which can lead to overheating. When drilling any metal other than brass or cast iron, lubricate the drill bit frequently to cool it. Use oil for steel, kerosene for aluminum and turpentine for glass. When drilling glass, make a small pool of lubricant around the hole in a putty ring.

Thin metal should be clamped to a wood backing when being drilled to reduce distortion and keep the drill from jamming as it breaks through to the other side. A piece of thin sheet metal revolving with a drill is extremely dangerous.

Circular saw attachments

These fittings for power drills are very popular, because sawing by hand is a time-consuming business.

Two types are made: *hand-held* ones consist of a 5 in. or 6 in. blade, a frame to hold drill and blade steady and allow it to be slid across the wood to be cut, and a fixed top and retractable bottom blade guard. There is also a *saw table,*

Fig. 7. *A hole saw cutting large circles out of an aluminum sheet, which is firmly clamped to a wooden backing.*

NELSON HARGREAVES

Fig. 11. *The rapidly passing slots allow you to see, through the disc, whether you are holding the blade at the right angle.*

Wire sizes required in extension cords for common power tool amperages, according to length of cord.

Feet of cord	Amperage of tool 4	6	8	10	12	
25	18	18	18	16	14	
50	18	16	16	14	14	Wire sizes*
75	16	14	14	12	12	
100	14	14	12	12	12	

*The smaller the wire size number the larger the wire. As extension cord length increases wire size must be increased for a given amperage lead.

Full load current ratings in amperes for typical single phase 120–125 volt A.C. motors. The lower powered motors are used in small power drills, the larger ones in ½ inch drills and other power tools.

Horsepower	Full load amperage
⅙	4.4
¼	5.8
⅓	7.2
½	9.8
¾	13.8
1	16.

Note: Full load amperages are given for the purpose of selecting extension cord wire size. With less load or no load on the motor, the amperage is lower. Be sure you know actual operating amperage.

Power drill speeds for typical operations

Under 1000 r.p.m.	2000 r.p.m.
Drilling:	
ceramics	Drilling steel
masonry	with high speed bits
Use of hole saws	Drilling aluminum
and power auger	Drilling brass
bits	Polishing and sanding
	Wood boring with spade-type bit

Note: precise drill bit speeds depend on drill bit diameter. Small drill bits must turn faster to attain the same cutting edge velocity. In critical work, make a test hole in scrap material to determine the speed to use in the final work.

Drilled hole sizes for common wood screw sizes

Screw size	Body hole		Thread hole (hardwood)		Thread hole (soft wood)	
	nearest fraction size	number or letter size	nearest fraction size	number or letter size	nearest fraction size	number or letter size
4	7/64	32	1/16	52	3/64	55
6	9/64	27	5/64	47	1/16	52
8	11/64	18	3/32	40	5/64	48
10	3/16	10	7/64	33	3/32	43
12	7/32	2	1/8	30	7/64	38
14	1/4	D	9/64	25	7/64	32

Note: The thread holes (called lead holes) are sized to assure that the wood screw threads have a firm bite in the wood without so much friction as to risk breakage of the screw. Lead holes are not needed in soft wood for screw sizes smaller than No. 6.

or bench-mounted circular saw, where the blade projects upward through a flat table top.

Of course, neither machine is intended for heavy types of work. Partly, this is because a circular saw requires a great deal of power to drive it—in fact, rather more power than can be provided by even the largest drill motor. There is, therefore, a possibility of overloading the motor to such an extent that it burns out. A drill-driven saw should only be used for light work such as battens, moldings and plywood and not kept too constantly in use.

Another reason is that, on many jobs—for example, repair work, or where you are fitting architraves or other moldings—it is necessary to keep removing the drill bit and swapping it for a saw-blade. Often this slows you down to the point where a hand-saw would be quicker.

So if you intend to do carpentry on a major scale, a better alternative is to buy a proper *bench saw* with an adequate motor, which will do everything that a drill-driven saw table can do—and more—and do it faster. The use of this tool is fully described in CHAPTER 41. It is used in exactly the same way as the smaller, less powerful saw table, so only the use of the hand-held type of saw is described here.

Using a hand-held circular saw

Saw attachments are used chiefly on medium- or large-sized drills, and then only for work that is unlikely to overstrain the motor. The 5-inch blade gives a depth of cut of about 1½ in. and the 6-inch blade one of around 1⅝ in. when set up normally on the machine, and this is about the maximum depth of cut the motor will stand. If the motor shows signs of slowing down or jamming, stop work immediately or you may burn it out on the spot. It is essential that the motor should be kept running at a high speed all the time to keep it from being damaged. Do not press the saw forward too hard, and always start the motor before the blade touches the wood, so that the speed of the motor stays up. You need a straight edge, and some practice, to bring in the blade at exactly the point where you want to cut; sighting straight down the blade will make it easier.

Saw cuts can be kept straight by nailing a batten to the wood you are cutting, and running the saw along it; or by using the adjustable *rip fence* on the saw, which guides it parallel to the edge of the wood.

Four types of blades are available: the *rip blade,* with coarse teeth, for cutting along the grain; the fine-toothed *cross-cut;* the *planer blade,* which gives an extra-neat result; and the most useful type, the *combination blade,* which cuts at any angle to the grain. Sharpening and maintaining the blades is fully described in CHAPTER 41.

These blades will not cut metal, so when using the saw on old wood it is essential to remove all nails and screws. To prevent the blade catching on anything underneath the wood, and to reduce the strain on the motor, set the depth gauge of the blade to only slightly more than the thickness of the wood

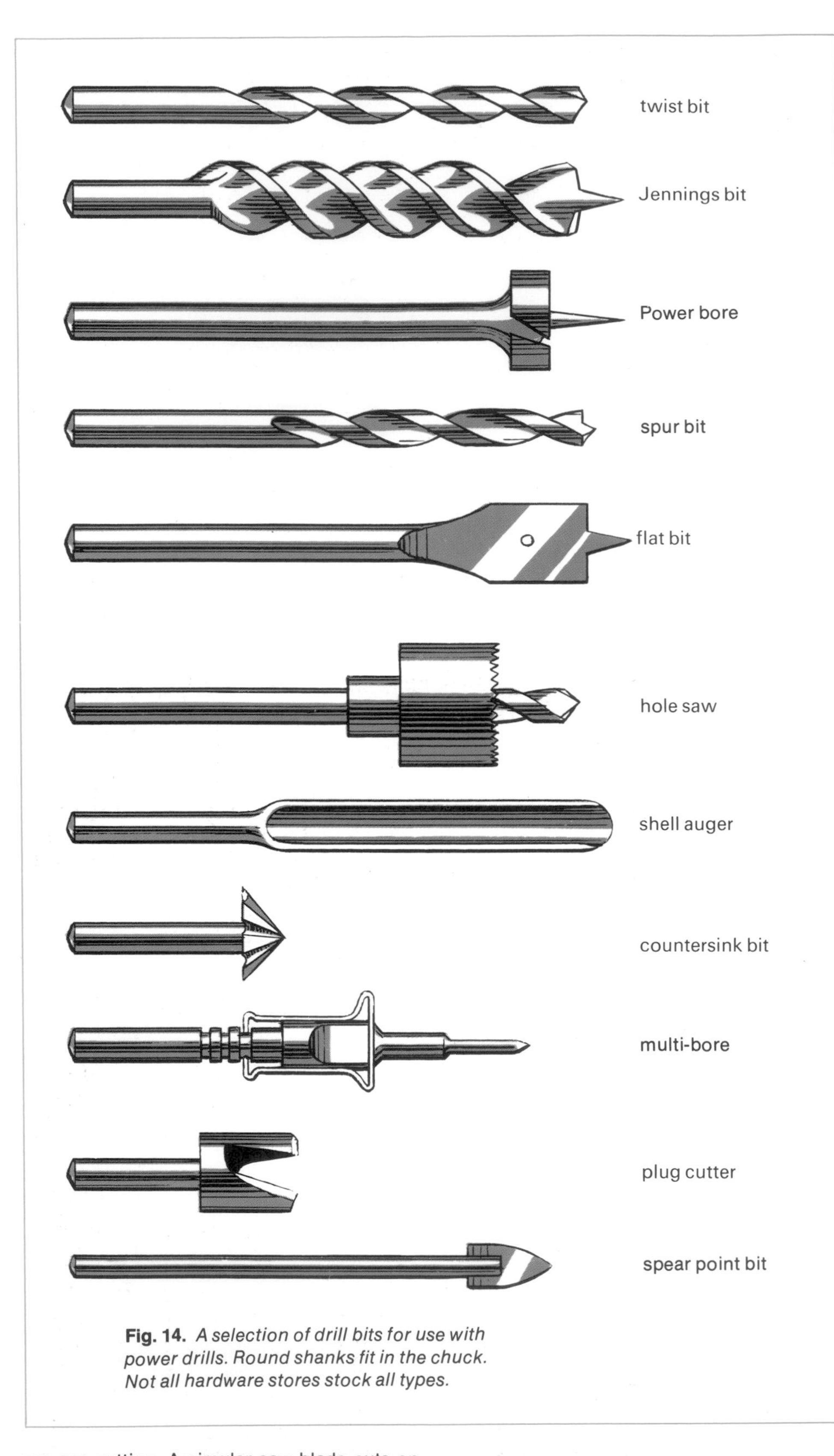

Fig. 14. *A selection of drill bits for use with power drills. Round shanks fit in the chuck. Not all hardware stores stock all types.*

you are cutting. A circular saw blade cuts on the upstroke, so setting the blade as shallow as possible gives a neater result by flattening the angle at which it cuts.

If the blade of the saw wanders off the cutting line, do not twist the saw to straighten the line. This may jam the blade in the cut, with disastrous results. After shutting it off, take the saw out of the cut, go back a few inches and cut along that section again.

Sabre saws

A power-driven sabre saw is used in the same way as a hand-held coping saw—that is, for cutting curves and complex shapes. Its blade is small and pointed and moves rapidly up and down with a stabbing motion. Various types of blades are available for cutting wood, plastic and sheet metal, but it will not cut very thick boards or sheets. It can manage a 2 in. thick softwood board, or hardwood half as thick.

Sabre saws should not be pressed forward too hard, or the highly tempered blade may snap. But they should be held firmly down on to the material they are cutting to reduce vibration.

Most blades are narrow enough to cut ½ in. radius curves but will not turn a right-angled corner. They can, however, be started in the middle of a piece of wood by tilting the machine forward on its nose and gradually lowering the blade into the wood until it is upright (see Fig. 2).

Sabre saws are available both as power drill attachments and as integral tools, hand-held or bench-mounted with the blade pointing upward.

Sanders

Several types of sanders can be fitted to a power drill. The most commonly used is the *disc sander.* A flexible rubber disc is mounted in the chuck of the machine and an abrasive paper disc is fastened to it with a recessed central screw.

The sander is used at an angle, so that only one side of the disc touches the surface being sanded (see Fig. 4). If the disc is laid flat against the surface it produces circular marks called *swirl marks,* which may be deep and difficult to remove. Even with the disc used at the correct angle, slight swirl marks are unavoidable.

A special type of disc called the "Swirlaway" reduces these marks to a minimum. The disc is made of metal and is flat and completely rigid. To give it flexibility in use, the shaft on which it is mounted has a ball joint in it.

The *drum sander* consists of a wide revolving drum made of stiff rubber, with an abrasive belt fastened around its perimeter. It makes no swirl marks, but can only be used for sanding small curved objects. On large, flat surfaces it tends to give an uneven result.

The *orbital sander,* on the other hand, can be used to give a perfect finish to any surface. It has a large, flat sanding pad covered by an abrasive sheet. This moves to and fro in a small circle without revolving, so swirl marks are negligible. Orbital sanders are available both as attachments and as integral tools.

The abrasive discs, belts, and sheets for all these tools are available in coarse, medium and fine grades as well as special types such as "wet-and-dry" and "finishing" for rubbing down paintwork.

Other attachments

Many highly specialized attachments are available for power drills. These include *rotary files* for finishing the edges of metal sheets, *polishing pads,* made of lambswool, that fit over the rubber sanding disc, *wire brushes* for removing rust from metal, *screwdriver attachments,* useful when a large number of screws have to be driven, and even *paint stirrers* and *hedge trimmers.* Among the most useful are *grinding wheels,* which can save a lot of time in sharpening knives, chisels and plane blades. Special extra-tough wheels are made for sharpening masonry drills.

CHAPTER 34

Building a gabled roof — English style

The styling of a gabled roof often reflects not only the traditional design, but the nature of the weather in the part of the world where it is built. In Britain, the pitch is usually steeper than in the United States. And, as shown in the photo below, exposed purlins as well as rafters are frequently a feature of the roof's interior. The lumber dimensions, the construction methods, and the terminology are those used by home builders in Britain, where all of the homes shown were built. If you'd like to capture some of the English atmosphere when you build, you can do it with American materials keyed to your local building code.

A gabled roof is a pitched roof with vertical ends, as opposed to a hipped roof, which has sloping ends.

The angle of slope or *pitch,* of the gabled roof, can be between 20° and 60°. The steeper slopes are more efficient in dispersing rainwater, but are generally used for purely decorative effect. Some angles of pitch will spoil the look of the roof. An angle of 45° somehow looks odd to many people, whereas any angle between 50° and 60° seems normal.

The design of the roof will depend on a variety of factors, the most important being the building regulations that apply in your case. In Britain, for example, the Building Regulations lay down standards for all types of construction and these must be adhered to exactly. Local authorities may also insist on new erections conforming to certain other requirements—your gabled roof may have to conform with the design of others on your street, for example. In historic towns the local regulations are likely to be particularly restrictive.

This open gabled roof is not supported by collar ties or joists. The supports that are there—purlins, rafters and struts running to the chimney breast are, therefore, of heavier timber than would otherwise be needed.

Components of a gabled roof

The wooden framework of a complete gabled roof for a brick house is shown on the following pages. Some of the terms you will encounter when planning your roof are:

Wall plate: in a brick house, a piece of timber, usually 4 in. x 3 in., built into the top of the brickwork along the long walls. It is bedded in cement mortar and supplies a fixing point for the foot of the rafter on either side of the roof. In a timber house it is the top component of the timber frame, usually 4 in. x 2 in.

Common rafter: one of the main support timbers of the roof, which run in pairs between the wall plates and the ridge board and at right angles to them. The size depends on the spacing between rafters, and the span and loading of the roof.

Ceiling joists: timbers that run parallel to the short walls of the house, horizontally from wall

PAF INTERNATIONAL

plate to wall plate. The feet of the rafters are fixed to the ends of the joists as well as to the wall plates. Their size depends on their spacing and span, and the load they carry.

Ridge board: the highest point of the roof, this is a piece of sawed timber which runs the full length of the roof. It is usually 1¼ in. thick and its depth is sufficient to accommodate the angled ends of the rafters plus a bit standing proud above the rafter tops. The rafters are nailed to the ridge board.

Purlin: an intermediate support timber that is as long as the wall plates and runs parallel to them. Purlins are nailed to the inside edge of the rafters, usually about halfway up their height.

Collar ties: pieces of timber that run horizontally between pairs of rafters parallel to the short walls. They are nailed to about every fourth rafter.

Struts: supports that run diagonally from the purlins to a suitable surface, such as an internal dividing wall, to provide extra support. On this surface, the struts are nailed to pieces of timber called *sole plates.*

Fascia: a length of prepared timber approximately 6 in. x 1 in.; the size depends on the size of the rafters. It is fixed to the foot of the rafters to provide a finish here. The rainwater gutter is fixed to the fascia.

Soffit: a timber, plywood or asbestos cement covering which is fixed to the underside of the rafters where these protrude beyond the walls of the building. It provides a watertight seal and a neat finish.

Bargeboards: a pair of boards like fascias except that they are sloping and are fixed to the last rafters at the gable end of the roof.

Eaves: this describes the whole area of the ends of the rafters.

Tilting fillet: a piece of timber with a wedge-like cross-section fixed near the foot of the rafters. Its top edge is flush with the top of the fascia board. It lifts the ends of the roofing tiles slightly to provide a good fitting and firm ends.

Battens: 2 in. x 1 in. sawed timbers which are nailed to the top edge of the rafters parallel to the ridge. The roofing tiles or slates are fixed to the battens. Their spacing depends on the size of the tiles or slates.

"Lookout" rafters: timber supports for the end rafters of a roof that oversails (projects beyond) the gable walls. They are fixed at right angles to the common rafters. They do not run the whole length of the roof but are nailed to the side or underside of the first rafters inside the gable walls.

Hanging beams or runners: used to brace ceiling joists, especially over long spans. They bind the joists together and prevent them from twisting. Hanging beams run parallel to the long sides of the roof, and are fixed to the joists by small vertical *hangers* which are placed on alternate sides of the hanging beam. Hangers are often 2 in. x 1 in.

Sarking felt: bituminous coated felt nailed to the rafters under the battens. It adds waterproofing to the roof and aids heat insulation.

Roofing tiles: the traditional finish for gabled roofs used to be slates, but these have generally been superseded in Britain by concrete tiles with a colored finish. The tiles are fixed to the battens by hook-like lugs called *nibs* and sometimes by nails passed through molded holes in the tile. Each tile overlaps the one below it by half its length, so that the roof is two tiles thick.

Eaves tile: a shorter tile to start off the roof covering. Its size is that of the other roofing tiles minus the overlap. The area of each ordinary roofing tile visible when the roof is complete is known as the *margin.*

Ridge tile: a special tile shaped to cover the apex of the roof over the ridge board, and seal the gap between the topmost tiles on either side of it.

Approaching the job

Building a gabled roof is relatively straightforward if approached in the correct manner. It is, however, a long and heavy job which is hard to do single-handed. A great deal depends on correct positioning of the timber and it is particularly important that the ridge board is in square. You should have at least one helper.

You will be working at a considerable height

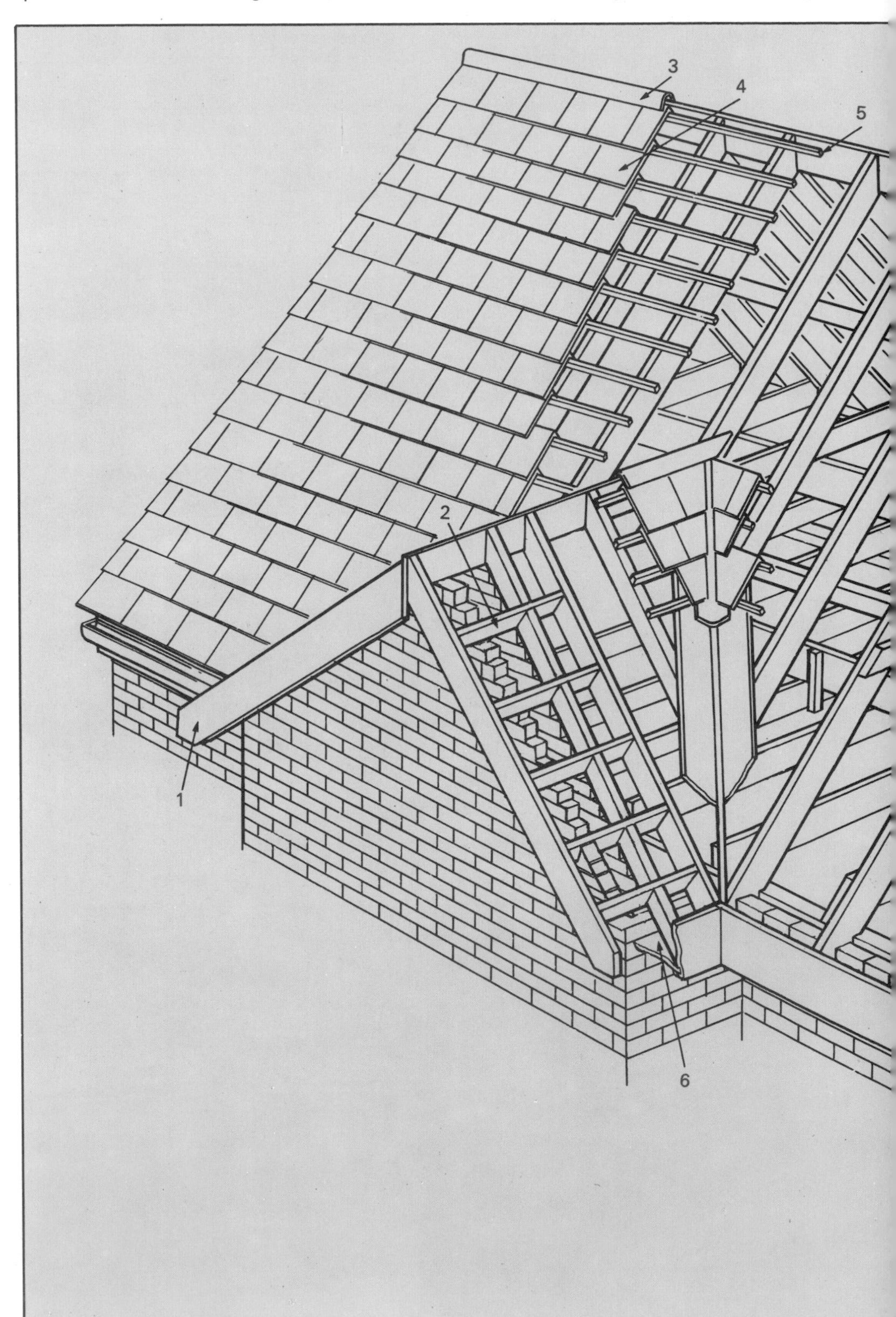

from the ground in most cases. Building a gabled roof on to a garage, however, will not present too many difficulties in this respect and a work platform constructed of two step-ladders and a sturdy plank may be sufficient if it is properly braced. The garage, however, may be too high to allow you to use this make-shift support, and if you are building the roof on a two-story house you will need a higher and more efficient work platform.

A mobile scaffold, which can be rented, will enable you to reach the job. This is a tower made of prefabricated scaffolding-like sections that fit together to reach any reasonable height. At the top there is a good size platform made up of planks. The whole structure is mounted on wheels so that it moves around easily. The wheels can be locked to hold the tower still. The tower's work platform is reached by an ordinary ladder.

Setting out the wall plates

Once planning and building permission have been obtained and the walls of the building (but not the gable ends) are complete to the top, mark out the positions of the joists on the wall plates. By the time you come to start the roof you will know the size and the spacing of the joists. Mark their positions on the wall plates by laying the two wall plates together on the ground with their long edges butting. Draw lines to indicate where one edge of the joists will meet the wall plate. On one side of these lines make a pencil cross. This will ensure that you do not nail some of the joists on the wrong side of the pencil lines.

The wall plates can now be fixed in place on the brickwork. Be careful to position them cor-

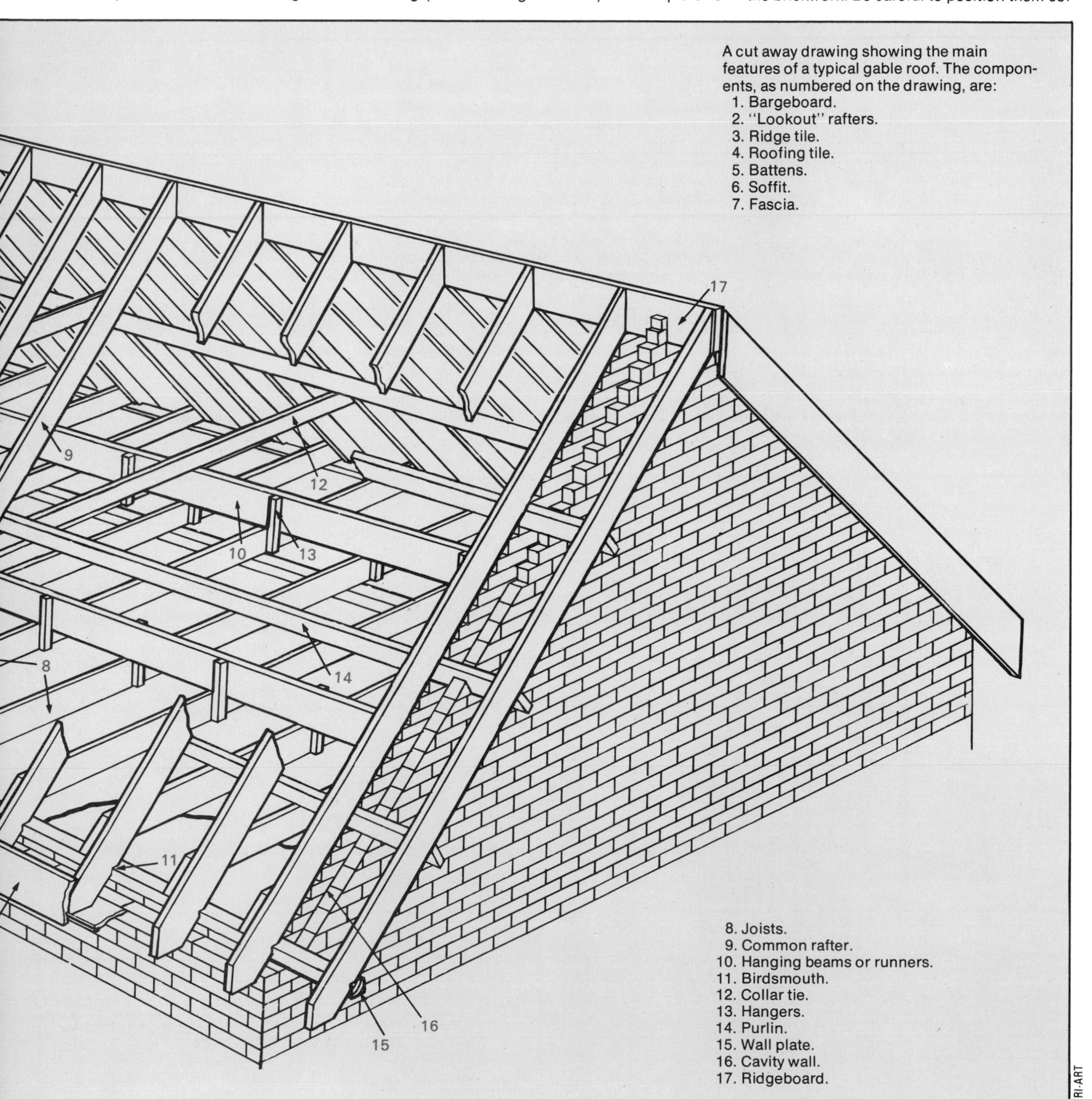

A cut away drawing showing the main features of a typical gable roof. The components, as numbered on the drawing, are:
1. Bargeboard.
2. "Lookout" rafters.
3. Ridge tile.
4. Roofing tile.
5. Battens.
6. Soffit.
7. Fascia.
8. Joists.
9. Common rafter.
10. Hanging beams or runners.
11. Birdsmouth.
12. Collar tie.
13. Hangers.
14. Purlin.
15. Wall plate.
16. Cavity wall.
17. Ridgeboard.

rectly—if the roof is to oversail the gable walls all the parts of the wall plates that protrude must be of equal length. The wall plates are fixed to the long walls of the building because the joists, which meet the plates at right angles, usually run the shortest distance across a roof. In a brick cavity wall, the wall plate is bedded in cement mortar on top of the inside course. Figs. 1 and 2 show a cross-section of the eaves of a gabled roof and the position of the wall plate. If there is an internal dividing wall running parallel to the ridge of the roof, top it with a sole plate the same size, and fixed in the same way as the wall plate. This will provide a midway support for the ceiling joists and reduce their span and—hence, the size of timber needed.

The joists

In Britain the sizes of ceiling joists are listed in the Building Regulations. Four standard spacings are listed, but the most useful and most commonly used is 16 in. The chart given below lists sizes of joists spaced at 16 in. centers. It will enable you to make a reasonably accurate dimensional drawing to take to your local authority. It is only a rough guide, but close enough that you will not have to completely re-design.

Size of joists (in.)	Maximum span between the wall plates
3 x 2	6 ft. 2 in.
4 x 2	8 ft. 3 in.
5 x 2	10 ft. 4 in.
6 x 2	12 ft. 2 in.
7 x 2	14 ft. 3 in.
8 x 2	16 ft. 3 in.
9 x 2	18 ft. 4 in.

The length of the joists should be sufficient to allow a small overhang at the two long walls to enable the rafters to be nailed to them at a later stage. Cut all the joists to the required size—any slight variation in their length when attached will not be noticeable when the roof is finished. Skew-nail the joists in place at the positions marked on top of the wall plates. (In a timber house, for reasons given below, the rafters *must* be immediately over the studs, so the joists must be just beside these.)

Setting out the roof

The most difficult part in building a gabled roof is to set out the various components correctly. The task requires precise measurement and marking—a trial and error approach to this sort of job will be costly and frustrating.

When you have designed your roof make a scale drawing of it. Use a scale of 12:1—1 in. on the scale drawing will thus equal 1 ft. in the roof. This will enable you to calculate the span and the height of the roof accurately, which you will need to do to execute the most difficult job—setting out the rafters.

Size of rafters

In Britain, the size of timbers used for roof rafters is governed by the Building Regulations. These list sizes of rafters for roofs with a pitch of between 35° and 40° and for roofs with a pitch of 40° or over. The chart below gives rafter sizes and lengths for roofs pitched at 40° or over, for a 16 in. spacing of rafters. For roofs pitched at between 35° and 40° the permitted span is 2 to 3 in. shorter. Use the chart as a guide to draw up your dimensional plan to present to the building inspector.

Rafter size (in.)	Maximum span across the building
1½ x 3	7 ft. 3 in.
1½ x 4	9 ft. 9 in.
1½ x 5	12 ft. 2 in.
2 x 3	7 ft. 11 in.
2 x 4	10 ft. 7 in.
2 x 5	13 ft. 2 in.

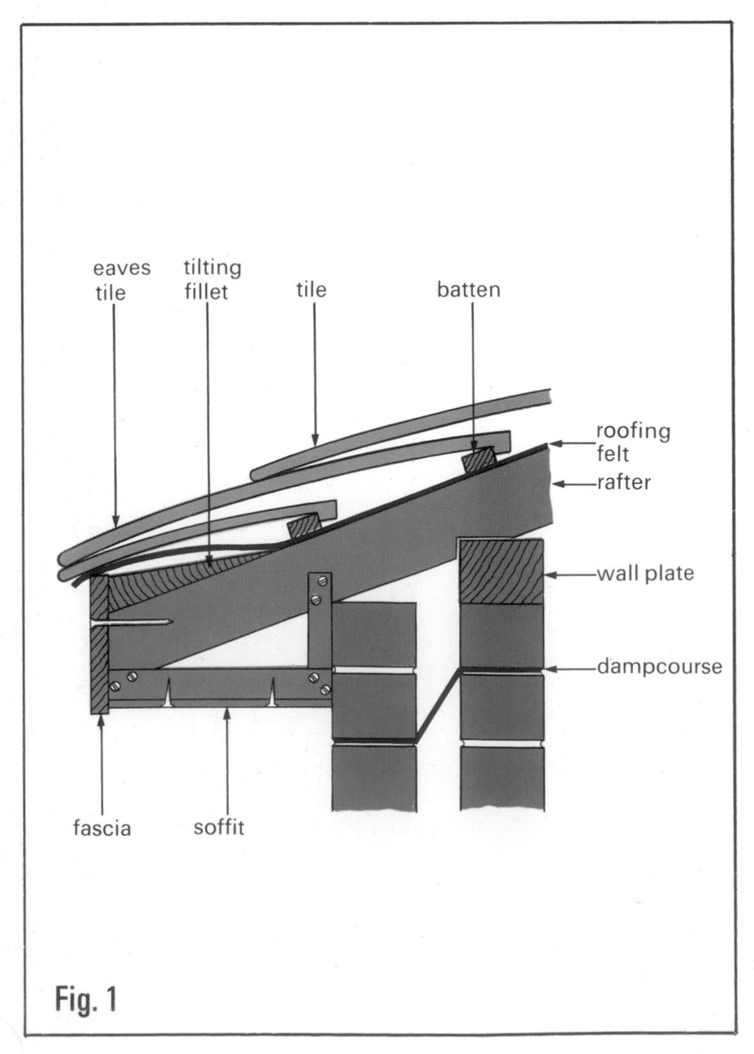

Fig. 1. *A cross-section of the eaves of a gabled roof. Here the rafters oversail the wall. An angled cut made at the bottom of the foot of the rafter provides a fixing for the soffit board.*

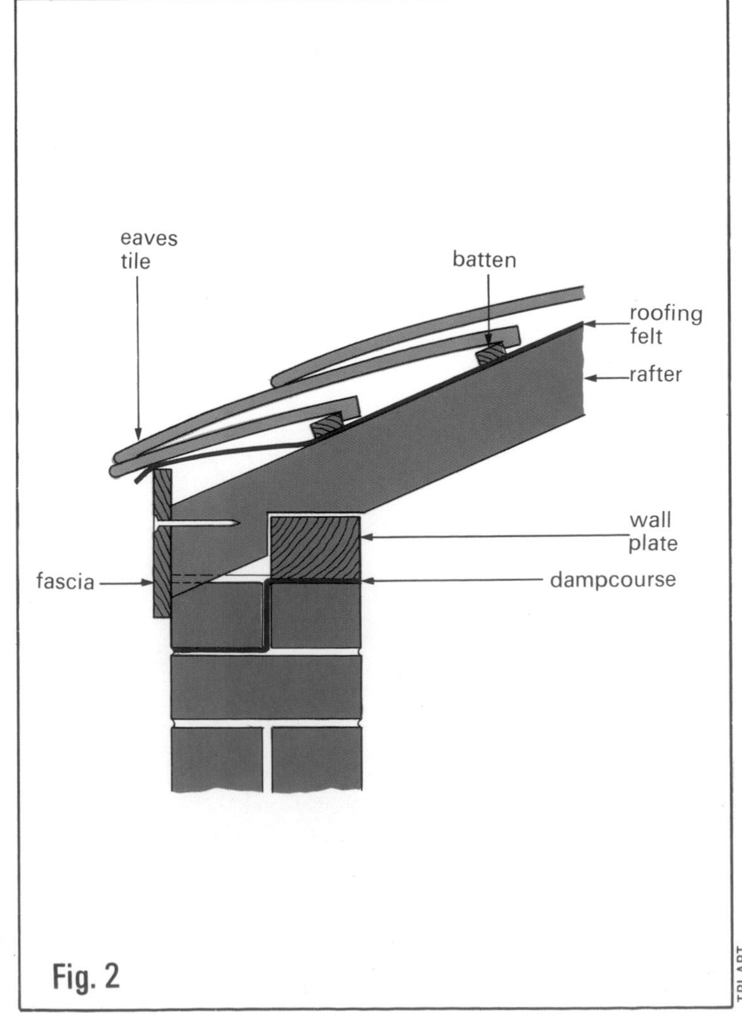

Fig. 2. *An alternative finish at the eaves. The rafters end flush with the outer skin of the wall. A soffit is not needed as the fascia seals the rafter ends against the weather.*

Setting out the rafters

The method of setting out a rafter is shown in Fig. 3. The method is the same whether you want the roof to oversail the long walls or whether you want it to end flush with these walls. The only difference is the shape of the foot of the rafters. (See Figs. 1 and 2. In the case of the former, the foot will have to be cut so that it has one vertical surface to take the fascia and one horizontal surface for the soffit or its supporting frame. In the case of the latter, the foot of the rafter has a single angled cut made so that the fascia can be fastened on vertically. More details on fascias and soffits and their fixing are given in CHAPTER 31.)

To set out the rafters, lay a long piece of timber (marked "A" in Fig. 3) on a level piece of ground and position wooden pegs along its edges so that it stays in position. Near one end of this piece make a pencil mark. From this mark, using your scale drawing to obtain the correct dimensions, measure and mark a distance equal to half the span of the roof. The span of the roof is the distance between the outside edges of the two wall plates. Next take a second piece of timber, "B," a few feet longer than the height of the roof, and peg it to the ground at right angles to "A." Position it so that its end butts the second marked line on timber "A." One edge of "B" should be inside the second line by a distance which equals half the thickness of the ridge board.

The next step is to measure and mark, on these pieces of timber, the position and shape of the wall plate and ridge board. Then take a piece of the timber you intend to use for the rafters. Cut it to a length a couple of feet longer than that needed—remembering any overhang that you may require at the eaves. You can judge the length by reference to your scale drawing or by measuring the distance between the marked positions of the ridge board and wall plate on the two pieces of timber that are pegged to the ground.

The rafters of a gabled roof fit over the wall plate by means of a *birdsmouth,* a triangular-shaped cut, made near the foot of the rafter. This ensures that the weight of the rafter presses directly downward on the top surface of the wall plate and not on its outside corner. The birdsmouth should never be cut deeper than one-third the width of the rafter.

Now take the over-length rafter and lay it

diagonally across the two pieces of timber pegged to the ground. One end of this rafter should run to and just pass over the marked position of the ridge board. If you are using normal ridge tiles on your roof, there will be a U-shaped gap under their centers to allow the ridge board to project 1–2 in. above the tops of the rafters without lifting the ridge tile off them. Allow for this projection when fitting the rafters to the ridge board; the deeper you can make this board the better, because it will add to the strength of the roof.

Near its other end, the rafter should pass over the marked position of the wall plate. The part of the marked position of the wall plate that is covered should not be more than one-third the width of the rafter, because of the limitation on the depth of the birdsmouth. When you have correctly positioned the over-length rafter, lightly nail it to the two pieces of timber.

Then mark out where the birdsmouth will be and the angle at which the rafter meets the ridge board. Use a straight edge to do this, sighting what can be seen of the previously marked areas as a guide. If your roof is to oversail the walls at the eaves, you will also have to mark the bottom corner at the foot of the rafter so that a surface can be cut that will be parallel to the joists in the finished roof. If these eaves are to be flush with the long walls there is no need to cut the foot of the rafter to shape. Ensure that the bottom end of the rafter is cut slightly over-length—the excess wood can be trimmed off when all the rafters have been fixed in place. The method of doing this is described in CHAPTER 20.

The next step is to remove the rafter from the pieces pegged to the ground and cut out the birdsmouth and the angle at the top of the rafter and, if necessary, the surface at the foot of the rafter to which the soffits will later be fixed.

Now make another rafter, marking it by laying the first rafter on the top of it. Keep the *top* edges of the two boards flush so that, even if there is a variation in the depth of the boards, the distance above the birdsmouth will be equal. Mark the second rafter using the first as a guide and then cut it to shape.

Now try these two rafters on the walls of the building. Place a piece of scrap wood the same thickness as the ridge board between the tops of the rafters. With lengths of battens, temporarily prop the rafters in place in the same manner as is used later for fixing the first set of rafters. This is described below and shown in Fig. 7. Check that the rafters slope at about the pitch you require—it is unlikely that the angle will be perfectly accurate. If the angle is badly off, the two rafters have been cut to the wrong length and you will have to do the setting out process again.

If the angle is right, take the two rafters down. Mark the first one "Pattern" and use this to mark out all the other rafters. These can now be cut to shape and placed in convenient positions on the ground around the building. Remember when using the pattern to have its top edge flush with the top edge of the rafter being marked out. Then, even if the width of the rafters differs, the height above the birdsmouth will be the same.

Setting out the ridge board

The length of the ridge board depends on whether you want the roof to oversail the end walls or not. If you do, it should oversail the walls by the same amount as the wall plates. The depth of the ridge board must be at least the depth of the angled cut made on the top ends of the rafters, and preferably more, so that the top of the ridge board projects over the tops of the rafters.

It will involve less work if the ridge board is one single length of timber. If you cannot get a timber long enough, the pieces that make the ridge board should be scarf jointed together before you set it out (see Fig. 5 for how to make this joint).

Setting out the ridge board involves marking on it the positions the tops of the rafters will occupy in the finished roof. You must work accurately but, since the rafters are in opposite pairs, you only have to mark one side of the board.

First mark out the positions of the rafters on the wall plate. The placing of the joists will determine the rafter positions. One face of each rafter, when fixed, touches one side of each joist and is nailed to it. To be equally spaced, the rafters must later on be nailed to one side (the same side) of each joist. Mark a line on the wall plate away from the side of the joist to a distance equaling the thickness of the rafter. Do this on the same side of each joist. Make a cross between the marked lines and the side of each joist as shown in Fig. 4.

Now mark out the position where the rafters will meet the ridge board in the finished roof. To do this, lay the ridge board at right angles across the joists just over the wall plates. Position the ridge board accurately, and lightly nail it to the joists. Place a straight edge across the ridge board lining up with the side of each joist from which the position where the rafters meet the wall plate was marked. Draw pencil lines to indicate the sides of the joists on the ridge board. On the side of the pencil line *away* from the joist, make a cross (see Fig. 4). When the ridge board is erected the top ends of the rafters will cover these crosses and one edge of each rafter will run along the pencil lines.

Erecting the roof

The most difficult part of this operation is to fix two pairs of rafters and position the ridge board correctly at the same time. The two pairs of rafters should be those at the ends of the roof, but *inside* the house walls. (If the roof is to oversail the end walls, do not fix the rafters on the tips of the wall plates first. These are nailed in place later.)

To fix two pairs of rafters and the ridge board in place, first lay one pair of rafters on the ground in roughly the position they will occupy in the finished roof, leaving a gap wide enough for the ridge board between their top ends. Nail a piece of batten across each rafter near the top so that the rafter tops and length of batten form a triangle. The batten should be placed far enough from the apex of the triangle so that the ridge board, when slotted into the gap, will occupy its desired position when resting on the batten. Use one nail only to nail the batten to each of the rafters. This will allow some play in the assembly during the positioning. Repeat this process with the other pair of rafters.

Lay one pair of rafters on the joists at one end of the roof so that the apex of the triangle they form points towards the opposite end wall. At the other end of the roof, position the rafters on the wall plates on the marked positions and nail them lightly in place. You will need some assistance to do this. Next, nail a

Fig. 3. *To set out the first rafter, peg two lengths of timber to a level piece of ground. Mark on them the shape and planned position of the wall plate and ridge board. Lay a piece of the timber you intend to use for the rafters across these two points. Mark the position of the wall plate and the ridge on the rafter.*
Fig. 4. *Mark the points at which the rafters will meet the ridge board before you fix the ridge in place. Do this by laying the ridge on the joists just over the wall plate. Mark a pencil line at a point that indicates one side (the same side) of each rafter.*
Fig. 5. *The ridge board can be jointed along its length with a scarf joint and gang nails.*
Fig. 6. *The method of fixing one pair of rafters to the ridge board. The first rafter is nailed in place from the opposite face of the ridge. The second rafter is skew-nailed in place.*

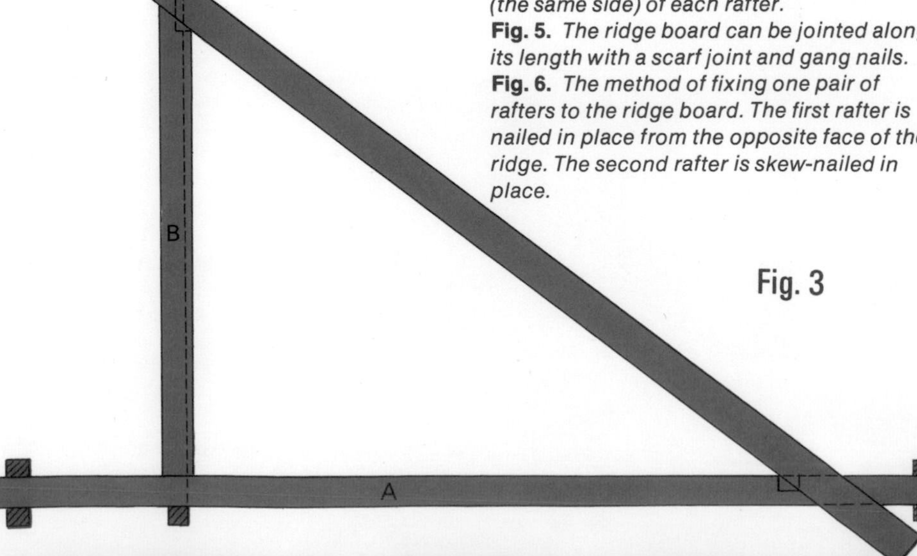

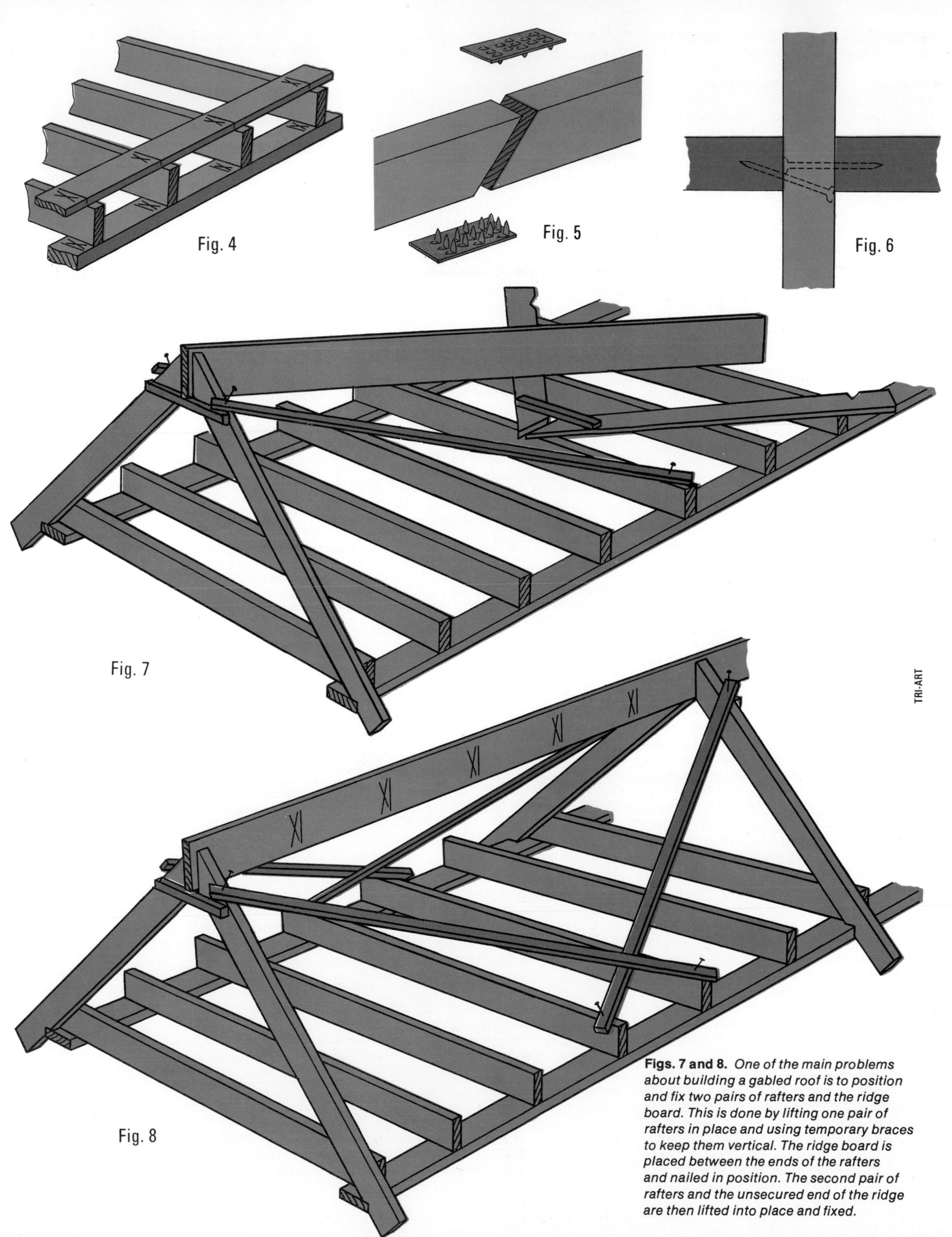

Figs. 7 and 8. *One of the main problems about building a gabled roof is to position and fix two pairs of rafters and the ridge board. This is done by lifting one pair of rafters in place and using temporary braces to keep them vertical. The ridge board is placed between the ends of the rafters and nailed in position. The second pair of rafters and the unsecured end of the ridge are then lifted into place and fixed.*

long temporary support brace to each rafter near their tops and to each end of one of the joists some distance away (see Fig. 7). The birdsmouth cut in the rafters should fit over the wall plates.

Lift the ridge board into place so that one end is between the top ends of the rafters that have been temporarily fixed in place. The other end should rest on the joists at the other end of the roof.

Now lift the second assembly of rafters up and secure it in position with support braces. Slide the ridge board along the channel formed by the ends of the rafters and the battens until it occupies the correct position. Once the ridge board is in place, skew-nail the feet of the four rafters firmly to the wall plates and to the adjoining joists. Then nail the tops of the rafters to the ridge board. Now remove the battens and the support braces and check that everything is straight.

Once two sets of rafters and the ridge board have been fixed in place, the fixing of the other rafters is simple. Lift each one to the job and nail it in place on the position marked on the wall plate and ridge board. One rafter of each pair can be nailed to the ridge board from its opposite face. The second rafter of the pair is skew-nailed to the ridge board (see Fig. 6).

The next step, if the roof is to oversail the end walls, is to provide some additional support for the rafters which are positioned outside the end walls. In a brick house, this is done by nailing lengths of timber to the sides of the first two rafters inside the end walls, and passing out through notches in the top of the brick gable—which must, of course, be built up to its full height at this stage.

To build a brick gable end, the outline of the brickwork must first be determined. To do this, tie a length of string to the end of the ridge board and take it to one edge of the wall that runs parallel to the ridge. Tie the other end of the string to a masonry pin and knock it into a convenient mortar course. Do the same at the other side of the roof. The lengths of string now show the required outline of the brickwork at the gable ends.

When the lines are in place you can start to build up the brickwork in the usual way. The end bricks of each course have to be cut so that they follow the slope of the lines. You can cut bricks with a bolster and club hammer. If the gable ends are to be finished with a tile verge the bricks should be cut carefully to ensure as neat a finish as possible. If you intend finishing the gable with a bargeboard, though, you do not have to be so careful—any small irregularities in the outer profile of the bricks can be filled with mortar to the string line.

In a timber house the gable ends are filled in by 3 in. x 2 in. framing, the top members of which carry the short lookout rafters.

Use timber the same size as the rafters for these pieces, space them at the same intervals as the rafters, and nail them to the rafters inside the end wall. Place them so that each is supported by one of the brick steps in the end wall.

In all roofs, the tops of the end brick walls have to be covered so at least one rafter has to be outside the end walls. If the roof is not to oversail these walls the inside face of this rafter is butted against the outer surface of the end wall.

Other supports

The next step is to provide greater internal bracing for the roof structure. This is done by means of purlins, collar ties and struts. The size of purlins is governed in Britain by the Building Regulations but if you plan your roof for 9 in. x 2 in. purlins you will not be far wrong. Purlins run the whole length of the roof and are nailed to the underside of the rafters. Usually one purlin is used for each side of the roof and they are positioned about centrally along the length of the rafters. At their ends, purlins are built into the inside leaf of the cavity wall of the brick gable, so they give extra support to the end rafters. Where they run into the brickwork they must be wrapped in building felt as a precaution against rot.

Next, install the collar ties and diagonal struts. Collar ties can be the same size as the rafters and nailed horizontally between pairs of rafters. The maximum spacing is between every fourth pair of rafters, but you may prefer to put in more than that for extra strength.

If the house has a central internal dividing wall running parallel to the ridge board, struts should be fixed to the collar-tied rafters running from the intersection of collar-tie, purlin and rafter to a 4 in. x 3 in. sole plate bedded in mortar along the top of the wall exactly like a wall plate. The struts should be firmly skew-nailed to the sole plate.

If no partition wall is available to take the ends of the struts, these are nailed to hanging beams of runners, fixed to tops of the joists and running at right angles to them. They should be bigger than the joists—exact sizes involve a lot of engineering calculation, but design for 9 in. x 2 in. and you will have a starting point—and set on edge. Most roofs that require these supports have two, one in each half of the roof. Fasten them to the joists with hangers, usually 2 in. x 1 in. pieces of timber, which are nailed at the corners of the hanging beams and the joists on alternate sides of the hanging beams.

The fascias and soffits

The methods of fixing and the materials used for these components of a roof are described in full in CHAPTER 20. But note that, in a timber house, the soffit support boards are nailed to the ends of the rafters and to the studs—this is why the former must be above the latter.

At the verge of a gabled roof two types of finish can be used. The first, mentioned briefly above, is to cut the end bricks of each gable course carefully and smooth them over with mortar prior to tiling. An alternative which is more common in older British houses, is to fit a bargeboard. These are lengths of timber like fascias, except that they run up the slope of a gable rather than being fixed horizontally. The bargeboards are not part of the roof support—they simply give a weatherproof finish at this point of the roof. Bargeboards and their fixings are dealt with in CHAPTER 31.

The tilting fillets should also be installed at this stage. These run the whole length of the roof and are fitted in the V-shaped intersection between the top of the fascia board and the feet of the rafters. Not all roofs require a tilting fillet, but their purpose in those that do is to lift the eaves tiles or slates slightly and to provide a firmer support for these tiles.

Once the basic components of the roof are in place, it is advisable to treat all the timbers with a preservative to prevent attacks of dry rot and woodworm. Some local authorities insist that roof timbers be treated in this way.

The roof covering

The gabled roof must, of course, be covered with some type of waterproof material. Slates were once in widespread use but they are less common nowadays and tend to be expensive. Tiles made from cement and colored a dull green or red or dark brown are more usual. To supplement the waterproofing qualities of both tiles and slates a bituminous roofing felt is nailed to the rafters before the tiles or slates are applied.

Nail the strips of roofing felt across the tops of the rafters, working from the eaves toward the ridge board so that the overlaps point "downhill." Lay the first strip so that it overhangs the long wall a little way and overlap the joints between the strips of felt on all the successive strips. To apply the felt you will need to build a cat ladder.

A cat ladder allows you to clamber up and down the slope of a roof in reasonable safety. It is simply a ladder made of square timber with rungs about 15 in. apart. It has a block of wood 4 in. x 2 in. nailed to it at the top so that it can be hooked over the ridge. The cat ladder is secured by tying it to the main ladder—the one you used for access to the roof—or, if possible, by lashing it to a convenient chimney. It is a useful piece of equipment for all roofing jobs.

When the felt has been applied, fix battens to the rafters, parallel to the long wall. Both slates and tiles are fixed to these battens which are 2 in. x 1 in. pieces of timber. They can be joined along their length if necessary, providing you do this over a rafter. The spacing of the battens depends on the size of tile or slate used; both materials are fitted so that they overlap. The space between the lowest batten and the fascia board is narrower than the space between one batten and another. This is because the eaves tile is narrower than the other tiles.

Slates are fixed to the battens with zinc nails—two to each slate is usual. Tiles hang on the battens by means of nibs, small protruding pieces on the underside of the top of the tile. Nail holes are provided in each tile. When covering a roof with tiles or slates work upward from the foot of the rafters, fixing the eaves course first. This will ensure that the overlap points "downhill." The waterproofing qualities of the tiles are enhanced if the joints between them that are at right angles to the wall plates are staggered from course to course. At the ends of the roof (nearest the gable walls) a tile 1½ times the length of the ordinary tiles is used.

CHAPTER 35

Shelves and how to fit them

If you are pressed for storage space in your home, extra shelves should prove a godsend. They take up far less floor space than cupboards, and can hold almost as much.

Shelving is the most versatile of all storage methods. Shelves can be built with any degree of strength to support the heaviest objects. Adjustable shelves can be moved up and down in a few minutes to take objects of any height. At the same time, things stored on shelves are always visible and accessible, and do not have to be hunted for as they would if stored in drawers or cupboards.

Shelving can be an important decorative feature, too. Room-dividers in modern houses are often made entirely out of shelves, on which ornaments, books, or hi-fi equipment can be arranged. Glass shelves fastened to walls make an ideal setting for the display of prized possessions, particularly since the glass allows the objects to be viewed from below.

Many kinds of shelving units are available ready-made, in a wide range of sizes and strengths. They take only a few minutes, and almost no skill, to set up and are instantly adjustable. In most systems, the brackets that support the shelves can be moved up and down metal tracks that are attached to the wall or occasionally freestanding or pole-lamp style.

Ready-made shelving systems, however, do have serious drawbacks. For one thing, nearly all of them are fairly expensive. For another, in most of the systems the shelf brackets and frames are large. Furthermore, most ready-

Below. *A simple white-painted shelving unit can set off your dishes or ornaments to their best advantage.*

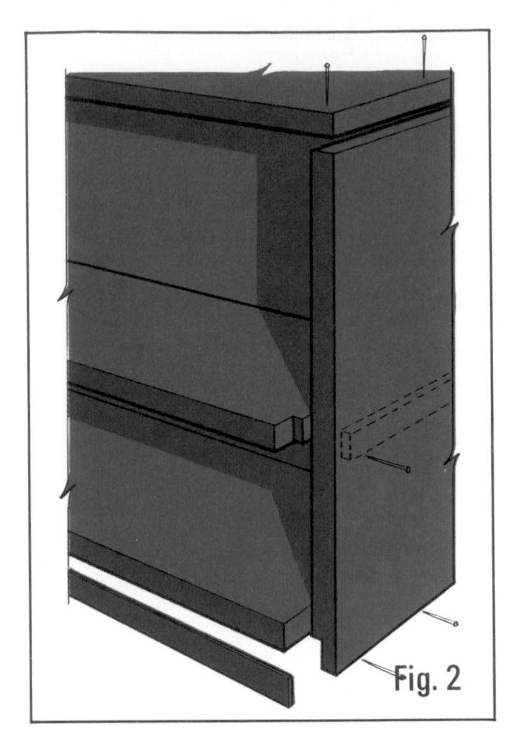

Fig. 1 *(top left). Boxed-in shelves.*
Fig. 2 *(above). Stopped dado joints are often used to make this kind of shelving.*

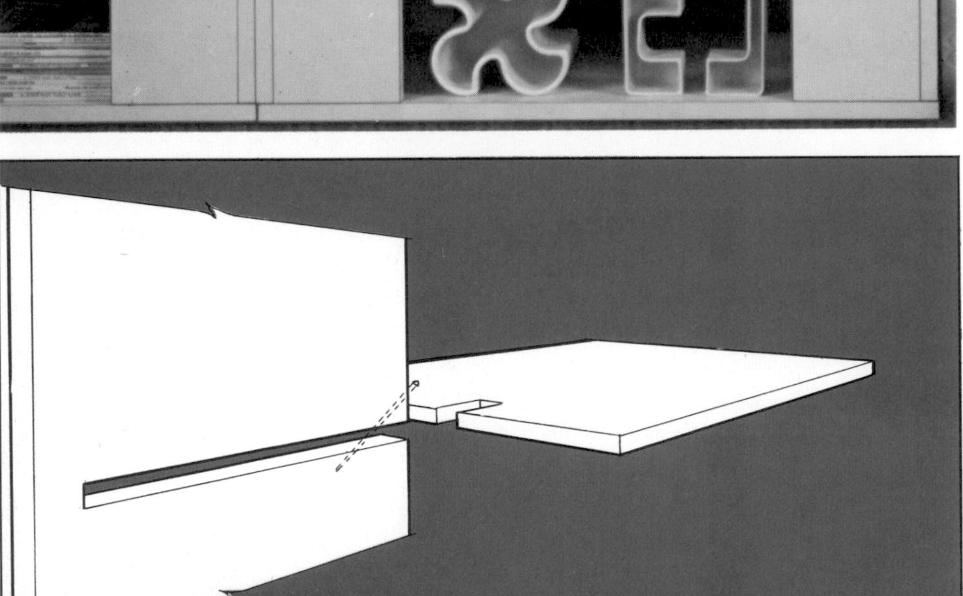

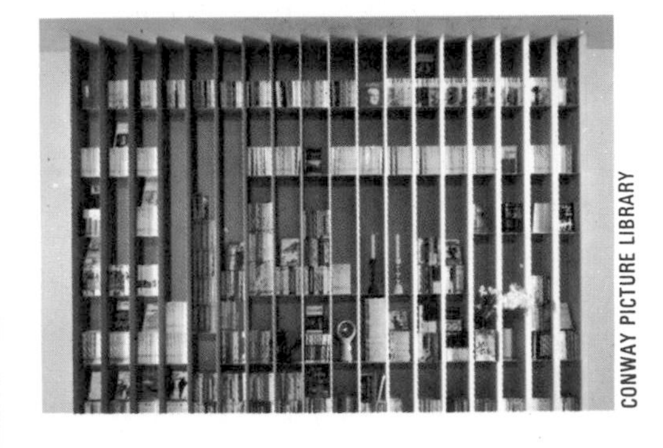

Fig. 3 *(middle left). Cheap, light "egg-box" shelving made of plywood can give a dramatic effect, as in* **Fig. 4** *(above).*

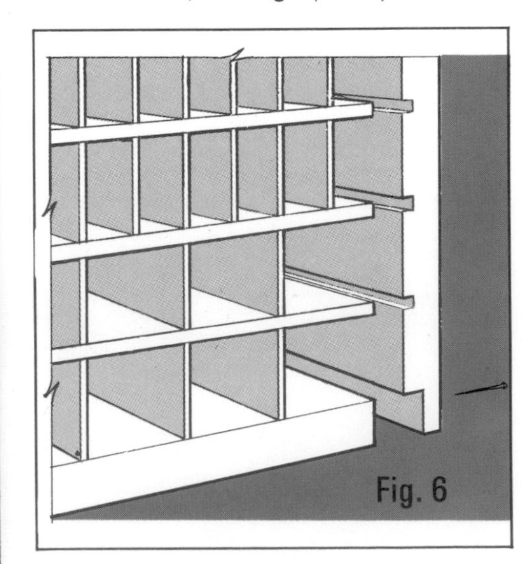

Fig. 5 *(bottom left). Vertical spacers strengthen shelving and give it variety.*
Fig. 6 *(above). The shelves are supported by dado joints, the spacers nailed.*

CINDY CASSIDY/PHOTO: JOHN VAUGHAN

Fig. 7 *(top left). Proprietary shelving brackets support specially-made shelves.* **Fig. 8** *(below). Details of the support.*

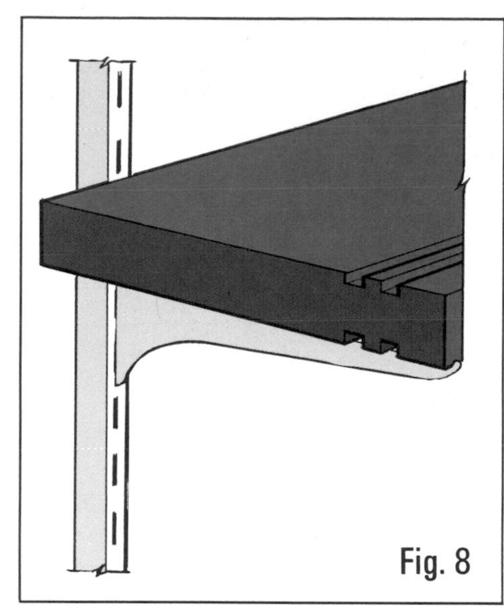

BRIAN MORRIS

Fig. 9 *(center left). Glass shelves are ideal for ornaments.* **Fig. 10** *(below). They can be mounted invisibly on clear plastic buttons.*

JOHN BETHELL

Fig. 11 *(bottom left). Hanging shelves supported by roof beams are strong enough to take heavy loads—and there are no awkward brackets.* **Fig. 12** *(below). How it works.*

Fig. 10

made shelving will carry only light or medium loads, and the stronger it is, the more massive.

It might be well at this stage to define what is meant by light, medium and heavy loads Consider a light load on a shelf 10 lbs. per foot length or less. An example would be light ornaments or a shallow stack of magazines stored horizontally. A medium load, such as that created by paperback books or large glass storage jars, might run to 20 lbs. per foot. A heavy load may be as great as 40 lbs. per foot, and might be made up of large hardback books or records stored on edge. These items are much heavier than you might think.

Unless you intend your shelving for a particular use that never changes, it is probably not a good idea to buy or make light-duty shelving. It is all too easy to set up a lot of flimsy shelves with the intention of storing a few magazines on them, and gradually build up the height of the stacks until the shelves collapse. On the other hand, the sturdy metal framing of a heavy-duty ready-made system looks out of place in many living rooms.

The way out of this quandary is to build your own shelving. It can be as strong as you like, and in many cases the brackets or fasteners holding up the shelves can be made completely invisible. Where this is not possible, careful design of the brackets can at least give a "clean" appearance.

Home-made shelving is normally not adjustable. This, however, is less of a drawback than it would be for a bought system, for you can design it for your particular needs and then it will not need to be adjusted. In practice, most people who buy adjustable shelving never alter the settings anyway.

Invisible supports

There are two types of "invisible" supports for shelves: those that are unseen from above, used for shelves below eye level, and those which are genuinely invisible from any angle. The first kind is easier to build.

Shelves that run the full length of a wall or alcove, and which can be supported at each end by other walls, are much easier to mount unobtrusively than freestanding shelves with open ends. Small glass shelves in alcoves can be held up by virtually invisible clear plastic supports that plug into holes in the wood or plaster at each end (see Fig. 10).

A wooden shelf mounted below eye level can rest on horizontal wood strips bolted or screwed to the wall. (see Fig. 18). Provided these strips do not come right to the front of the shelf, and their ends are trimmed at an angle, they will not be noticeable. A long shelf can also be given extra support by running a strip along the back. If it is more than 6 ft. long and carries heavy loads, one or more brackets may be needed in the middle of the shelf to stop it from twisting and sagging. These brackets can also be hidden—see below.

A shelf running between solid walls and mounted above eye level can be fixed in such a way that nothing shows at all. The supporting strips at each end of the shelf are made either of very narrow wood strips (narrower than the thickness of the shelf) or of aluminum angle strips. Instead of resting on the strips, the shelf is grooved at the ends and slides on to the strips, effectively concealing them (see Fig. 19). If the grooves are "stopped" (do not reach all the way to the front) nothing will be visible.

Grooves in the ends of wood shelves may be cut with a plow plane, if available. The end of a piece of particleboard is hard to plow neatly but may be sawed. This can be done with a hand saw, but better with a power saw with its fence set to locate the cut. Use successive cuts, re-positioning the fence, to get the desired cut width.

(If you don't have a plow plane, grooves can be made just as well with a narrow mortise chisel the same width as the groove.)

This method is very suitable for nominal 2 in. lumber shelves, because they are quite thick and will conceal a wide, strong supporting strip.

Fig. 13 *(left). A highly original shelving idea: three wooden shelves drilled at the corners and hung from thick, knotted ropes. The ropes are fastened to the wall at both top and bottom.*

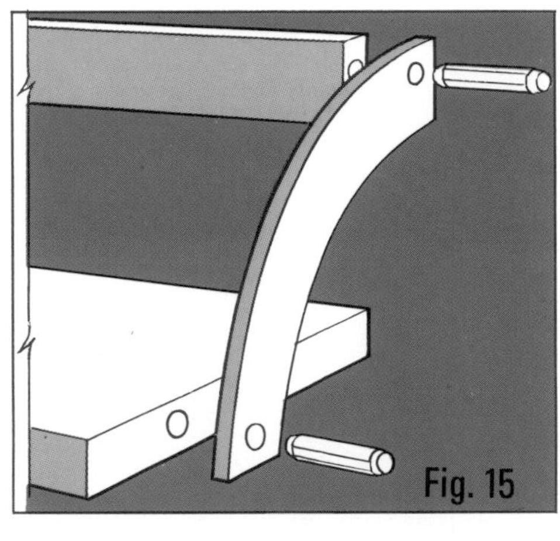

Fig. 14 *(left). If you can't hide brackets, why not make a feature of them?* **Fig. 15** *(above). The curved plywood brackets are held to the shelves by dowels.*

Fig. 16 *(above). Shelves look good in alcoves, and are easy to install there.*

Fig. 17 *(above). One of the simplest masonry mountings: screw-eyes and wall plugs.*

All these techniques can also be applied to shelves mounted around inside and outside corners (see Fig. 20). The shelf on one side of the corner acts as a brace for the shelf on the other side. However, both ends must be firmly attached to their mounting and not just resting on it, for the force exerted by pressing down on one end of a corner shelf may lift the other end. A shelf resting on wood strips can simply be screwed to them. One slotted into invisible mountings is harder to attach, unless the mounting strips are very wide and can have a screw passed through them from above. In masonry walls plug and screw the shelf to the wall instead. A small pocket cut out of the less visible side of the shelf does it as in Fig. 21. In wallboard or plastered walls use toggle bolts or mollies, instead.

Freestanding shelves

Shelves which do not have their ends supported by a wall are harder to mount unobtrusively. Most people putting up this kind of shelf settle for small, neat brackets, but if you feel you must have invisible mountings, there are solutions to the problem.

The simplest method is suitable only for timber-framed stud walls—common in most rooms of the average home, but not in basements. Large lag screws—at least 9 in. long and $\frac{3}{8}$ in. thick—are screwed into the studs, which are generally 16 in. apart. The shelves for this type of mounting should be at least "$\frac{5}{4}$" thickness—a lumber term meaning $1\frac{5}{32}$ in. Once they are firmly screwed in, their heads are sawed off. Holes the diameter of the screws are then drilled into the shelf from the back edge and the shelf is then slid on to the screws (see Fig. 23). This method is very strong, provided the screws project about two-thirds of the width of the boards.

Another similar method for narrow shelves on brick or masonry block walls uses steel angle brackets (also called corner braces). The horizontal parts of the brackets are slotted into the shelf as the coach screws were. The only difference is that the slots in the shelf are rectangular, rather than round. These slots are made by drilling several holes and cutting out the wood between them with a long, narrow chisel. It is easier to make the slots too wide and insert narrow pieces of wood as a wedge to hold the shelf firm to the brackets.

The other half of the bracket is harder to hide. One solution is to screw it to the wall studs and hide it with a backboard. If there are three or more rows of shelves and a backboard runs behind them all, it can be quite a decorative feature.

A more satisfactory way of hiding the brackets is to recess them into the wall $\frac{1}{4}$in.–$\frac{1}{2}$ in. and cover up the recess with spackle. A neat, shallow channel can be cut in a plaster wall with an ordinary carpenter's chisel, though you will have to sharpen it afterwards. The blade should be held at an angle so that its cutting edge always points toward the center of the channel. Then, if the blade slips, it will damage the plaster *inside* the channel instead of making a long gash in the wall. Once the channel is cut, coat the brace with oil paint to stop rust bleeding through the spackle.

Brackets should not be placed too near the ends of shelves or they will make it liable to sag in the middle. A proper position for a pair of brackets under a freestanding shelf is roughly one-quarter and three-quarters of the way along it. With this arrangement, neither the middle nor the ends are very far from a solid support. A shelf should always be fixed rigidly to its brackets to stop it from tipping up if the end is pressed down.

Featured supports

If you cannot hide the supports of your shelves, the best thing to do is to bring them out into the open and make a feature of them. One way of doing this is to buy a ready-made shelving system, which has the great merit of being adjustable. If this doesn't appeal to you, there are many good-looking shelf brackets you can buy or make yourself.

If you have period furnishings, plain wood brackets are probably the most suitable. A typical design, intended to be made out of ¾ in. thick hardwood, is shown in Fig. 22. The shape can, of course, be varied to taste, provided that the vertical depth of the bracket is at least half the width of the shelf. The bracket is best fastened to the back plate with a mortise-and-tenon joint as shown, but can be screwed on from the rear if you prefer.

One of the simplest ways of holding up shelves, and one that looks particularly good with modern furniture, is to run vertical boards up the ends of the shelves to turn them into a wall-mounted box (see Fig. 1). All the shelves except the top one should be attached to the vertical boards by stopped dado joints, and the top shelf rabbeted at each end (CHAPTER 3 describes how to cut both these joints). This construction, which makes the shelves look like a bookcase, is very strong. (For light loads shelves may be supported by just the strength of nails driven into the ends.) A backboard, even of hardboard, fitted behind the unit makes it even more stable, by acting as a brace. In a long unit, short vertical spacer boards can be put between the shelves to hold them apart. An attractive random effect can be created by placing these boards at irregular intervals.

There are many other methods of giving your shelving an interesting appearance. For example, the method illustrated in Fig. 13 uses heavy rope to hang the shelves from rings at the top of the wall. The rope is knotted under each shelf; the height of the shelves can be adjusted by reknotting the rope. This method is too flexible for fragile ornaments to be put on the shelves, but it is ideal for books or magazines.

One of the most convenient types of shelving consists of large open-sided boxes that can be stacked on each other. As long as the boxes are not stacked too high, they provide a strong, stable storage space that can be rearranged to any shape. If all the boxes are made the same height, but some are twice or three times as wide as the others, an enormous variety of arrangements can be made to suit any use.

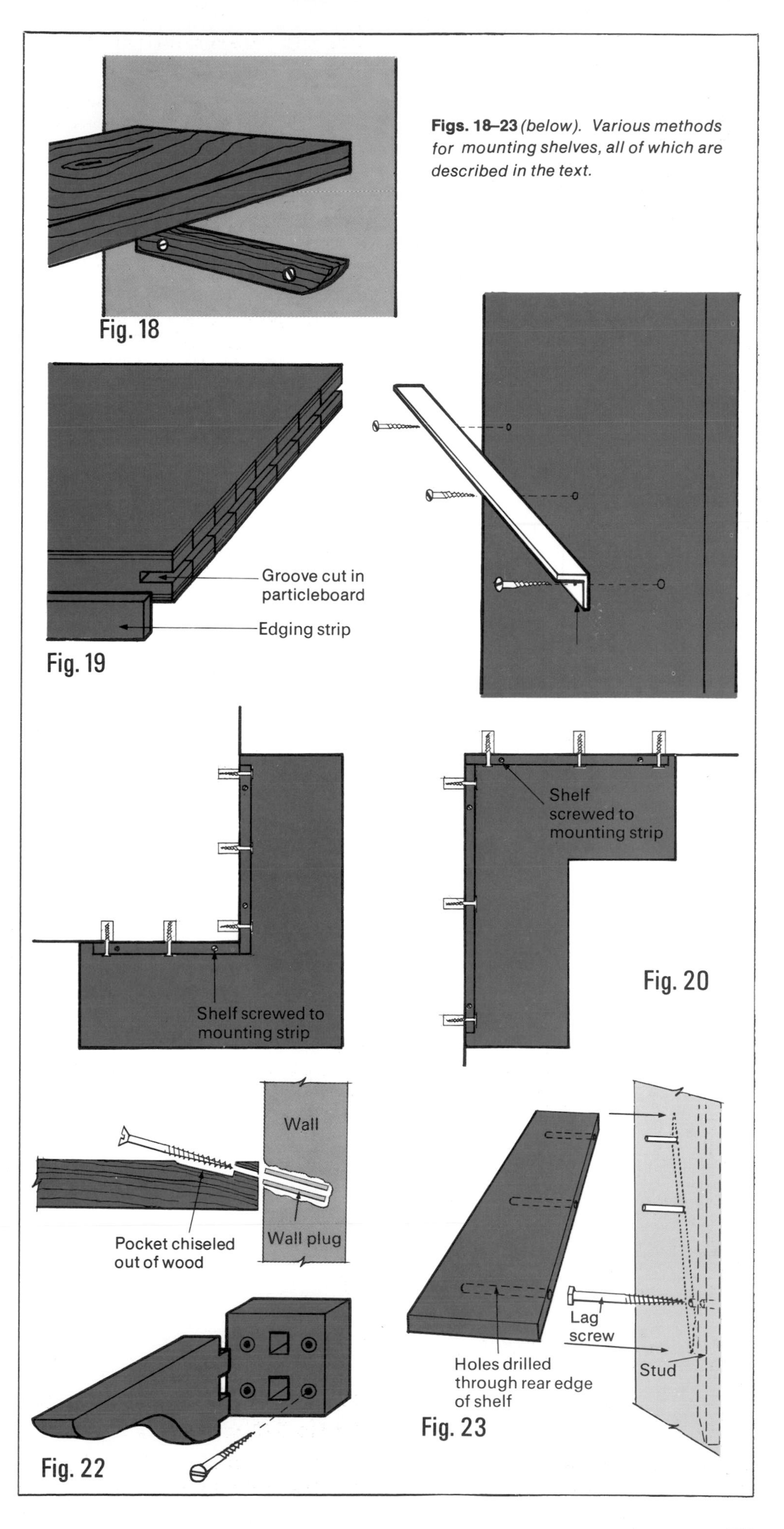

Figs. 18–23 *(below). Various methods for mounting shelves, all of which are described in the text.*

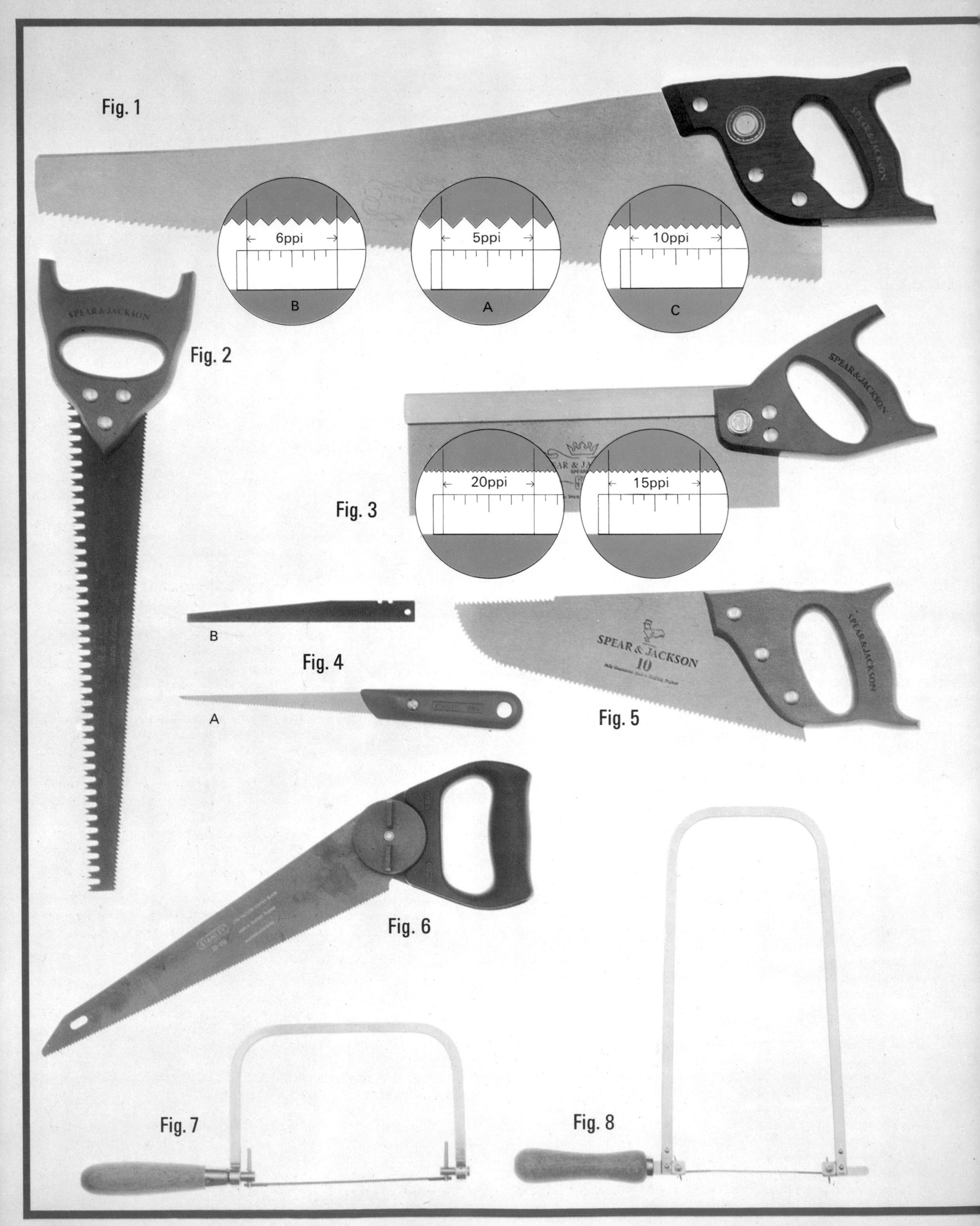
Fig. 1
SPEAR & JACKSON
6ppi
5ppi
10ppi
B
A
C
Fig. 2
SPEAR & JACKSON
SPEAR & JACKSON
20ppi
15ppi
Fig. 3
B
Fig. 4
A
SPEAR & JACKSON
10
SPEAR & JACKSON
Fig. 5
Fig. 6
Fig. 7
Fig. 8

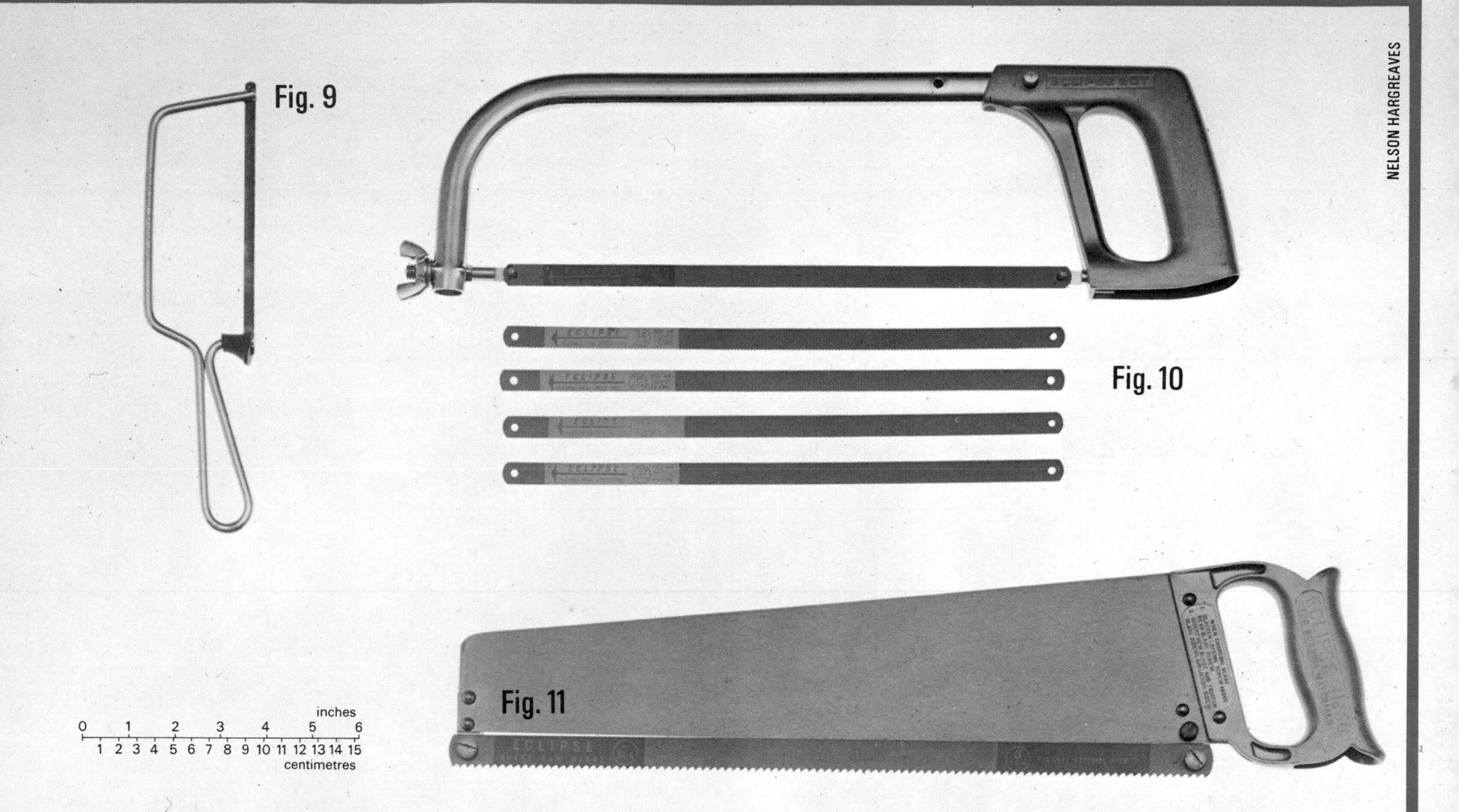

DATA SHEET

Saws

Efficient and easy cutting is a prime requirement of all do-it-yourself jobs. There is a wide range of saws available to enable you to do just this—and this DATA SHEET tells you all about them.

The type of saw you use for a job depends on the material you are using, how accurate the cutting needs to be and where the job is located. Knowing which saw to use will simplify all do-it-yourself jobs.

Knowledge of a few technical terms will help explain the DATA SHEET:

Points per inch (ppi) refers to the number of saw teeth to the inch along the saw blade. Woodworking saws with a small number of ppi make a rough cut but cut fast. Those with a larger number cut smoothly, but not as fast. (Points per inch is often shortened simply to "points.")

The *kerf* is the name given to the width of the saw cut.

The *gullet* is the distance between one saw tooth and the next. The gullet carries sawdust out of the kerf to make the task of sawing easier. Saws suitable for cutting cord wood have deeper gullets and larger teeth for fast, rough cutting.

Fig. 1. Hand saw. There are three types of hand saws: **A.** Rip saw. 26 in. long with 5 or 5½ ppi. It is used for cutting *with* the grain. The teeth are chisel-edged to chip out the wood. The large gullet carries the sawdust out of the kerf. **B.** Cross cut saw. 24 in. to 26 in. long with 6 to 10 ppi. The saw is used to cut across the grain on hardwoods and softwoods and for working with the grain on very hard woods. The knife-point shaped teeth give the sharper cut needed when working across the grain. **C.** Toolbox saw. 12 in. long with 10 ppi. The toolbox saw is small enough to fit a carpenter's toolbox. It's for general use, smooth cutting.

Fig. 2. Double-edged pruning saw for cutting green wood. One side is fine-toothed for cutting slender branches and the other has large open gullets to carry away sawdust when cutting larger ones.

Fig. 3. Back saw. 12 in. to 14 in. long usually with 13 to 15 ppi. It is used for jointing and for cutting across the grain on small pieces. The back may be brass or steel. A narrower type saw with 15 ppi for cutting dovetails is called a dovetail saw. Its blade is thin to give greater accuracy.

Fig. 4. A. Pad-handle keyhole saw. **B.** Metal keyhole saw blade. Both are used for cutting along straight or curved lines.

Fig. 5. Flooring saw, 6 to 10 ppi. The rounded nose allows you to cut into floorboards without damaging adjacent boards.

Fig. 6. General purpose saw. One of several variations. It is used for cutting wood laminates, plastic. With hardened blade, it cuts metal. It is a handy odd job tool but is not recommended for accurate work. This handle is adjustable to enable work in awkward places and positions.

Fig. 7. Coping saw. It has very fine teeth and is used for cutting tight curves. Tension is applied to the replaceable blade by tightening the handle.

Fig. 8. Fret saw (deep frame). It is similar to the coping saw but is deeper to allow work with larger boards. There are many types of blades available, the choice depending on what material you wish to cut.

Fig. 9. Junior hack saw. General purpose saw for light metal work. Fits small toolboxes.

Fig. 10. Adjustable frame hack saw. It can take 10 in. to 12 in. blades. Blades are available in range of ppi from 18 to 32.

Fig. 11. Metal-cutting hand saw. This is available with 20 in. blade with 15 ppi for cutting non-ferrous metal like aluminum roofing. Special purpose saws like this vary in form, and are not stocked by all hardware stores. If the saw you want isn't available ask your hardware dealer for a substitute that will serve the same purpose. It may have to be ordered from the manufacturer.

CHAPTER 36

Using veneers

The price of wood gets higher every year, particularly of expensive and decorative hardwoods. As a result, more and more articles that would once have been made of solid mahogany or teak are now built from cheap softwood and covered with a decorative hardwood veneer, giving much the same effect as solid wood if it is properly done. Some decorative woods are too fragile or unstable to use solid, and these have to be applied in veneer form. Sooner or later, you will probably want to apply some veneer yourself. Here is how to do it.

You can apply veneer successfully by hand without much specialized equipment, even though special heated presses are used by many professionals. But try to avoid veneering anything except flat surfaces to begin with; special formers are usually necessary with curves such as those found on a number of old cabinets. If curves must be veneered, you can achieve reasonable results on mild curves by using special contact adhesive made for veneering.

Buying and handling veneer

Veneer may not be available in your locality, but is usually sold by suppliers in larger cities. The ideal place is, of course, a cabinetmakers' supply dealer, but there are very few of these even in large cities. If you can't find one, check the yellow pages under "plywood and veneers." Some large lumberyards also handle veneer, including exotic types. Albert Constantine & Son, Inc., 2050 Eastchester Rd., Bronx, N.Y. 10461 and similar cabinetmakers' supply houses are mail order sources. Antique dealers who restore their own furniture sometimes have a few sheets of veneer around and may agree to sell you some. And failing all these, larger art and handicraft shops sell small pieces of veneer for marquetry.

A number of veneer-like products are also sold in some hobby shops. Some have a backing that gives greater stability to the extremely thin veneer and others a coat of glue on one side. Always follow the manufacturer's instructions carefully when using these proprietary products to ensure the best possible results.

Cabinetmakers' supply shops generally require a certain minimum, such as 3 square feet, for veneer orders. Check on the lengths and widths when you buy. They are often "random." Veneer is very fragile and particularly easy to split, so you may have some difficulty taking a piece this size home on a train or bus. It should be carefully packaged for protection, the usual thickness is 1⁄28 in. Some antique furniture has much thicker veneer, and if you are matching this you will be wise to show the dealer a sample piece. It is completely unrollable— though slightly less likely to split.

Methods of veneering

Veneer can be stuck down with hide glue, urea resin glue (often called plastic resin glue), polyvinyl (white) glue, resorcinol resin glue, or contact adhesive. Hide glue is a solid substance when cold and only becomes liquid and adhesive when heated after soaking. As it cools, it soon becomes too thick to flow around easily, so it has to be kept hot during use. In contrast, water based resin adhesive flows easily at room temperature and in normal conditions allows ample working time. Contact adhesive makes both surfaces stick fast on contact but requires careful pre-alignment. Because of the differences between these adhesives, the method of applying the veneer differs in each case.

The principle of the *traditional hide glue method,* sometimes called *hammer veneering,* is to spread hot, liquid hide glue on both surfaces, lay the veneer in place, and work excess glue out from underneath it by pressing the surface by sliding a special *veneer hammer* over it, as in Fig. 7. This produces a smooth, flat finish. The use of the veneer hammer is very important, because otherwise the glue layer may be uneven, causing a rippled surface. A veneer roller often replaces the hammer in modern veneering. If the glue dries before you have finished smoothing the veneer, it can be reheated by ironing the surface of the veneer with a heavy iron. The glue holds the veneer firmly in place from the moment it is applied, but takes some time to gain its full strength.

This method is not as widely used today, as other adhesives make the job simpler. Always use hide glue, however, when re-veneering part of an article already veneered with hide glue—if resin adhesive and hide glue are used together the bond may be weak.

The *white glue and C clamp method* is best for small areas no more than 2 ft. wide, and for curly, wavy or burr veneers, which are hard to lay flat. The base wood and the veneer are both coated with adhesive. The veneer is put in position and a thick, flat board clamped in place on top of the veneer. Since this type of glue flows easily, the excess will be forced out by the pressure of the clamps, and as long as the clamped-down board is really flat, the veneer will have a perfectly smooth surface. The clamps are essential to prevent the veneer lifting during the fifteen or so hours the adhesive takes to dry; they must be arranged so as to exert an equal pressure all over the board. But as an even pressure can be exerted for only a distance of about 1 ft. from a big clamp with a block under it, the greatest width that can be adequately veneered by this method is about 2 ft. You can veneer long strips of this width by using several clamps. If the surface being veneered is uneven; as on old furniture being restored, place a layer of soft, fibrous wallboard between it and the clamp-down board. This "gives" enough to exert an even pressure on most such uneven surfaces.

Removing old veneer

If you are re-veneering an old piece of furniture, remove the old veneer before re-covering it. First take any remaining polish off the surface by roughening it with coarse sandpaper or steel wool, or scratching it with a chisel. Lay hot wet rags all over the surface and apply a very hot iron to boil the water. Continue until the veneer begins to peel, adding water as necessary. When it does, take an old carving knife or chisel and work it under the edges. The veneer will be easy to lever off. Finish the preparation by scraping the softened glue from the base wood. Wash it with methylated spirits.

Hide glue method

You will need to work as quickly as possible when veneering by this method, so have the following tools and materials ready and within reach.

1. Sufficient hide glue. The glue may be in the form of a solid sheet or small flakes, or in a semi-liquid state.

2. A glue pot. You can use two different-sized tins; one should fit inside the other with at least 1 in. space all round it. Fill the larger pot with a few inches of water and drop a few bolts or stones of equal size into the bottom. Better, make a wire support for the smaller tin. Put the hide glue in the smaller tin and place it in the larger tin on top of the bolts. This allows the glue to be heated without any risk of burning.

3. A source of heat. Any type of stove will do; the double boiler described above prevents overheating.

4. A veneer plane (illustrated in Fig. 1) or coarse sandpaper wrapped round a flat block.

5. A veneer hammer (illustrated in Fig. 1) or roller.

6. A sharp knife, a chisel, and a metal rule.

7. A clean rag and hot water.

8. Various grades of sandpaper down to finishing grade.

When you have assembled all these things, clean the surface to be veneered. Then, if veneering a piece for the first time, roughen it with a veneer plane or coarse sandpaper, both used at an angle of 45° to the grain, to give a good key for the glue. If the surface is particularly absorbent, for example in the case of very coarse-grained wood, size it with watered-down glue—this will prevent so much glue being absorbed that the veneer does not stick properly.

Now match the pieces of veneer to give the most attractive grain pattern. Use the knife and metal rule to cut them so that they are about 1 in. oversize around the edge of the surface to be covered. If two or more pieces are being laid next to each other, lay them in place with the edges overlapping by an inch. Then cut through the center of the line of the overlap along a steel rule, cutting through both pieces at once. Be sure to hold the knife vertical. Discard the offcuts and stick the pieces edge-to-edge with veneer tape (Fig. 4).

Use a double boiler to heat the hide glue according to the manufacturer's instructions. This is usually preceded by overnight soaking. Add a little water as required for easy flowing consistency. Spread the glue evenly but as thinly as possible over the adjoining surfaces of both the wood case and the veneer. Work quickly as the glue soon thickens. Lay the first sheet in place. Take the veneering hammer and draw it over the veneer parallel with the grain but with the blade slanted at approximately 45 degrees to it. Ways of using the hammer are shown in Fig. 11. Always use a sliding movement with the veneer hammer or roller and never a tapping movement, as this would mark the surface. Try to keep an even pressure on the hammer at all times. If you use a veneer roller, simply draw it over the surface with

Opposite. *A chessboard with a difference made out of squares from offcuts of contrasting wood veneers.*

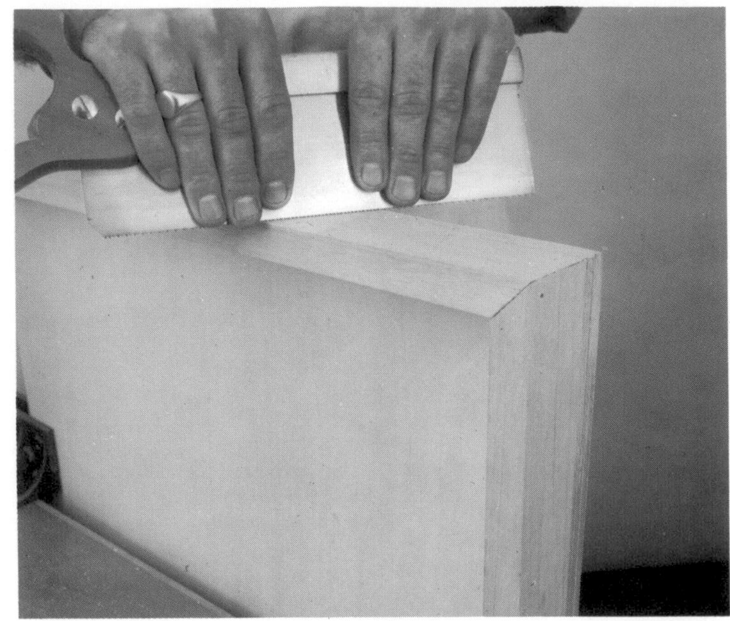

enough pressure to smooth the veneer. Then clean off any lumps of excess glue around the edges with a hot wet cloth.

The hide glue may set before the veneer has been pressed completely smooth, but you can soften the glue so that it can be worked again by ironing it with a warm iron. Always keep the surface slightly damp to allow the veneer hammer to slip easily over it and to prevent the veneer from being scorched by the iron. But try not to reheat the glue more than necessary, as you may dry out the veneer and cause contraction cracks.

When you have smoothed over the whole piece, check the surface for blisters or ripples. To remove these faults, slit the blister in the direction of the grain with a razor blade or modelling knife. Work glue into the slit with a thin knife. Then wet the blister and press on slit half down at a time with a flatiron at medium clothes-pressing temperature. Then stick a piece of veneer tape over the cut. This prevents it from opening up and air being drawn back beneath the veneer. Place a

Fig. 1. *A veneer plane (bottom) has a finely toothed blade for slightly roughening the surface to be veneered. Alternatively you can use medium sandpaper. The veneer hammer (top) used to smooth the veneer in place has a brass smoothing edge. Some hammers have a hardwood edge, but the edge must be smooth.*

Fig. 2. *A back saw can be used instead of a veneer plane to roughen the surface.*

Fig. 3. *Two sheets of veneer are matched with the grain running in the same direction. The edges are overlapped by approximately 1 in. and a sharp knife and metal rule used to cut through both pieces down the center of the overlap to give a neat joint.*

Fig. 4. *Sheets of veneer can be bought cut from the same section of tree and then matched together (or quartered) to produce a very attractive pattern. Here the edges have been taped before cutting to prevent these delicate veneers from splitting.*

Fig. 5. *Hot, molten hide glue is applied with a brush to the edge of the piece—work as quickly as possible before the glue dries.*

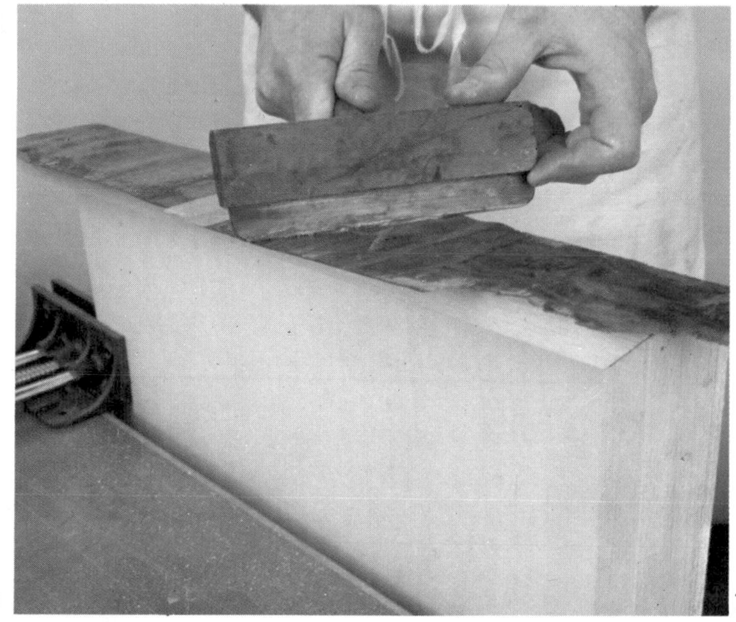

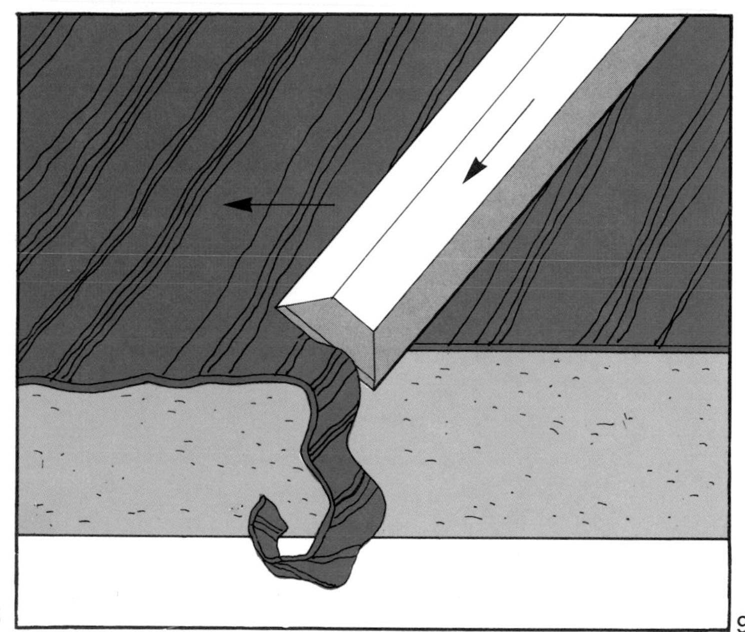

Fig. 6. *Here white glue is brushed as evenly as possible onto the face of the wood.*
Fig. 7. *Here an edge is veneered by the hide glue method. The veneer is smoothed in place by even strokes with the veneer hammer: an even pressure must be applied at all times. Excess glue will be forced out at the sides. If the glue sets before the veneer is smooth, remelt by ironing with a steam iron—but keep the surface moist.*
Fig. 8. *Veneer stuck with white glue can be ironed in place with a steam iron at 120°. The pressure of the iron smooths the veneer and the heat hastens the setting process of the adhesive. Keep the iron moving, and work from center outward to smooth off bubbles.*
Fig. 9. *The overlapping edges of the veneer are trimmed with a chisel; note the combined downward and sideways movement.*
Fig. 10. *Battens provide a more robust edge than a strip of veneer and can make a nice contrast with the top. The battens are stuck, bradded in place, and trimmed with a plane before veneering the top.*

weight on the repair until the glue is firm. Then remove the weight and the tape.

Leave the veneer to dry thoroughly, according to adhesive manufacturer's instructions. Then trim the edges, being careful not to damage the surface. If the veneer overlaps a corner, use a chisel to trim it to size as shown in Fig. 9. Remove any veneer tape stuck to the surface. Sand the whole surface lightly with the grain with very fine sandpaper. If you are re-veneering part of an old article, stain the new veneer to match the old. French polish the surface to finish the job, if that was the original type of finish.

The white glue and C clamp method

Have the following ready: white (polyvinyl acetate) adhesive, a mixing pot, a sharp handyman's knife, a chisel, sandpaper in various grades, a large sheet of brown paper, a flat, thick, smooth board large enough to cover the veneer, and sufficient C clamps.

Remove the existing veneer (if any) and clean the surface with sandpaper, but do not roughen it as described above. Match the veneers and cut them out oversize, but do not wet them. If you are laying two pieces side by side, cut the adjacent edges along a steel rule with their edges overlapped during the cutting so that they can be butted accurately against each other.

Spread the glue evenly over the surfaces to be joined—keep it off the outer veneer surface—and place the veneer in position. Use "veneer pins" to hold it in place. Drive one at each corner, leaving the tips up for easy removal after the glue dries. Use a veneer roller to smooth the veneer.

Lay the sheet of brown paper over the veneer and put the flat board on top of it. Clamp the board in place with at least one clamp every 2 ft. The brown paper prevents the veneer from sticking to the board. Tighten the central clamps first, then the outer ones, so as not to create a bubble. Leave to dry for approximately 24 hours, then remove the clamps and board. Trim the veneer, sand and finish as before.

The contact cement method

If you use one of the contact type veneering adhesives, like Constantine's Veneer Glue, the procedure is much like that of applying plastic laminate to a counter top, except you apply veneer to furniture. The important point to keep in mind is that once the adhesive-coated surfaces are brought together with this contact-type adhesive, they cannot be separated or adjusted. The bond is instantaneous and permanent. So the veneer must be perfectly aligned when it is set in place.

To begin, prepare the surface on which the veneer is to be placed, and cut the veneer to size, as described earlier. If the veneer is in several pieces the joining edges should be trimmed to fit exactly and taped together on what will be the exposed surface. This way, the joined pieces can be applied as a single piece, which can be slightly oversize to make the job easier.

Next, coat both meeting surfaces (the back of the veneer and the surface on which it will be applied) with the veneer glue (contact type). Allow the coating to dry completely, which usually requires about 20 minutes. Then lay a "slip sheet" of wrapping paper on the surface that is to be veneered. This sheet should be slightly larger than the surface on which it is laid. It won't stick to the dry, adhesive-coated surface. (With this type of adhesive, both surfaces must have an adhesive coating in order to stick together—and you *do not* apply any adhesive to the paper slip sheet.) With the slip sheet in place, lay the veneer on it, adhesive side down—against the paper. Then align the veneer carefully with the surface under the slip sheet. You can flex the edges of the paper to see what you're doing, and assure alignment.

For the final step, it's handy to have a helper. Let the helper hold the veneer in position while you pull the slip sheet out about half an inch or so in one direction, so about half an inch along one edge of the veneer comes into contact with the wood surface to which it is to be applied—adhesive to adhesive. Starting from the center, and working outward, press this edge of the veneer in firm contact with the adhesive coated wood to hold the sheet in alignment. Be *sure* alignment is correct before you do this, as you can't adjust or separate the joined parts after they're in contact. They're bonded together for good.

Gradually slide the slip sheet all the way out, smoothing the veneer into firm contact as you go, working from the center outward to avoid trapping bubbles or ripples. You can use your hands for this, or better, a veneer roller. That completes the job except for trimming. The bond is final. No waiting for glue to dry. Apply the finish as with any other veneering method.

If you try this contact adhesive method on a small job first you can avoid errors when you tackle a larger one. But use only an adhesive made for veneer. Once you get the knack this is probably the easiest veneering method.

Fig. 11. *The veneering hammer in the hide glue method is run over the surface in the general direction of the grain but at 45 degrees away from it. When quartering, work from the center toward the edges.*

Edge and back treatment

When two surfaces at right angles to each other—such as the edge and face of a table top—are being veneered, one piece of veneer must overlap the other. This joint will later be sanded smooth, so it will not be very obvious. But you should make the surface that is seen most overlap the surface that is seen least—for example, the veneer on the top of a table should overlap that on the edges. In this case veneer the edges first, allowing the veneer to overlap both sides of the edges.

Leave the job to dry for the glue's required hardening time and then trim the veneer to size (see Fig. 9) so that the top of the table is smooth all over. Then veneer the top of the table and leave it to dry thoroughly.

An attractive method of finishing the edges of a table is to fix strips of solid wood matching the veneer to the edges with glue and the very thin brads which are called *veneer pins.* This method gives a more robust result, since the edges of a veneered table are most easily damaged.

If veneer is applied to one side only of a board, the veneer will create considerable stresses on that side, causing the wood to *bow* or warp. To prevent the board from doing this, cover the other side with a veneer of equal thickness. To keep the cost low, a cheaper veneer can be used for the backing.

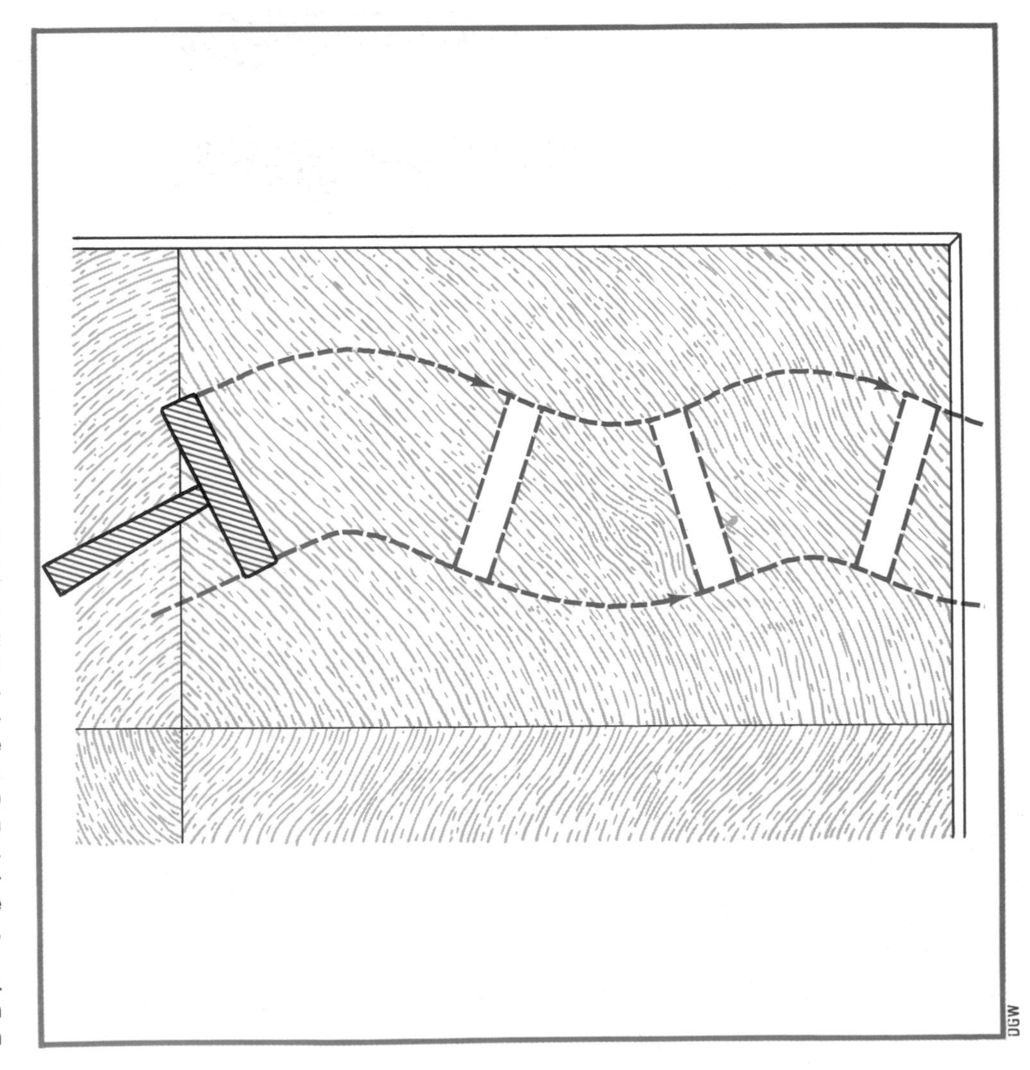

CHAPTER 37

Versatile gateleg table

A gateleg table is an especially useful item of furniture in modern homes where rooms may have several uses, or where the dining area is on the small side. When the end leaves are folded down, the table takes up a minimum of space and can be stored against a wall. But with one or both leaves opened up there is ample space for as many as eight people to sit at it.

The gateleg table shown here is easy to build and has been designed to meet as many different seating requirements as possible. Since the top can be altered to three different sizes, any number of people from two to eight can find it exactly the right size for them. The center section is wider than on many gateleg tables; this allows the legs on either side of the table to be spaced far enough apart for you to sit with your feet between them. An extra bonus is that the single crosspiece between the legs in the center of the table (part H in Fig. 1) is designed not to get in the way of your feet.

Design and materials

The design of this table is most attractive, but at the same time thoroughly unusual; it is intended to be made out of plywood throughout, including the frame. This apparently strange choice of material turns out to be very practical for two reasons. One is that plywood is far less likely to warp than solid lumber and so, because there is no need for rigid framing under the leaves, the construction is much simplified. The other is that the whole of the table is designed so that it can be cut from one standard-size sheet of ply. There is no attempt to hide the striped edges of the ply in this design; instead, a feature has been made out of these edges, so that they contrast with the plainness of the flat surfaces.

If you have no power bench saw, however, and want to avoid sawing all the frame pieces to their finished width by hand, you can make the legs and framework from solid lumber. This alternative will also save you trouble if you are going to paint the table or apply laminate or veneer sheets with edging strips, so that the striped effect is lost.

In the table shown here, the footrest is made out of round dowel to suit the contours of the feet. If you prefer, however, you can make this piece, too, out of two thicknesses of ply cut 1½ in. wide. There will be enough ply to spare in

Below. *This versatile gateleg table makes a useful-sized breakfast table when one leaf is extended. Made from birch ply, it has been covered with a high gloss oil base to give a bright finish.*

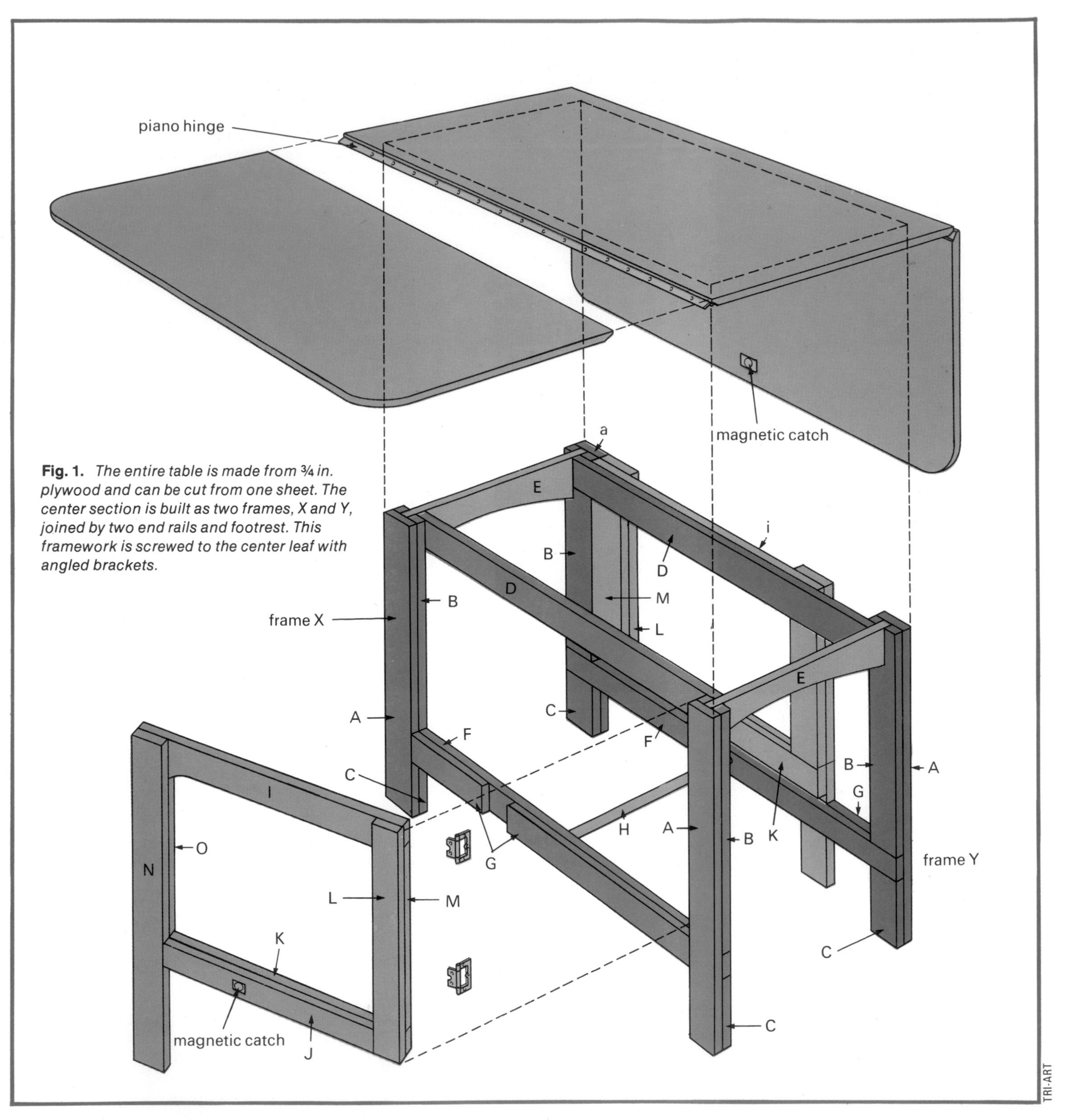

Fig. 1. *The entire table is made from ¾ in. plywood and can be cut from one sheet. The center section is built as two frames, X and Y, joined by two end rails and footrest. This framework is screwed to the center leaf with angled brackets.*

the large sheet for the two extra parts.

The joint between the end leaves and the center piece always presents a problem in gateleg tables. Sometimes the edges that meet are left square, but this results in an unsightly right-angled groove between them when the leaves are down. In this design, the hinged edges are cut at an angle of 45° (see Fig.1) and screwed together with lengths of piano hinge. This produces neat external corners when the leaves are down, and there is no groove visible. When the flap leaves are extended, there is a groove on the underside between the center piece and the two leaves, but as this is below eye level it is hardly noticeable.

The leaves and center section are attached to the piano hinges so that the folding edge of the hinges is about level with the upper side of the table top. As a result, when the leaves are dropped, they will be supported by their outer edges, and so will hang slightly out of the vertical. Again, this is hardly noticeable, but you could correct the slight error by fitting magnetic catches to the inside edges of the leaves and the crosspieces of the gatelegs (see Fig. 1). Small circular magnetic catches can be dropped directly into drilled holes, and hardly show at all. Or use a small standard type.

The end leaves in this design have been left square, with only the corners rounded off. But with skill and patience it would be possible to mark and cut them to an oval shape—make sure, however, that this oval is at no point smaller than the arc through which the leg pivots when it is opened.

Details of construction

The top of the table consists simply of two

leaves hung from the center piece. The framework beneath consists of four legs, two hinged gates and a number of crosspieces, all of which are joined with lap joints, most of them just glued together, though a few need screws as well. The framework is assembled as one piece and then screwed to the top with metal angle brackets.

All the legs and crosspieces are made out of two thicknesses of ¾ in. ply stuck together. (A useful hint: knock a few headless panel pins almost the whole way into one of the surfaces to be glued together so that their tops bite into the other piece and prevent the two pieces from slipping around when they are clamped together.)

All the pieces of ply could be glued together before the lap joints are marked out, and the joints then marked and cut in the conventional way. The method described here, however, is simpler, since almost no joints have to be chiseled out. The lengths of ply are cut out as in Fig. 3 and glued together as in Fig. 2 below, so that the correct shapes for the lap joints are preformed and no cutting has to be done. The only exception to this is with the crosspiece on the main frame, which is made out of pieces F and G. Here, it is advisable to cut out the two pieces G in one length and glue them to piece F. Then, when the whole frame has been put together, the joint to take the gateleg can be cut by direct measurement to ensure accuracy.

Choice of lumber

All the parts for this table can be cut from one standard-sized 4 ft. x 8 ft. sheet of ¾ in. plywood by arranging them as shown in Fig. 3. You will find that buying this one large sheet is cheaper than buying a number of smaller, different pieces. Your nearest lumberyard stocks this common standard size.

Plywood can be bought in various woods and grades: the grade denotes the number of flaws in the surface. Birch ply is good for outer veneer; it has a plain, light-colored surface that will take a good polish. But there is no reason why you should not use inexpensive fir plywood. The "appearance" grades for the plywood you are likely to use are as follows.

You may use either interior or exterior plywood for your table. The exterior type is highly resistant to moisture—important if the table will be used out of doors or in a very damp place.

Grade A-A has high quality veneer on both surfaces, so is suited to uses where both surfaces will be visible.

Grade A-B has high quality veneer on one surface, slightly lower quality on the other, for uses where only one surface will show. Both are smooth and solid.

Grade A-D is similar to the one above, except that a lower quality veneer is used on the back. (The quality of the surfaces decreases with the letters from A onward.)

Grade B-B has two smooth sides suitable for painting but not usually used for a clear or varnish finish. If you are using plywood with face veneers other than fir (birch, for example), your supplier will usually stock grades suitable for a clear finish. There are numerous other plywood grades, but those listed above will cover most requirements for this table.

Plywood should be stored flat and not on edge; otherwise it will twist and the edge may be damaged. Check that the sheet you buy is flat and that the edges are sound.

Materials and tools

The equipment needed to make this table can be very simple. The minimum requirements are white glue, a number of C clamps, some finishing nails, various grades of sandpaper, some wooden wedges as described below, a back saw and a rip saw. It will simplify the assembly if you also have three furniture clamps which will open to the width of the frame (that is across the narrow side) plus about 6 in.—this adds up to about 22 in. But you can do without them in a pinch by holding the frame together with tightly wound string and wedges as described in CHAPTER 38.

A power bench saw for cutting the leaves to shape and beveling their edges at 45° will save you a great deal of time and effort, and it is strongly recommended that you use one.

Lengths of piano hinge should be used for joining the end leaves to the center piece, since they support the leaves along the whole of their length and don't have to be recessed into the wood. The hinge for this table should be about 1 in. wide when opened out flat.

Each gateleg should be hung from two flush hinges (see Fig. 2); these, too, can be screwed in place without cutting them in. If you plan to have a circular-section footrest between the legs, you will need a length of 1⅛ in. closet pole.

Marking and cutting

Choose the best side of the sheet of plywood for the top surface of the table. Find the best two adjacent edges, ensure that they are at right angles, and measure from these to mark out the first piece, as shown in Fig. 3. Alternatively, use a large steel square to mark out two straight lines at right angles to each other, and measure and mark from these. Then cut along the marked lines by one of the following methods.

A *power bench saw* is the simplest means of cutting the 3 ft. sides on the leaves of the table. It should be fitted with a fine-toothed "plywood" blade, which gives an accurate straight edge that can easily be finished with sandpaper. The 45° edge needed where the leaves meet can also be cut accurately in this way.

Using a *hand saw* takes more time, but in the absence of a bench saw or sabre saw it is the only alternative. When you have marked out the first piece, rest the sheet on edge on the floor in front of your work-bench and grip it in the vise with the cutting line vertical. Use a crosscut saw to cut down the waste side of the

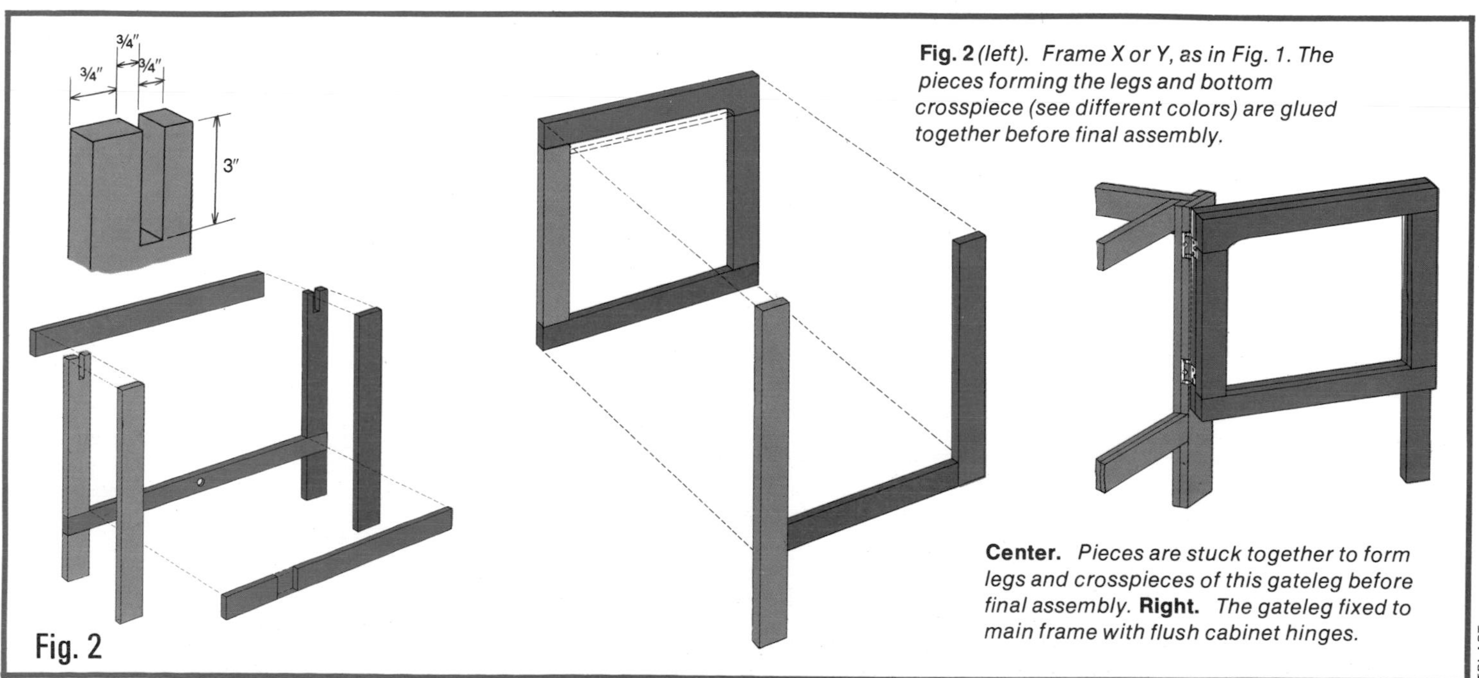

Fig. 2 *(left). Frame X or Y, as in Fig. 1. The pieces forming the legs and bottom crosspiece (see different colors) are glued together before final assembly.*

Center. *Pieces are stuck together to form legs and crosspieces of this gateleg before final assembly.* **Right.** *The gateleg fixed to main frame with flush cabinet hinges.*

line, keeping a straight line on both sides of the wood. The curved edge of pieces E and I can be cut with a keyhole or pad saw, or you can modify the design and cut a straight line instead.

To obtain the 45° beveled edges for the leaves, first cut them square and then plane them to size using a bevel gauge with an angle of 45° set as a guide.

The three main pieces of the table top should all be arranged so that the grain on their surface runs the same way. But only mark out one piece at a time, using a pencil, and then saw along the waste side of the pencil line. If you achieve a perfectly straight edge, you can use that as the edge of the next leaf; otherwise you should plane it straight first. If you marked all the pieces out at once, the thickness of the saw cuts would reduce the size of each piece when cut out. Cut out all the pieces in this way, then finally saw the notches in the top of the four corner legs as shown in Fig. 2. Use a back saw and coping saw.

The footrest stretches between the two horizontal crosspieces that lie across the width of the table and brace each pair of legs. Each of these crosspieces is made from the thicknesses of ply, and the footrest should be set into a hole cut in the inner thickness only. Whether this takes the form of a square or a 1⅛ in. diameter circle depends on whether you are using a square or circular-section footrest. Anyway, cut the hole at this stage, using a chisel or hole saw as appropriate.

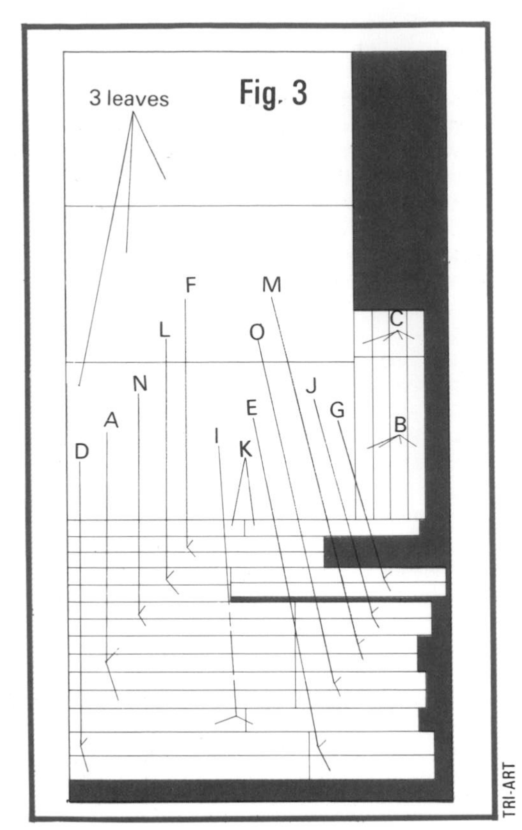

Fig. 3. *The pieces for the table (see cutting list alongside) can be cut from one sheet of plywood size 4 ft. x 8 ft.*

CUTTING LIST

All pieces are made of ¾ in. or plywood, except where stated.

	inches
Main frame	
legs:	
4 pieces (A)	28¼ x 2¼
4 pieces (B)	18¼ x 2¼
4 pieces (C)	6 x 2¼
top struts:	
2 pieces (D)	29 x 2¼
2 pieces (E)	16½ x 3
bottom struts:	
2 pieces (F)	32 x 2
2 pieces (G)	27½ x 2
Gatelegs	
top struts:	
2 pieces (I)	22 x 3
bottom struts:	
2 pieces (J)	17½ x 2
2 pieces (K)	22 x 2
inner legs:	
2 pieces (L)	20¼ x 2¼
2 pieces (M)	15¼ x 2¼
outer legs:	
2 pieces (N)	28¼ x 2¼
2 pieces (O)	14¼ x 2¼
Top	
3 pieces	20 x 36

Foot rest (H)

1 piece of 1⅛ in. closet pole, 16⅜ in. long.

"Doubling up"

There are only six pieces in the finished framework which are made of a single thickness of ply: the top crosspiece in each of the gates (I) and the four pieces at the top of the frame which are screwed to the table top (parts D and E in Fig. 1). Take these six pieces and keep them completely separate from the rest.

The remaining pieces of ply have to be "doubled up" by gluing two thicknesses together to give them extra strength. To do this, first match the two halves of a piece together, then take one of the halves and hammer three or four small finishing nails halfway into the inner face. Cut their heads off fairly close to the wood with a wire cutter. Provide yourself with a clean paint brush and spread white glue over the inside face of both halves. Press the pieces together. (If necessary, place a piece of wood on top of them and hammer it to make the headless finishing nails sink into the other piece. They will hold the pieces firmly in place while the clamps are set.)

Wipe off any surplus adhesive immediately with the rag. Any adhesive left on the surface would only become noticeable if you varnish the surface, but by then it will be too late to do anything about it. Sandwich the pieces between a couple of bits of scrap wood and hold them together with C clamps; wipe them again with the cloth and leave them overnight to dry. Repeat this with the other pieces; the more C clamps you have, the more you can do at once.

When the adhesive has set, clean up all the edges. If you have cut the pieces accurately in the first place, you can probably smooth them down with sandpaper wrapped around a wood

NIGEL MESSETT

Left. *When both leaves of the gateleg table are extended it becomes a good-sized dining table—and there is plenty of space to sit with your feet between the center legs.*

block. But if they are far out, you will need to plane them. Try to remove as little wood as possible—plywood is such unpleasant stuff to plane that you will need no encouragement to do this.

Assembling the framework

The rectangular framework under the center of the table consists of two flat frames (marked X and Y in Fig. 1) joined by the footrest and the two top crosspieces.

The two frames X and Y consist of two legs and a bottom crosspiece (each made from two thicknesses of plywood) and a top crosspiece (one of the single pieces of plywood previously set aside).

Trial assemble the four pieces in each frame flat on the floor with the joints in the legs facing upward; check that the top crosspiece does not cover the housing in the top of the leg. This top crosspiece should be screwed to the two legs, to hold the pieces firmly in position. Drill screw holes through the crosspiece and part of the way into the leg.

Now apply adhesive to all the joints, assemble the frame and clamp it until the adhesive is dry. Remember to wipe off any excess adhesive at once and check that everything is square. Repeat this with the other frame.

Once both the frames are assembled and dry, take them, the two top end rails (E) and the footrest (H) and make a trial assembly with the frame upside down, so that the four top members are resting on a flat surface to hold them steady.

Dismantle the pieces, apply adhesive to all the joints, and reassemble them clamping the joints with furniture clamps, or alternatively use some suitable improvised clamps. Check that the whole frame is square, using a piece of string to check that the two diagonals in every rectangle in the frame are equal in length.

The gatelegs

Take the four parts which make up each gateleg, that is, the two legs and the top and bottom crosspieces. Glue the joints and assemble the frame in the way previously described for frames X and Y. The legs of each frame should be laid on a flat surface with the cutouts for the lap joints uppermost.

When the adhesive has set, screw the gatelegs to the main frame as shown in Fig. 2. Make sure that the top and bottom of each gate is level with the top and bottom of the frame. Don't try to line up the horizontal crosspieces so that they meet—they have been designed to have clearance to prevent them from sticking. Just screw the hinges directly in place; there is no need to cut them into the wood.

Now mark the position where the longer upright leg of the gate crosses the bottom crosspiece of the main frame on each side, and cut halfway through the crosspiece with a back saw to form the sides of a recess for it. It should then be fairly easy to form the recess by separating the two thicknesses of plywood with a chisel inserted between them. Use the chisel, held bevel side up, to clean out any debris in the recess.

Final assembly

Place the center piece of the table top and the two end leaves side by side on a flat surface with the angled edges towards each other. Half open the piano hinges and lay them in the two V-shaped grooves. The hinges should go as far into the V groove as possible while the edges of the leaves continue to meet. Mark the position of the screw holes in the hinges on to the edges of the leaves. Start the holes with an awl, then screw the hinges in place. The screws must not be too long, and must be inserted at the correct 45° angle, or they may break through the table top.

Take the main frame and place it upside down on the center piece. Measure carefully around it to make sure it is set straight and exactly the same distance from the sides and ends. Then fix it in place with screws and angle brackets, again being careful that the screws don't break through the surface of the table top.

Round off the bottom edges of the legs slightly with sandpaper or a file to avoid any danger of them splitting when the table is dragged along the floor. Screw a small block of wood to the underside of each leaf in the right position to act as a stop to prevent the gateleg from opening too far. Hold the end leaves and carefully turn the table on to its feet, but be careful not to strain the piano hinges in the process.

Finish the table with varnish or enamel or veneer or plastic laminate. The edges of plywood are notoriously hard to finish neatly so if you are varnishing or painting the surface, be particularly careful to sand the edges of the ply very smooth. The flat surfaces should also be sanded with fine sandpaper to give a silken-smooth finish.

Below. *With both gatelegs closed and the leaves down this gateleg table becomes an extremely stable and useful-sized occasional table.*

NIGEL MESSETT

CHAPTER 38

Elegant tea trolley

This richly elegant tea trolley in oiled teak will be an attractive addition to your dining room. With its trays covered in hard-wearing plastic laminate it is also extremely functional. The real beauty of the trolley, though, is that it is very easy to make with its few basic joints.

A tea trolley is easy to build and all the joining involved is straightforward. The trolley in the photograph is made from teak, but you may wish to use a different wood to match the rest of the furniture in your kitchen or dining room. Mahogany is also very attractive, and more readily available at lumberyards. It is also easily worked.

The trays of the trolley are made of plywood and covered in plastic laminate. The color of the laminate used is a matter of taste, but the very dark slate green on the trolley in the photograph blends attractively with the color of the teak. The trays are edged with teak.

Preparing the trays

The first step is to cut the plywood for the trays. These are 32 in. x 18 in. x ¾ in. Clamp the two trays together in a vise and cut the sides and ends exactly square and to size. Check that the trays are square by measuring the diagonals.

The teak side members of the trays are joined to the long sides of the plywood sheets by means of a ½ in. x ¼ in. plywood fillet. This is glued into grooves cut in the edges of the plywood and the inside face of the side members.

Cut the grooves in the long sides of the plywood with a combination plane or table saw. The groove runs along the center of the long edges and is ¼ in. wide and ¼ in. deep.

Applying the laminate

The trays of the trolley are covered in plastic laminate. The underside of the trays is also covered, since if you covered the top only the trays would be likely to warp and twist. The pieces of laminate on the bottoms of the trays are called "balancers" in the cutting list. A good quality laminate should be used for the top "working" surface of the trays; a cheaper quality can be used for the balancers.

Cut the four pieces of laminate to size. The balancers only are glued in place at this stage. Spread contact adhesive over the surfaces to be glued and let it go tacky. Be *absolutely sure* the laminate edges are aligned with the plywood edges before you bring them together. Then put the pieces together and smooth any "bubbles" off toward the edges. No clamping is required.

With the laminate in place, carefully plane the edges of the balancers flush with the edges of the plywood. Do this with the two trays held together in a vise. Try to avoid removing any plywood from the edges. If you do accidentally remove any, make sure that you have not put the edges out of square, and check the groove to make sure that it is still ¼ in. deep.

Buying the wood

Half a panel of plywood (4 ft. x 4 ft.) is all you need for the trays of the tea trolley. In some lumberyards you may find "handy panels" that are 2 ft. x 4 ft. A pair of these will do. Either way, just cut them to the size shown in the drawings.

The solid wood for the legs and side pieces and for the crosspieces that take the casters, is of nominal 1 x 2 material, which has the actual dimensions shown (¾ x 1½). The strips of ¼ in. plywood for the fillets can be cut from a handy panel of plywood of that thickness if no scrap material is available. The ½ in. material for the end strips on the trays can be cut from ½ in. thick lumber, which is stocked by most lumberyards in widths from 2 in. up.

Left. *Good looking and efficient, this tea trolley will add interest to your dining room and simplify the job of serving meals.*

The side members

The strips of lumber that run along the long side of the trays have a finished size of 32 in. x 1½ in. x ¾ in. They are 1 in. longer than the plywood panels to allow for strips of edging ½ in. thick to be inserted between their ends to cover the short edges of the tray.

The side pieces have a groove cut in their inside face. This holds part of the plywood fillet that is used to join the plywood trays to the side members. The ¼ in. x ¼ in. groove begins ⅜ in. up from the bottom edges of the side pieces. Cut it with a combination plane, or table saw, its fence set to ⅜ in.

The next step is to glue edging strip to the short sides of the trays. Use pieces of 1 in. x ½ in. lumber for this—they can be cleaned off flush with the shelf top and with the bottoms of the side pieces later. The ends of the edging strip should be cut slightly over length to protrude ⅛ in. beyond the plywood at each end. Glue the strips in place, clamp them and allow the glue to dry.

You can now apply the top sheet of laminate. Glue it in place over both the plywood and the teak end pieces, and trim its edges off flush with the sides and ends of the tray. Then apply a thin coat of wax polish to the laminate. This prevents excess glue from sticking to the surface of the trays later on.

Now cut the overlap off the edging strip so that its ends are flush with the grooved sides of the tray. Then continue the groove cut in the plywood through the ends of the edging strip. Do this carefully with a saw and a chisel, not with a plane, which might rip off the edging strip.

Joining the sides to the trays

A ½ in. x ¼ in. plywood fillet is used to join the side pieces and trays together. The fillet is glued into the groove cut in the plywood tray, glue is applied to the groove in the side piece and the components are pressed together.

Before you do this, clean up the inside edges of the side pieces and sand them smooth. These surfaces will not be so accessible when the side pieces are fixed.

Cut the plywood fillet 1½ in. shorter than the length of the trays. Spread adhesive on the fillet and in the grooves. Push the fillet into place so that its ends are ¾ in. from each end of the tray. Then push the side pieces and the tray together. Clamp the assembly and place it on a level surface while the adhesive dries.

The next step is to cut 1 in. x ½ in. x ¼ in. plugs of teak, or whatever wood you are using. These fill the small gap left at the ends of the side pieces by the groove, and also add strength to the edging. Taper the ends of the plugs slightly to the edging. Taper the ends of the plugs slightly and chamfer one corner a little, as you would do for a dowel. Then spread adhesive on the plugs and tap them into the small gaps with a mallet. When the glue has dried, carefully saw off and plane down any parts of the plugs that protrude, and sand the assembly smooth.

The trolley legs

The finished dimensions of the trolley legs are given in the cutting list. The sides of the trays and the legs are joined together by means of lap joints. The position of these joints is shown in Fig. 11. The legs at the back of the trolley are joined to a handle that runs above the surface of the top tray, so they protrude above the top tray. The front pair of legs end flush with the top of the tray sides. A cross rail is joined to the bottoms of each pair of legs.

Cutouts for the lap joints must be made in the sides of the trays and the legs. The ones in the sides are 5/16 in. deep and 1½ in. wide, and start 3½ in. from the ends of the side pieces. Those cut in the legs are ¼ in. deep and 1½ in. high. Mark out the cutouts in the two shelves while they are held together in a vise. They should be "marked tight," to allow for any reduction in the width of the legs caused by cleaning them up. Mark and cut according to the techniques described in CHAPTER 5.

To avoid wobbles when the trolley is finally assembled, process all four legs together through as many stages as possible. Start by cutting them to exact length. Lay them together in a vise, with ends squared. Their finished length should be 28 in. for the back legs and 24½ in. for the front legs. All measurements for the positions of the joints must be made from the squared ends of the legs. Fig. 11 shows the positions of the joints on the back and front legs. The bottom of the lap joint on each leg is 6 in. from the bottom of the leg. The top lap joint in the rear leg corresponds to that of the front leg. Make the cutouts with a back saw and chisel.

The handle

The legs at the back of the trolley protrude above the top shelf and are joined at their tops to the handle. This runs between the two legs above the surface of the top tray. The joint used is a *box* joint (a near relative of the dovetail, but easier to make) which provides a greater gluing area than most other joints and adds an attractive touch to the finished trolley.

The box joint is shown in Fig. 7. The handle end of the joint consists of two tongues; these fit into two slots cut in the top of the leg so as to leave three tongues of wood protruding. Thus, the whole joint consists of five tongues; these should be of about equal width for the sake of appearance. In this case, the tongues on the handle are 5/16 in. wide and ¾ in. deep. The tongues on the legs are slightly thinner—9/32 in. approximately—and the same depth.

Cut the handle just over length. Square lines near the ends to indicate the finished length. From these squared lines, measure back toward the center of the rail a distance of ¾ in. Square lines through these points.

Use a marking gauge to mark out the box joint, as in Fig. 8, working from these squared lines toward the ends of the timber. Carefully saw down the lines made with the marking

1

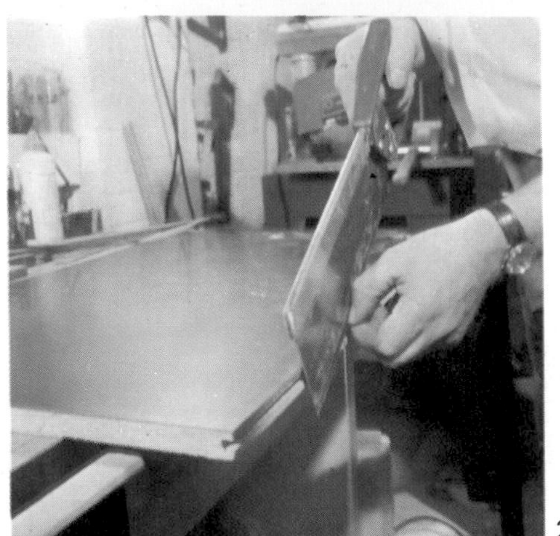

2

3

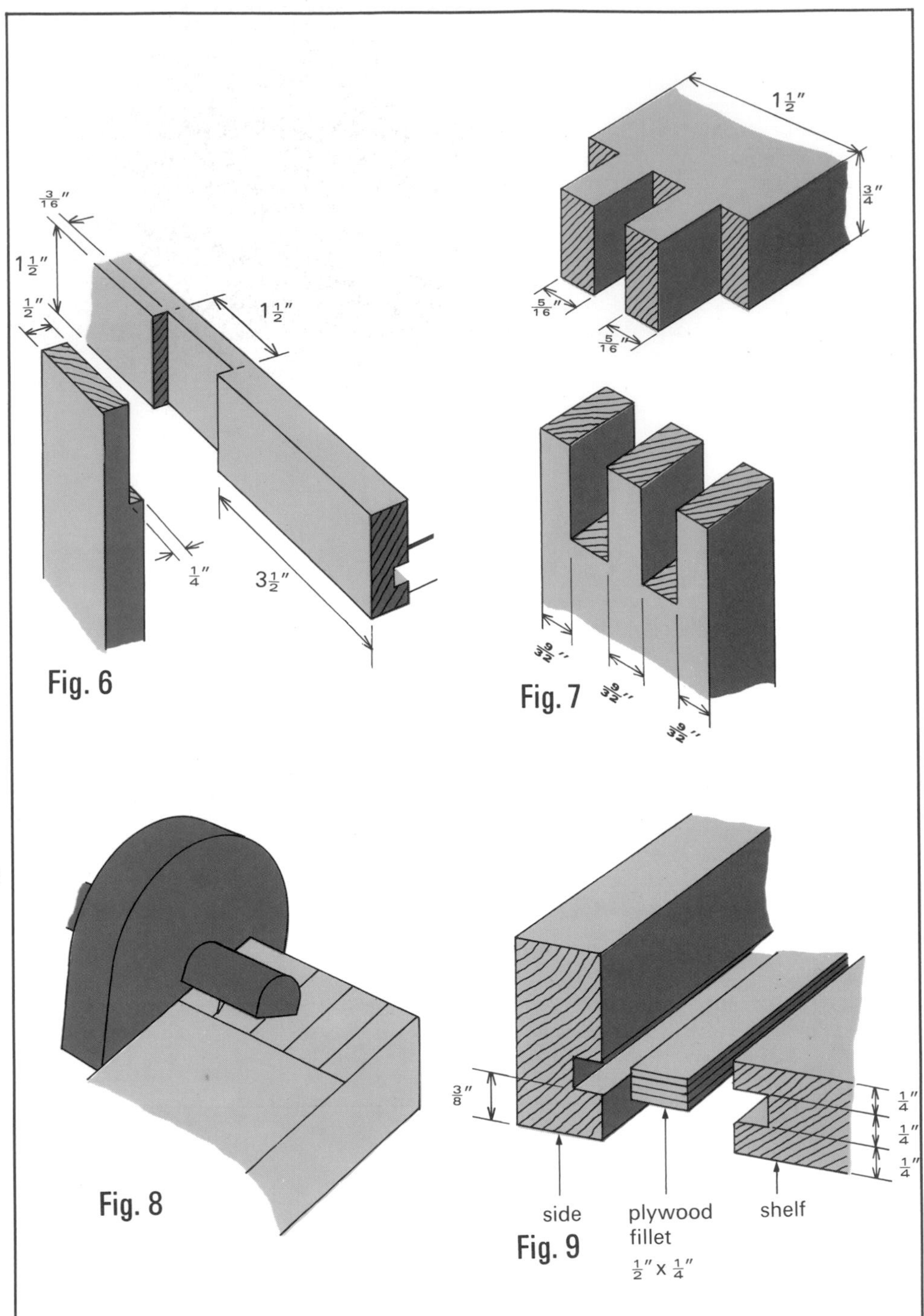

4

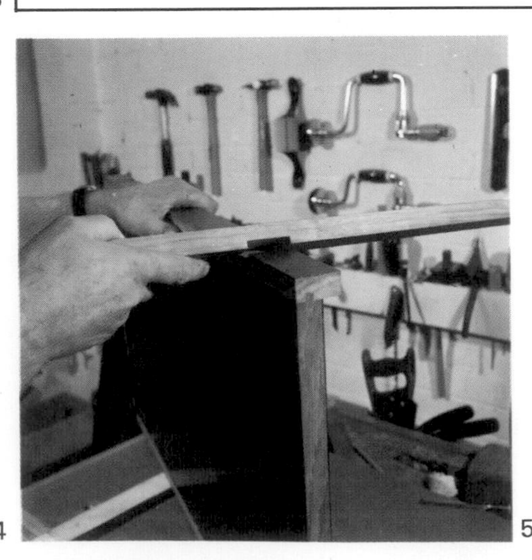

5

Fig. 1. *Applying the laminate to the tray.*
Fig. 2. *Trimming the edges of the laminate.*
Fig. 3. *Cutting the groove in the tray side.*
Fig. 4. *Fitting the side piece to the tray with a plywood fillet which is glued in place.*
Fig. 5. *Fitting one of the back legs into the lap joint cut in the side piece.*
Fig. 6. *The dimensions of the lap joint that is used to join the side pieces of the trays to the front legs.*
Fig. 7. *The box joint between the trolley handle and the back legs.*
Fig. 8. *The box joint is marked out with a marking gauge. The wood is divided into five more or less equal widths.*
Fig. 9. *The dimensions of the groove cut in the trays and sides, and of the plywood fillet.*
Fig. 10. *An exploded view of the tea trolley.*

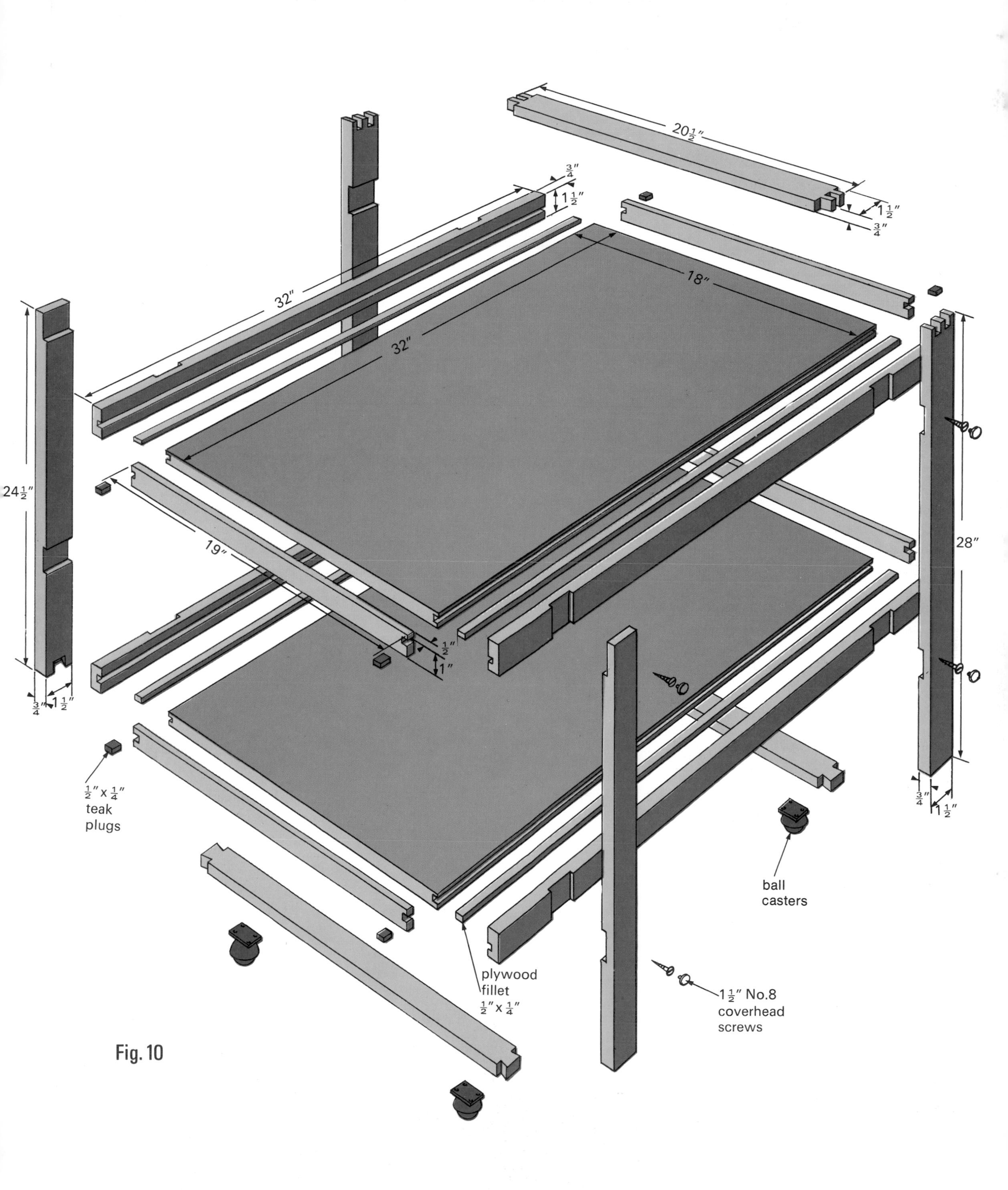
20½″
¾″
1½″
1½″
¾″
32″
18″
32″
24½″
19″
½″
1″
¾″
1½″
28″
¾″
1½″
½″ x ¼″
teak
plugs
ball
casters
plywood
fillet
½″ x ¼″
1½″ No.8
coverhead
screws

Fig. 10

gauge, using a back saw. Then use a coping saw to cut out the piece between the first saw lines. Instead of measuring twice, use the sides of the joint on the handle to direct-mark where the legs are to be cut. This is more accurate.

The cross rails

Both pairs of legs have cross rails connecting their bottom ends. The cutting list gives the dimensions of these rails. The legs and the cross rail are joined with a single dovetail joint, as shown in Fig. 12.

First cut the rails exactly to size. Square a line around the rail 9/16 in. from each end. From each side of the rail measure inward along this squared line a distance of 3/8 in. Mark these points. Set a bevel gauge (this tool is similar to a try square, but has one adjustable arm) to 10° and from the marked points draw two lines running outward toward the corners of the wood. Cut this joint to shape with a back saw.

Use the part of the joint cut on the cross rail as a template to mark out the leg. Cut this part of the joint with a sharp chisel.

Assembling the trolley

Assemble the trolley only after you have cleaned up all the inside edges of the trolley members. This should be done now because the pieces will not be so accessible when the trolley is assembled.

Apply adhesive to the lap joint cutouts on the insides of the legs. Push the legs and trays together. Clamp this assembly; this is best done with furniture clamps, but if you do not have any, bind the assembly with loops of strong twine. Force wooden wedges under the twine where it runs over the joints. Place paper under the wedges so they won't stick to the work.

Apply glue to the joints on the handle, cross rails and legs and push the pieces together. Check that the whole assembly is square by measuring all the diagonals; opposite diagonals should always be equal, or your assembly is out of square. If it is, clamp the entire assembly with furniture clamps—you will need at least six—or with the string and wedge method, after squaring.

Finishing the trolley

Clean up where necessary the outer surfaces of the trolley members by sanding. Sand off those parts of the box joint that protrude. Sand any rough parts of the trolley smooth.

Now, at the points where the legs are joined to the tray sides, drill holes for 1½ in. No. 8 steel screws with decorative heads. If coverhead screws are not available use oval heads. Fig. 10 shows this screwed joint.

Clean off the wax polish on the laminate surfaces of the trays with turpentine or a substitute.

Follow the manufacturer's instructions on whatever type of finish you use.

Fit casters to the lower crosspieces that join the legs, following the fitting instructions for the type you have. Ball casters are the most efficient and elegant, but you can use the cheaper wheel casters if you prefer.

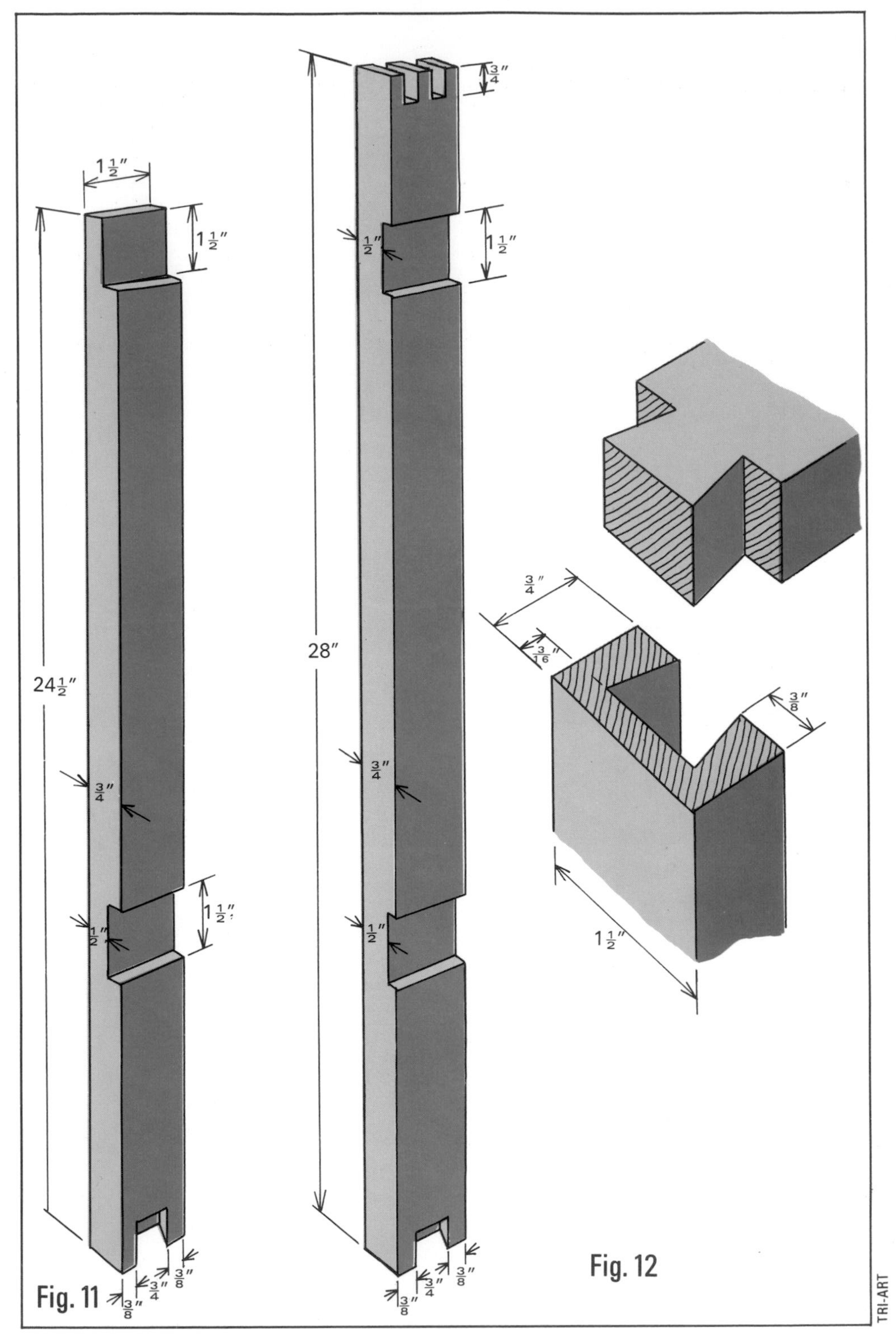

Fig. 11. *The front and back legs of the trolley showing the overall dimensions and the size and positions of the joints.*
Fig. 12. *The bottom cross rails are joined to the front and back legs of the trolley with a single dovetail joint.*

CUTTING LIST

(finished sizes)	inches	
4 side pieces	32 x 1½ x ¾	(nominal 1 x 2)
2 legs	28 x 1½ x ¾	(nominal 1 x 2)
2 legs	24½ x 1½ x ¾	(nominal 1 x 2)
2 bottom crossrails	20⅛ x 1½ x ¾	(nominal 1 x 2
handle	20½ x 1½ x ¾	(nominal 1 x 2)
4 edging strips	19 x 1 x ½	
8 plugs	1 x ½ x ¼	
Plywood 2 trays	32 x 18 x ¾	
Plywood 4 fillets	30½ x ½ x ¼	
Plastic laminate (sizes allow for trimming)		
Tray tops	32¼ x 18¼	
Balancers	32¼ x 18¼	

Also needed
8 decorative head screws 1½ in. No. 8, 4 castors, ½ pint contact adhesive.

The mansard roof—of many lands

It's sometimes called a gambrel roof in the United States, but it's basically the form known as mansard in Europe, Britain, and many other parts of the world. Not only is it an attractive style, but a very practical one that adds greatly to the space that might otherwise be a cramped attic. The one shown below was photographed in Britain. In keeping, the drawings, lumber sizes, and terminology that follow are also from Britain. As the mansard varies widely in the pitch of its two slopes, you can choose a version that best meets your particular building needs. Simply adapt its structure to the requirements of your local building code.

A mansard roof provides everything at once. Because of its height, for example, it makes the house look better. But the real trick is that it is higher even than it looks. The amount of space beneath a mansard roof is really surprising. You will be able to fit virtually another floor of spacious rooms within the roof structure.

Each side of a mansard roof has two slopes (see Fig. 2); first a steep slope up from the eaves, and then a more gradual slope to the top or ridge of the roof. The lower, steeper slope provides a large amount of space within the roof—sufficient for as many rooms as on the floor beneath, although they would be a little smaller.

Many large older buildings have this type of roof—because it gives an extra story without affecting the skyline too much—but it is equally suitable for the smaller and more modern home. Sometimes a mansard roof is built on top of a single floor, creating the appearance of a bungalow but giving the space usually associated with a two-story house. A variation of this roof is also used in some chalet bungalows.

A mansard roof creates stresses on the joists across the base of its upper section (see Fig. 1), so it is best not to use this roof for a span, from back to front, of more than 30 ft. It is, however, suitable for most medium-sized houses.

The mansard roof really consists of an upper and lower section. The lower section is a box-like structure stretching upward from the wall plates and ceiling joists of the house itself to the upper joists (see Fig. 1). The upper section is like a gabled roof with its own joists, ridge board and so on set on top of the lower section. Read CHAPTER 34 carefully—this gives instructions on how to build a gabled roof. The same components and methods used for constructing a gabled roof are used, or adapted, in erecting a mansard roof.

A mansard roof is generally designed so that the five "points" of the eaves, the ridge and the joints between the slopes fit on a semicircle (see Fig. 2). François Mansart, who developed this style of roof, thought this gave the best proportions.

If a mansard roof is drawn to fit into a semicircle, a roof with a 20 ft. span will give headroom of approximately 5 ft. 9 in.—but this depends on the size of the timber used. As a rough guide, the headroom is usually approximately two-sevenths of the span.

Note, however, that in the Building Regulations in Britain, "wall" means any part of a roof which slopes at less than 20 degrees to the vertical. Under these rules, a wall has to be constructed of different materials from those of a roof. Therefore plan your mansard roof so that the lower slope has a pitch of not more than 70 degrees from the horizontal. In other countries, check the regulations that cover roof building before designing the roof.

You will probably want to make maximum use of the room under the roof instead of treating it just as a larger-than-usual storage space. This means building in some windows. When designing your windows, ensure that they conform to any building regulations. In Britain, the building regulations specify that each room

Below. *This mansard roof adds character to the house, and there is more floor space than in a house with an attic and a gabled roof.*

must have a window space equal in area to at least one tenth of the floor area. Regulations in other countries may differ, so again check before you start. Then draw the position and structure of the window on the plans and work out how to keep it in style with the rest of the house. A typical dormer window suitable for setting into a mansard roof is shown in Figs. 4 and 5.

Draw a plan and elevation of the roof you intend to build, and then submit it for the approval of your local authority or building inspector (CHAPTER 34). If you are planning a conventional mansard, with the five points or "corners" planned to lie on the circumference of a semicircle, the two lower rafters between the eaves and the joints between the slopes must be the same length as each other, as must be the upper two in the gable section.

Fig. 2 shows how to set about designing a roof of this type. First draw a semicircle with a compass so that the diameter represents the roof span—set the two arms to a distance representing half the roof span using a scale of 1 in. to 1 ft., or if using metric measurements a scale of 10:1 or greater. Divide the semicircle into ten equal parts—use a protractor and measure steps of 18 degrees from one side to the other. The ridge is at the center; the joints between upper and lower sections can be placed three steps around the semicircle from here.

This straightforward way of drawing a cross-section of the roof ensures that the roof will look right. But if greater headroom is required, the lower slopes can be extended, or if greater width is needed they can be reduced in length. But do not move the lower slopes so much that they slope at more than 70° to the horizontal, otherwise the "roof" will become a wall!

Components of a mansard roof

The components of a typical mansard roof with vertical ends are shown in Fig. 1, and the function and sizes of most of them are described in CHAPTER 34. These are: wall plates, ceiling joists, rafters (but only those of the upper slope), purlins, ridge board, collar ties, struts, fascias, and soffits. A mansard roof can have the slope continued around the end, as with a hipped roof, as often done in town houses. Components of the mansard roof that have not been described previously include:

Pole plate: a piece of timber (there are four in Fig. 1) running the length of the house on top of the outside long walls, and along the joint between slopes. The lower pole plate rests on the lower joists and holds the bottom end of the lower rafters. The upper pole plate rests on top of the studs and holds the bottom end of the upper rafters.

They are usually the same size as the wall plates described in CHAPTER 34, that is 4 in. x 3 in.

Stud: this piece of timber supports the lower and upper rafters where they meet, and completes a triangle between each lower rafter and ceiling joist. They are usually the same size as the pole plates, and form the framework for the internal walls.

Upper joists: these timbers run in the same direction as the lower joists in the original ceiling, see Fig. 1, from front to back of the house. Their size depends on their spacing, span and the load they carry, as with a conventional gabled roof. Guidelines for the sizes are found in CHAPTER 34.

Lower rafters: these are the same size as the upper rafters and run from the bottoms of the upper rafters to the lower pole plates.

Scaffolding

Always have adequate scaffolding (see page 145). Remember that a mansard roof is taller than a comparable gabled roof and the scaffolding needs to be that much higher. Be especially careful when working on the upper section of the roof—there are many timbers beneath that could injure you if you fell. If a mansard roof is added to the top of an existing terraced house, you may need to seek specialist advice for adequate scaffolding.

Construction

The roof shown in Fig. 1 and described below is built from individual lengths cut and individually fitted in place. The upper and lower rafters form a continuous line seen from the front of the house; the tops of the lower rafters butt against the bottoms of the upper rafters. But they could be staggered slightly. Alternatively, the upper section could be made from prefabricated gable trusses fitted in position.

Lower section

The sizes of the wall plates and lower joists and how to fit them are covered in CHAPTER 34. But remember to mark the position of the joists on the wall plates before fixing the wall plates in position. With a cavity wall, the wall plates can be placed on top of the inner or outer skin as long as they are of brick or similar construction.

The method of erecting the lower section is a further application of the method of erecting a gabled roof described in detail in CHAPTER 34. The triangles formed by the studs and the lower rafters are erected on the lower pole plates along both the long walls and propped in place while the upper pole plates are fitted on top.

First, cut the timber for the lower pole plate to length. Do not mark the positions of the lower rafters on the pole plates as you can line them up later with the lower joists (the rafters lie directly over the joists and the studs to one side of the joists). Fit the pole plate in position on top of the rafters and skew-nail it in place.

To position the studs, measure from the plan the distance of the edge of a stud from the pole plate, and mark this distance from the pole plates along one lower joist at each end of the roof. Then use a straight length of timber to draw a line between these two points, cut the runner to size and nail it along this line. Again from the plan, measure the height of the studs and cut them to size.

The length of the lower rafters is measured in the same way as for the gabled roof (see

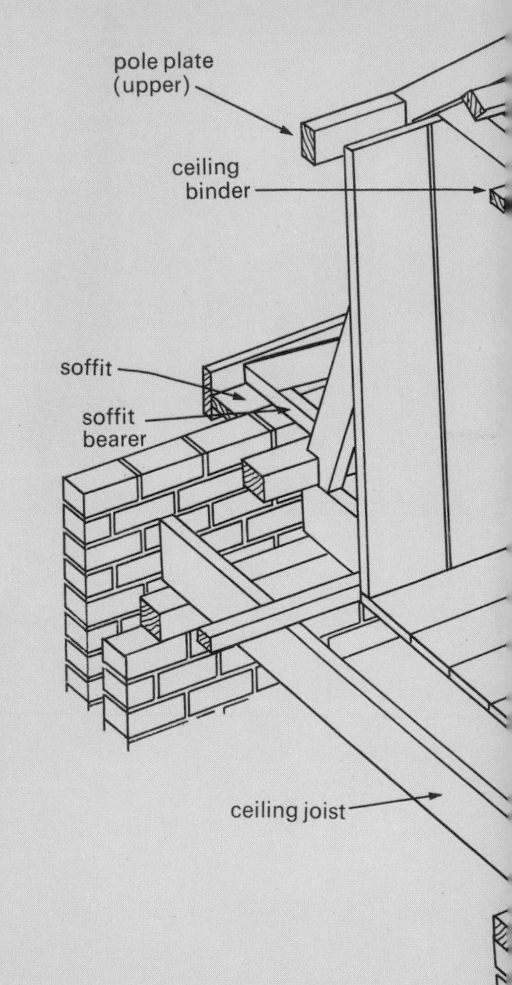

Fig. 1. *The framework for a mansard roof is shown here set on a cavity wall. The inner wall is brick and takes the weight of the ceiling joists. The roof is continued beyond the edge of the outer wall by nailing the soffit bearers and sprockets to one side of the lower rafter as shown, and gives a pleasing shallower slope to the end of the roof.*

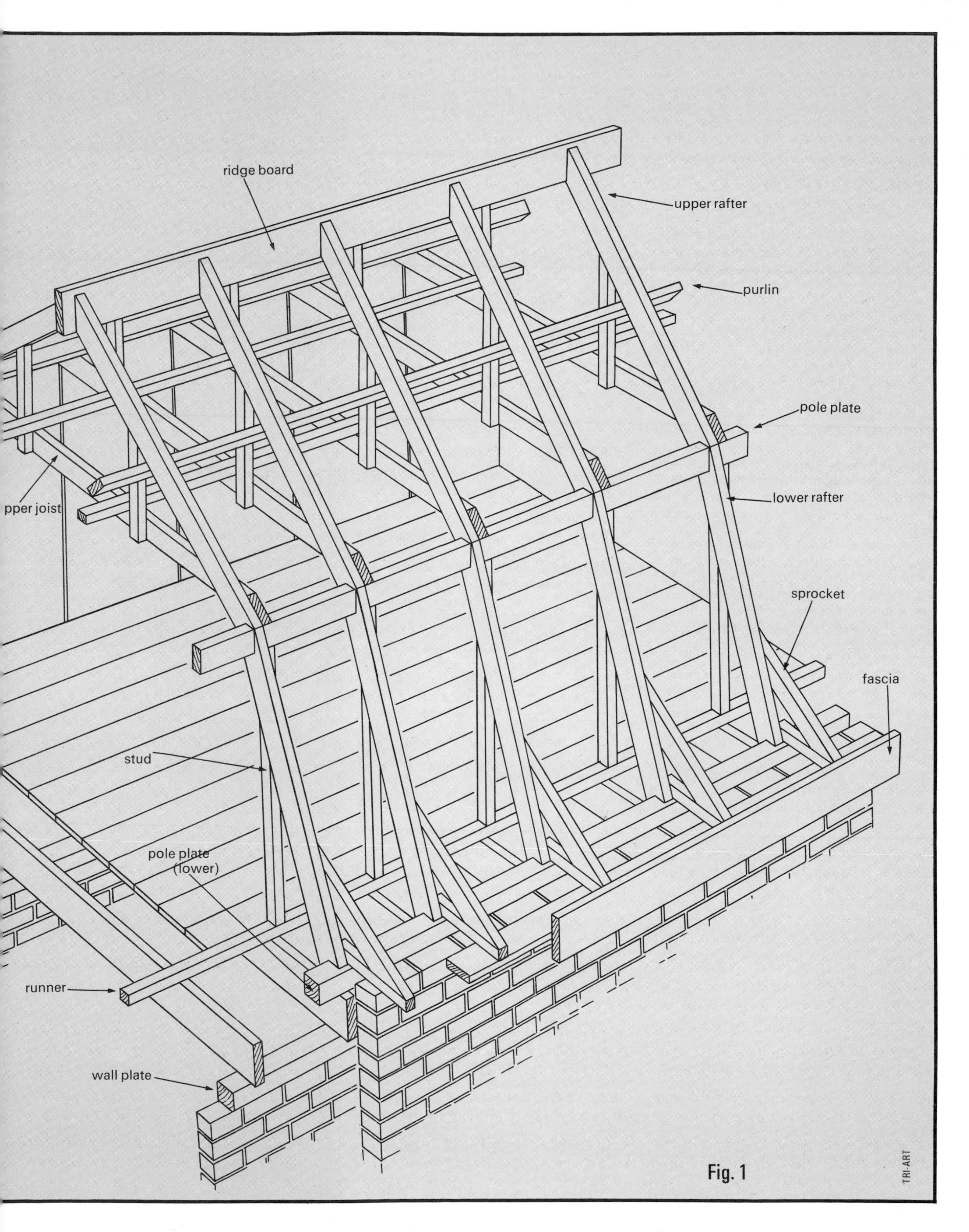
ridge board
upper rafter
purlin
pole plate
pper joist
lower rafter
sprocket
fascia
stud
pole plate
(lower)
runner
wall plate
TRI-ART
Fig. 1

CHAPTER 34). Fix a length of wood to the ground and from one end measure the distance of a stud from the wall plate. Butt a stud at right angles against the wood at this point and fix it to the ground. Place a short length of the timber used for the pole plates on top of the stud to allow for its thickness. Then take a piece of timber for the lower rafters somewhat longer than the rafter is to be —you can measure its length roughly from the plan—and place it to form the third side of the triangle. Mark and cut the rafter to shape, birdsmouthing (CHAPTER 34) both ends to fit over the pole plates. The rafter should be long enough to overlap the stud. Nail a short length of timber across the stud and rafter to make a triangle that will hold the frame in shape.

Take this frame onto the roof and check to see if it fits. Place the lower end of the rafter over the pole plate and the stud over the runner. The rafter will lie directly above the ceiling joist. Check that the stud is vertical and that the slope of the rafter is as designed. Alter your triangle if necessary. Then copy it to make the remaining triangles.

To erect these frames and the upper pole plate, simply use the same method as with the rafters on a gabled roof in CHAPTER 34. Put one frame in position at one end of the roof with the stud next to a joist, and the rafter on the pole plate and directly above the joist. Check that the frame is still upright and support it temporarily with a batten by nailing one end of the batten to the top of the stud, and the other end to a nearby joist. Rest another frame flat on the joists at the other end of the roof with its "feet" toward the end of the roof. Place the upper pole plate in position in the birdsmouth on the frame that is held upright by the battens. Then lift the other end of the pole plate, raise the second frame into position and place the pole plate in the birdsmouth on top of that frame. Check for squareness and nail both frames permanently to the joists. Then fill in with the intermediate frames.

The upper joists lie above the upper pole plate. They are generally laid on top of the pole plates and nailed in place alongside the rafter, as in Fig. 1. In an alternative arrangement, they can be butted against the lower end of the upper rafter as in Fig. 3 and held in place with a sheet of ply—approximately 12 in. to 18 in. long, 4 in. wide and ½ in. thick—glued and nailed to either side. This arrangement is found in many prefabricated trusses.

Nail one joist in position between the upper pole plates at front and back of the house. Check again that the entire framework is square, and then fill in the intermediate joists. Then build up any end walls to this level.

Upper section

The upper section of rafters, purlins and trusses is then cut and erected (see CHAPTER 34 for details). When this framework is completed, brick the ends up right to the ridge. Usually the roof extends over the top of the wall but, especially in a terraced property, the walls may be built to project above the roof line to prevent rainwater from draining on to a neighbor's property. In this case the top of the wall will require a coping to prevent water entering the brickwork. (See Conversion, below.)

Tiling

Cover the roof with roofing felt, nail battens in place on the rafters and tile the roof (as described briefly in CHAPTER 34). Start from the eaves and work toward the ridge. On the lower, steeper slope you should nail every tile or slate in place, and not every third tile as you can do on the upper slope.

After the last row of tiles is laid on the lower slope, a piece of lead or zinc should be nailed in place as a soaker or apron (see Fig. 3). This is to provide a watertight joint between the two different slopes of tiles. Nail it to the battens and rafters with galvanized nails at the top edge and mold the lower part around the lower row of tiles. Then nail the first row of tiles on the upper slope in place.

Fitting a dormer window

A dormer window is the most convenient way of providing the lighting necessary to turn the roof space into a room. The dormer window, as shown in Fig. 5, lies between the top and lower pole plates. Leave out enough of the lower rafters to leave space for the window. Check that the studs forming the inner and outer face of the window are vertical and that the "flat" roof shown here has a slope away from the main roof of at least one in 40 so that rainwater runs off into the gutter. The horizontal battening between the inner and outer studs should be halved or notched into them.

The flat roof and sides of the window should be boarded and covered with a minimum of two layers of roofing felt (see CHAPTER 34 for a brief description). Lay strips of lead or zinc in the corners where the dormer window and the roof join so that the water flows down into the gutter. The sides and top of the window can be tiled if you prefer; every tile hung on the sides of the dormer window must be nailed in place. In this case, the top of the window should continue the slope of the upper part of the roof instead of being flat; the junction of top and sides should be fitted with a soaker as described above.

Conversion on a terraced house

If it is possible to obtain planning permission—and the cooperation of your neighbors—a mansard roof is an ideal way of adding an extra floor to the top of a flat-roofed or shallow-roofed terraced property. If there is a large chimney stack with a dividing wall reaching high above the flat roof, it is possible to let the pole plates and ridge board into it without disturbing the neighboring property. However, it is normally necessary to build up the dividing wall to a sufficient height, and this will invariably mean disturbing the edge of any adjoining roof. Flashing will have to be built into the neighbor's side of the party wall to provide a watertight joint with the roof. Because of these potential difficulties check carefully that your local authority will grant you permission before you plan too far ahead.

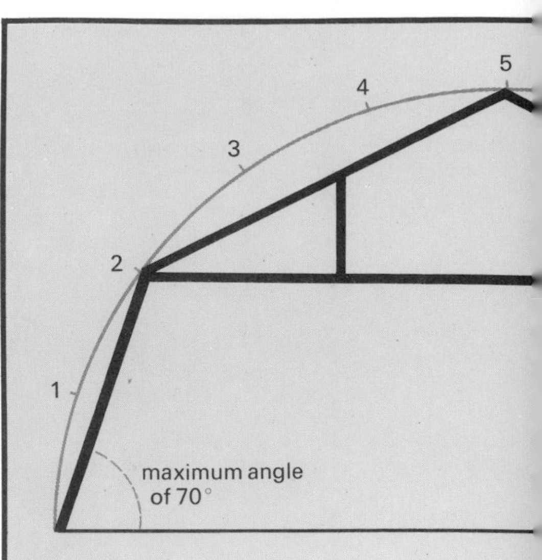

sarki

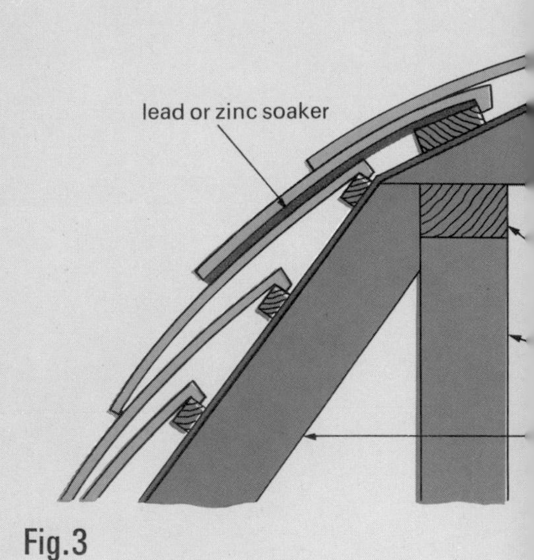

Fig. 2. *A mansard roof is traditionally designed to fit into a semicircle which is divided into ten equal sections.*
Fig. 3. *A lead or zinc soaker fitted between a couple of tiles makes a watertight joint where the lower and upper rafters meet. Here the upper joist is fitted by the alternative method: it does not extend over the upper pole plate as in Fig. 1 but is cut at an angle so that it meets flush with the bottom end of the upper rafter. A sheet of ½ in. plywood is nailed over both sides of the joint as found in many factory-made trusses.*
Fig. 4. *Side view of a typical dormer window. The flat roof must slope at least 1 in 40.*

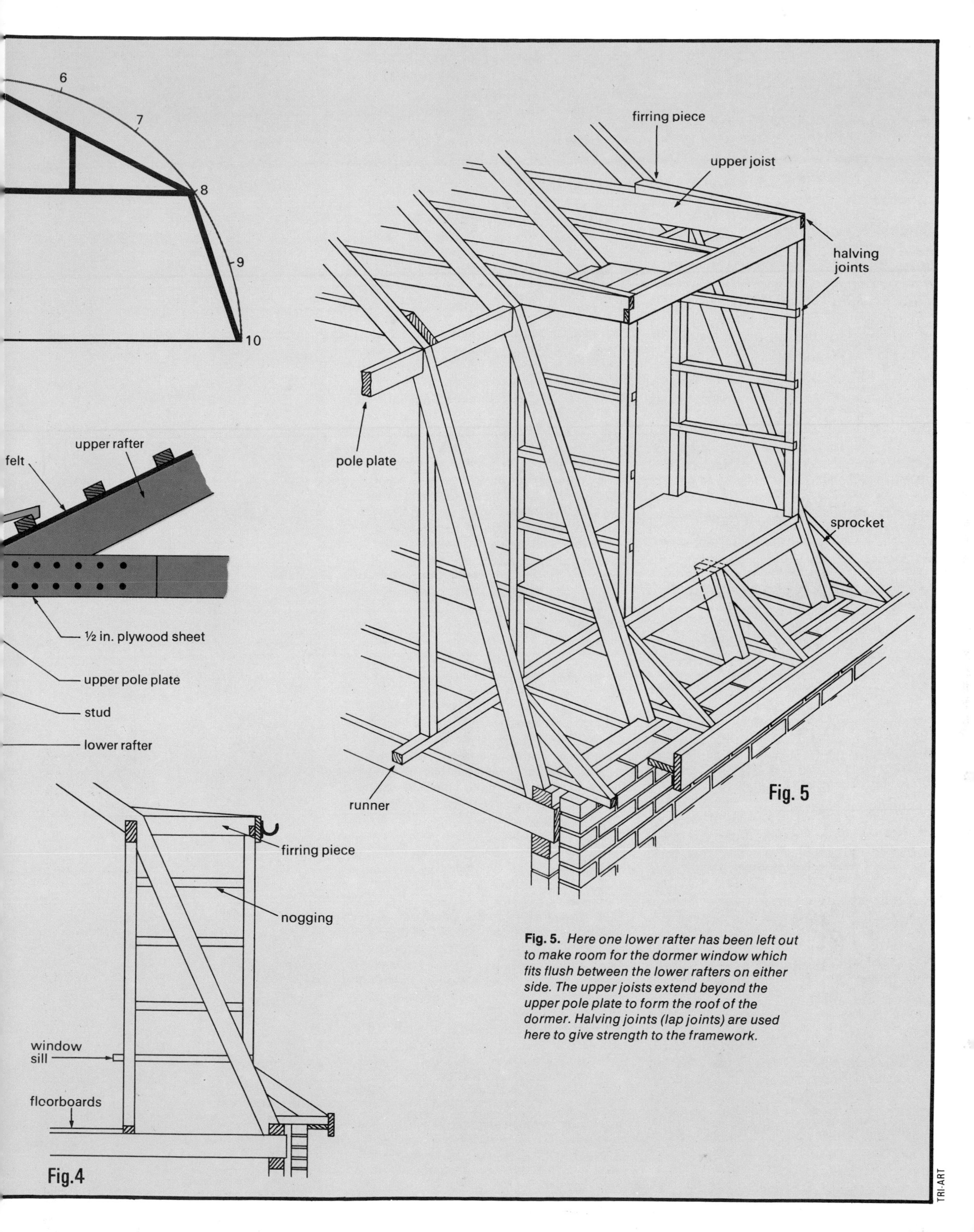

Fig. 5. *Here one lower rafter has been left out to make room for the dormer window which fits flush between the lower rafters on either side. The upper joists extend beyond the upper pole plate to form the roof of the dormer. Halving joints (lap joints) are used here to give strength to the framework.*

Easy-to-build country-style doors

Doors with a "country" look are usually seen outside the home where their strength provides necessary security on garages, sheds and other outbuildings. If constructed carefully and well designed, though, this type of door can add an interesting and attractive touch inside the home.

The basic construction of diagonally braced doors is simple. The face of the door consists of vertical lengths of timber which are butted together along their long edges. They are held together by cross members, horizontal pieces which are as long as the door is wide. Most of these doors also have braces, pieces of lumber that run diagonally between two ledges. These usually run *upward* from the hinged side, sometimes one upward, one downward; they give added strength to the door.

There are many designs for these doors, some of which are shown in Figs. 2–6. If the door does not have to be particularly strong—for a cupboard, say—you can do away with the braces. You can use only two cross members, one near the top and one near the bottom of the door.

Another variation is a framed and braced door. This has stiles and rails joined together, with the battens glued and nailed into a rabbet cut on the inside edges of the stiles and rails. The construction of this type of door is basically the same as that described in CHAPTER 23 for making Dutch doors though, of course, the door is not cut in two.

Typically, these doors are 1½ in. thick, including the nominal 1 in. vertical boards and the cross members and diagonals, all of which are actually ¾ in. thick. Doors on outbuildings, however, need to be fairly solid for security, and for one of these you may wish to increase the thickness. You can then use nominal 2 in. lumber for all the pieces. The width of the lumber depends largely on the design you choose but 4 in. is a common size for the purpose. Doors made from wider boards tend to look heavy and unattractive.

Tongue-and-groove boards are often used for the verticals. These help to provide a weathertight seal. Square-edged boards can be used but these are not as weatherproof, especially when they shrink and gaps appear between them.

Basic construction

The first step in the construction of most designs of this type door is to butt the edges of the vertical boards together. You will first have to decide how many you need to make up the door. You may have to cut some of them narrower than the rest. In doing this you must ensure that the arrangement is symmetrical. For example, if the door width is 34 in. and you intend to use nominal 4 in. boards, this width will be approximately bridged by nine 4 in. boards plus one 2½ in. board. The door would look odd if you simply put the 2½ in. board at one edge of the door and trimmed it to fit. In this case you should use eight 4 in. boards, cut two of them down to just over 3 in. and put one at each side of the door; trimming them equally to a final fit.

Cut the boards a little too long and the outside boards just over width. If you are using T & G boards, paint the tongues and grooves before you begin the assembly of the door. (But see "Designs with a difference" below.) Paint the edges of square-sectioned timber also. This makes the door more weatherproof.

The verticals should be fitted together tightly and the cross members nailed to them. To do this, you will have to clamp the boards together. Two or three sash clamps would do the job, but you may not have any. An alternative is to nail a long, reasonably straight piece of lumber to the floor parallel to the skirting. Do not drive the nails right home. The distance of this piece from the skirting should be the planned width of the door plus, say, 2 in.

Cut four wedges, each of which should be a little narrower than 2 in. (or whatever distance you have chosen). Lay the boards on the floor between the nailed-down strip of lumber and the skirting, then force them together with the wedges, used as "folding wedges" in two pairs (see Fig. 7).

Fixing the cross members

Now mark out on the vertical boards the finished length of the door and the position of the cross members. To do this, mark a line across the verticals in the middle of their length. Take all measurements from this line. On each side of the line, measure half the planned height of the door. Square lines through these points across the vertical boards.

Now mark the lines that will indicate the position of the cross members. If you intend to use three, the middle one should run across the center of the door. On each side of the central squared line mark a distance equal to half the width of the middle ledge. Square lines through these points. The top and bottom cross members can be between 1 in. and 3 in. away from the top and bottom of the finished door. They should not be farther away, or the ends of the vertical boards may become damaged when the door is in use. Mark out the positions of the top and bottom cross members from the central squared lines.

Before assembling the door further you will have to decide how the braces, if they are

Right. *An attractive, naturally finished country-style door. An interesting design variation is the use of narrower lumber for the diagonals than is used for the cross members.*

NIGEL MESSETT

to be used, will be joined to the cross members. If you intend to use simple butt joints you can go ahead and fix the cross members, as described below. If you want to notch the diagonals into them, as shown in Fig. 3, you will have to cut the cross members to shape before you attach them. The distance of the point where the diagonals meet the cross members from the ends of the ledges is a matter of design. Fig. 3 shows one design and indicates the positioning and size of the notches.

Once the cross members have been cut to shape, you can fix them to the clamped-up verticals. *Lightly* nail the cross members to the verticals in the marked positions. Use finishing nails to do this, nailing through the edges of the cross members into the vertical boards. Release the clamps, or knock out the wedges if you are using them, and turn the assembly over. Place rough pieces of wood underneath the ledges to hold them away from the ground.

Now nail through the vertical boards into the cross members. Use nails about ¼ in. longer than the combined thickness of the cross members and verticals. Nail through the door into the rough pieces with finishing nails. Then turn the door over, knock off the rough pieces and *clinch,* or bend over, the nail points with a nail punch. This pulls the cross members and verticals together tightly.

The door can now be cut to size. Saw along the marked lines that indicate the top and bottom of the door. Plane off the excess wood on the long sides of the outside vertical boards.

Fitting the diagonals

If you have not mounted the diagonals, these can now be cut to size and shape and nailed in place. Remember that these add to the strength of a door by acting as tie rods.

If the diagonals are to butt against the cross members, mark on the cross members the points at which the diagonal ends meet them. Cut the diagonals slightly over length and nail them temporarily *over* the cross members, so that they cover and protrude beyond the points marked. Near the end of the diagonal lay a straight edge and line it up with the edge of the cross member—the edge to which the diagonal will run. Draw a line along the straight edge. Repeat the process at the other end of the diagonal. Remove the diagonal from the assembly. With a back saw, cut down the lines marked at the diagonal ends.

Once you have cut the diagonal to shape and tried it for fit, cut the other one to length in the same manner. Lightly nail them to the vertical boards and follow the procedure outlined above to enable you to nail through the verticals into the cross members. These nails should be clinched, too.

Insert braces

If the cross members have a V-shaped notch cut into them, cutting the diagonals to shape will be a little more difficult. It involves reproducing the V shape on to the end of the diagonals, but at a different angle (see Fig. 8). To do this, mark a line along each cross member to indicate the maximum depth of the V, and across each cross member through the point of the V. Nail the diagonals lightly in place over the cross members and draw a line to indicate the edge of the cross member. Now mark on the diagonal the maximum depth of the V and the position of its point. The place where these two lines cross will be above the point of the V. Draw two lines from this point to the line indicating the edge of the cross member. Repeat this at the other end of the diagonal. Remove the brace and cut it to shape.

If you have cut the V shape in the cross members accurately, you can use the first brace to mark out the second. If in doubt, though, repeat the process outlined above. Nail the diagonals in place.

Designs with a difference

There is a wide variety of designs for this type of door. Many of the variations are small, but the less conventional designs can be attractive as well as functional.

A simple variation is to use two different widths of vertical boards for the door face, placed alternately across the door. The assembly is exactly the same as for a conventional door. Make sure that the arrangement of the boards does not spoil the symmetry of the door.

Fig. 1 shows a door with a diamond-shaped window in the center of the top half. The hole for the window is cut after the cross members are attached, but before the diagonals. The diagonals must be positioned so that they do not obscure the window. And the window should *not* cut *all the way* across any vertical board.

To make this type of door, lay the vertical boards together in the manner described above and clamp them together. Mark out the position of the cross members. Then mark out the diamond shape. Its size is largely a matter of taste, but it should not be too large or the door will lose some of its rigidity. An attractive size for a door 2 ft. 6 in. wide is 10 in. along the vertical and 6 in. along the horizontal.

Assemble the door in the manner described above. Mark out the diamond with a pencil. At the four corners, drill a ¼ in. hole. The diamond shape can then be cut out with a sabre saw or, if you do not have one, a compass saw.

An alternative is to clamp the verticals and mark out the position of the cross members. Do not attach them yet. Mark out the diamond shape and number each of the verticals. Release the clamps. The boards in the center of the door will have part of the diamond marked on them. Cut along the lines with a back saw. Reassemble the boards in their correct position, lining up the pencil marks that indicate the position of the cross members. Re-clamp the battens and assemble the door in the usual way.

If you intend to cut the diamond by this second method, and you are using T & G boards, you should not paint the tongues and grooves before the initial assembly. If you do, it will be difficult to get them apart again. They can be painted when you reassemble them after the diamond has been cut.

Once you have cut out the diamond, you can reassemble the rest of the door, including any

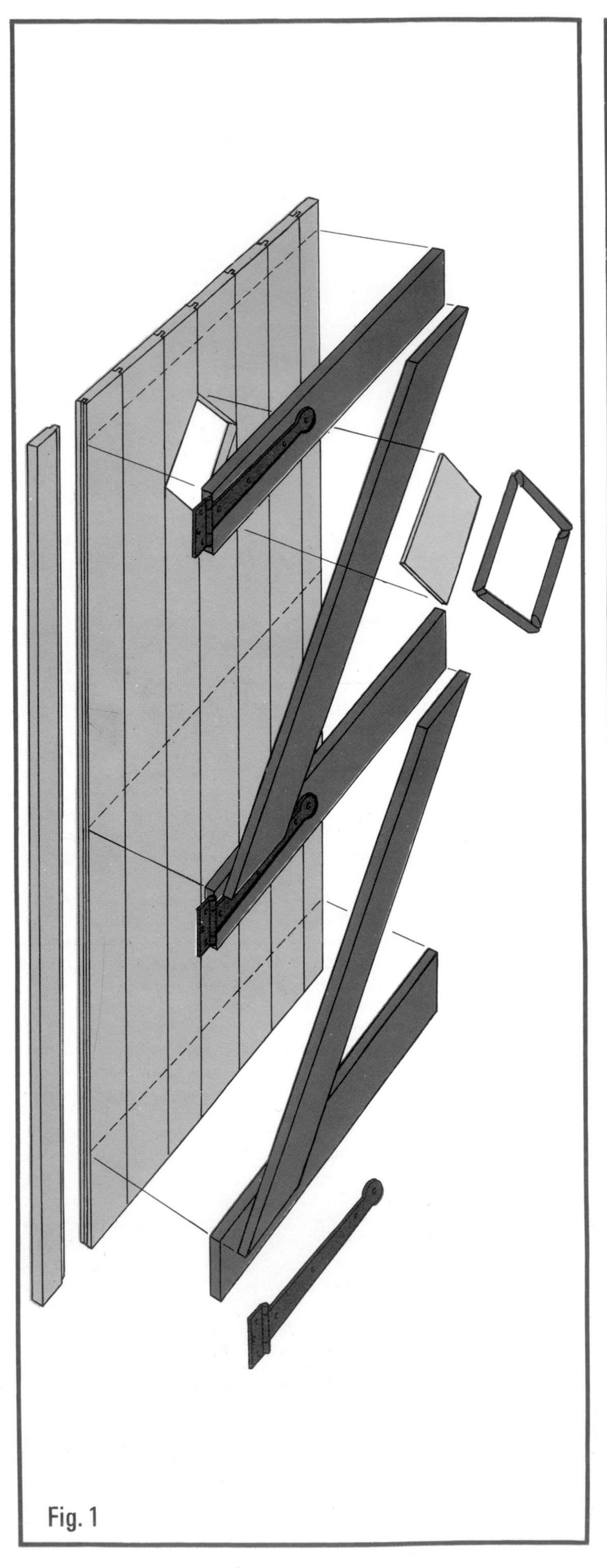

Fig. 1

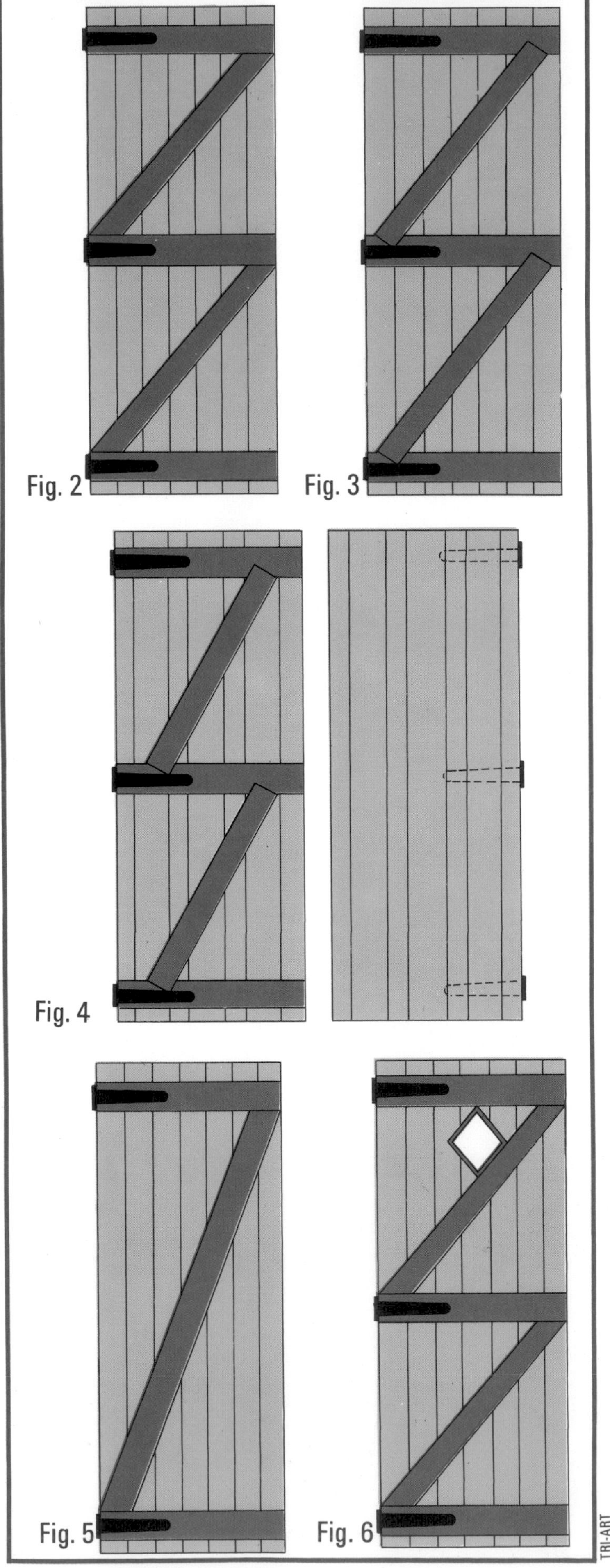

Fig. 2 Fig. 3 Fig. 4 Fig. 5 Fig. 6

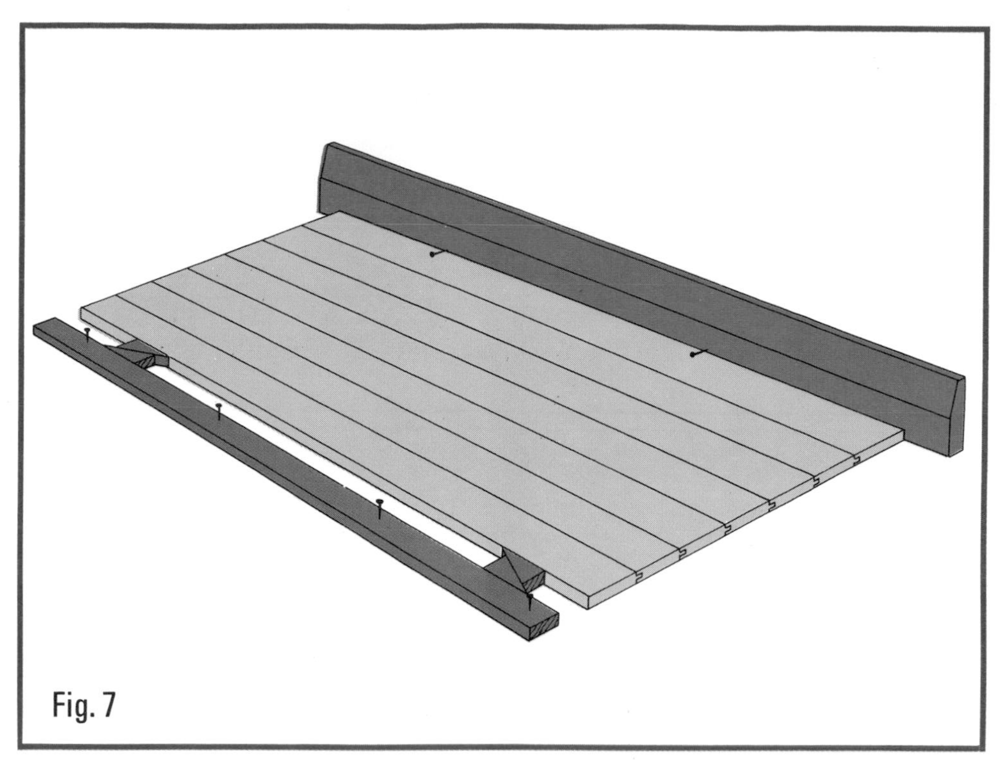

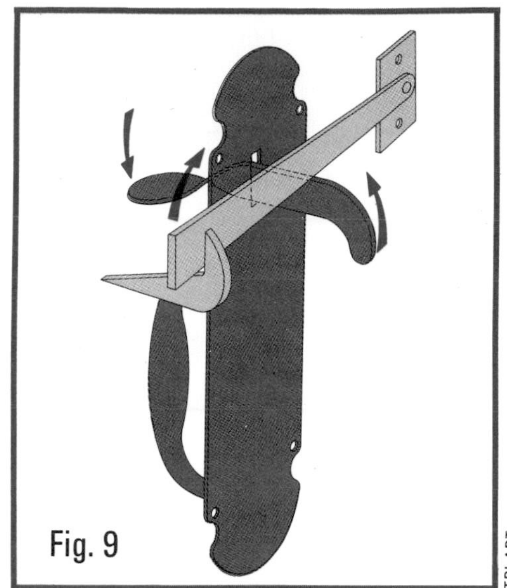

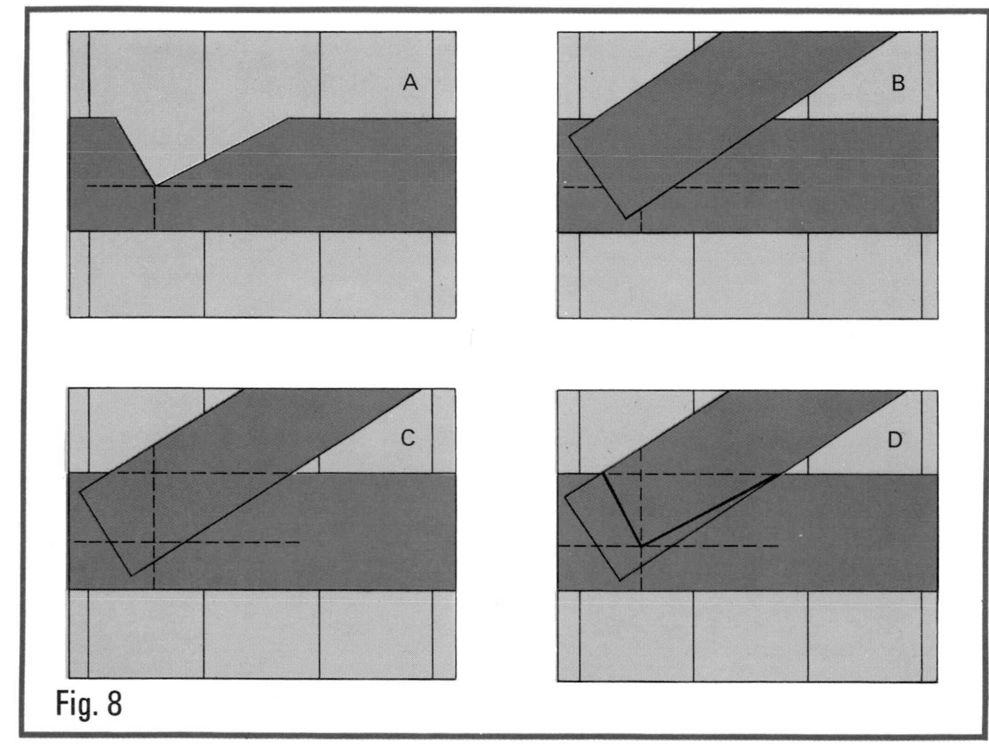

Fig. 1. *An exploded drawing of one design for a country-style door. The diamond-shaped window is held in place with putty and strips of molding. The diagonals can be positioned so that they do not obscure the window.*

Figs. 2 to 6. *Five designs for this type of door. The basic construction requirements—the braces must run diagonally, usually from the hinged side upward, and the door must be symmetrical—allow plenty of scope for designs with a difference.*

Fig. 7. *One method of clamping the door boards while you fix the cross members. A piece of lumber is nailed to the floor. Pairs of wedges are knocked in place between the lumber piece and the door edge. This pushes the door firmly against the baseboard of the room.*

Fig. 8. *If the diagonals are to be inset in the cross members, cut the cross member to shape and fix it to the door. Mark it as shown in A. Lay the diagonal over the cross member (B). Sighting the marks on the cross member, draw intersecting lines on the diagonal (C). Draw lines joining up this intersection with the points where the diagonal meets the edge of the cross member (D). Then remove the brace and cut along these lines.*

Fig. 9. *Where security is not really important a thumb latch will hold the door closed.*

diagonals. These do not have to run right to the ends of a pair of cross members, so you can avoid obscuring the cutout for the window. When the door is assembled and cut to size, fit a glass pane into the window.

You can cut the glass for the door yourself using a glasscutter and a straight-edge but your glass dealer can do the job for you. Fit strips of wood molding around the sides of the diamond cutout from one side of the door to give a firm mounting for the window. Then mount the glass in place with putty. If the sides of the diamond-shaped cutout have been irregularly cut then rabbeted molding will disguise this. The molding is nailed in place—the method of attaching the molding is basically the same as that described in CHAPTER 26 for fitting new panels into paneled doors.

Hanging the door

These doors are hung with tee hinges. They are shown, mounted in place on the door, in Fig. 1. The long part of the hinge is screwed to the cross members. If the door is fairly heavy and has been built from thicker lumber than that suggested above, a heavier type of hinge may be necessary. The method of hanging doors is fully described in CHAPTER 16.

You will also need to mount some kind of lock, bolt or latch to the door so that it closes securely. A barrel bolt is the most commonly used on these doors. These are made from iron, brass or bronze, but for purely functional purposes an iron bolt is sufficient. The length of barrel bolts varies but a 6 in. bolt is sufficient for most types of doors. The plate of the bolt is screwed to the door and the metal socket or staple is attached to the door frame. The bolt is pushed into this socket to secure the door. If the door frame is thick, the bolt can slide directly into a hole drilled in the frame.

A thumb latch will hold the door closed, but does not lock it. The components of a thumb latch are shown in Fig. 9. A rim lock is the most secure way of locking the door. These are easy to install, the main part of the lock being screwed to the door and the smaller part, to the door frame.

1

2

3

4

NELSON HARGREAVES

Power tools: 2

Anyone who has used both a hand tool and its drill-powered equivalent will appreciate the fantastic saving of time and effort that a power drill brings. Sooner or later, however, he will want the convenience of ready-to-use tools and will graduate to integral tools, as used by professionals.

Integral tools are designed in one piece for one purpose, instead of being adaptations. Many have more powerful motors than those of drills and as a result work faster and more efficiently than their equivalent drill attachments.

In many cases, an ordinary power drill with an attachment is quite good enough for amateur use. For example, the amateur carpenter would seldom think of buying a full-size integral drill press instead of his power drill. But with other tools, bolt-on attachments to power drills are often inadequate. This is particularly true with circular saws. A drill-driven saw will simply not provide adequate speed in cutting large pieces of hardwood. So the first integral tool that many amateurs buy is a bench saw.

The bench saw

Circular saws of this type have powerful motors—at least ½ hp—and large blades ranging from 6 in. to 12 in. in diameter. The blade turns much faster than the one on a drill attachment, which not only speeds up the saw's cutting rate, but also gives a cleaner result.

However, both integral bench saw and drill-driven saw table are used in exactly the same way, and if you have a drill-powered saw, you can use these operating instructions provided that you do not overtax the tool's limited power.

A bench saw consists of a flat table top through which the saw blade projects. Wood is slid over the table toward the blade, which is adjustable for depth and angle of cut and protected by a slide-away guard that reduces the risk of you cutting your fingers off—though this is still a tool that must be used with great care.

Opposite. *Some of the many ways of using a bench saw.*
Fig. 1 *(above, left). A push stick is used to protect fingers from the revolving blade. The saw is a lightweight drill attachment.*
Fig. 2 *(above, right). Cutting a miter on a more powerful integral bench saw, using its built-in protractor.*
Fig. 3 *(below, left). Cutting a bevel. A straight-edge is laid along the rip fence to make the cut more accurate. The push stick will finish the cut.*
Fig. 4 *(below, right). Using the same saw to make the second cut of a large rabbet.*

Wood is fed into the saw at the correct angle by using one of two *guide fences.* One of these, the *rip fence,* is parallel to the saw blade, but can be set any distance away from it by sliding it sideways in a groove up the edge of the table. Wood is slid along it into the blade for straight cutting parallel to the edge, generally along the grain. The other fence, the miter gauge, is for cutting across the width (and grain) of wood at any angle. It does not lock in position, but moves from the front to the back of the table by sliding in a slot parallel to the blade. On the front of the guide there is a protractor. This is set to the desired angle, and the wood is rested against it. Then wood and guide are pushed together toward the blade, which cuts the wood at the angle the protractor is set to.

Both guide fences can be removed entirely if necessary, so that large sheets of hardboard or plywood can be cut without anything getting in the way. It is as well to check the angle of the rip fence from time to time. If it is not exactly parallel to the blade, it may cause the blade to jam when making a cut in a long piece of wood. Some, but not all, bench saws have slip clutches that disengage the motor when this happens, to stop it from burning out.

Of course, the saw is not restricted to making cuts straight along or across wood. It is an extremely versatile tool.

Sawing techniques

Small pieces of wood

It is dangerous to feed small pieces of wood into the blade with your hands, because your fingers get uncomfortably close to the blade and the slightest slip may cause a serious accident. Wood is very likely to slip on a bench saw, because the tremendous torque of the blade tends to wrench it aside if you are not holding it firmly. This is a particular problem when using the crosscut guide at an angle.

Small pieces should be pushed toward the blade with a *push stick*—a short lath with a V shape cut out of the end so that it can hold the piece of wood firmly (see Fig. 1). It doesn't matter if the push stick gets cut, because you can make another in seconds. Fingers are not so easily replaced.

Miter and taper cutting

Miter cutting, or cutting wood at a 45° angle to make a mitered joint, as described in CHAPTER 30, can be done quickly and accurately. To cut a miter across the face of a piece of wood—as if making a picture frame—set the protractor on the crosscut guide accurately to 45°. Then lay the wood against the guide and slide the guide and wood together down the table into the saw blade (see Fig. 2).

As a general rule when crosscutting, you should hold the wood with both hands on one side of the blade, and allow the offcut to fall away freely. If you push from both sides, the pressure tends to close up the cut around the blade, causing it to jam and "buck" dangerously. If you must hold both sides, curve the wood slightly with your hands to hold the cut open.

To cut a bevel, or miter along the edge of a piece of wood—as if making a box with mitered corners—set the blade at a 45° angle. Many saws have built-in protractors here, too, to ensure accuracy. On some bench saws, the table tilts instead of the blade. Slide the wood and crosscut guide towards the blade in the normal way, but grip the wood extra firmly (see Fig. 3).

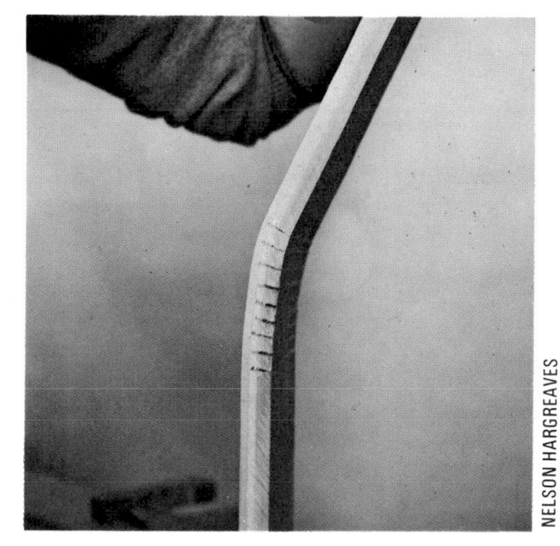
NELSON HARGREAVES

Fig. 5. *A bench saw is particularly useful for kerfing—bending a piece of wood into a curve by making rows of deep cuts in it.*

Taper ripping is cutting a very shallow taper on a long length of wood so that it is narrower at one end than the other. It is used, for example, in cutting table legs.

It is best done by making an adjustable jig out of two moderately long battens. Set them face to face and fasten them together by a hinge at one end and a slotted metal strip fastened with wingnuts at the other. By moving the free ends a distance apart and locking them at this distance with the strip and wingnuts, the jig can be set at any shallow angle (see Fig. 10).

In use, the jig is slid along the fence together with the wood to be cut. This method is particularly convenient when a large number of identical pieces have to be cut.

Dado joints

Dadoes and other grooves can be cut very simply (but also very accurately) by setting the saw blade to the required depth of cut and cutting the sides of the groove first, using the fence to keep them straight. Then slide the fence away and remove the wood between the cuts by repeatedly passing the wood over the saw blade. Mark the extent of the groove on top of the wood, or you may cut past the edges. Except with very wide grooves or housings, this method is faster than chiseling by hand, although for a stopped dado you will have to cut the last inch or two by hand, since the curved saw blade cannot reach the inside corner of the housing. It is also more accurate, because the depth of cut is constant all over the groove.

When cutting tenons, wood can be removed in the same way.

Rabbets

There are two ways of cutting rabbets on a bench saw. One way is to cut along one side of the rabbet, using the fence to ensure accuracy, and then turn the wood through 90° and cut the other side (see Fig. 4).

This involves two operations for each rabbet. A faster way is to mount the blade on "wobble washers"—a pair of angled washers that make the blade wobble from side to side as it revolves. As a result, the blade cuts a wide groove instead of a neat line. The width of the groove is restricted by the size of the slot in the saw table, because if the blade "wobbled" too far it would cut the table. But you can always make several passes to cut a wide rabbet or a groove.

When you have set the blade on its washers, fasten a piece of old battening to the fence to protect it and move it until the blade just brushes the battening at the apex of its wobble. Now any piece of wood that is slid along the battening will have a rabbet cut out of it the same width as the wobble of the blade—or narrower if you adjust the blade to cut farther into the temporary battening fence. The depth of cut can still be adjusted in the normal way.

Great care should be taken when using wobble washers because the oscillation of the blade makes it even more dangerous than an ordinary circular saw blade. At all costs, keep your fingers well away from it and use a push stick to move the wood you are cutting. Use the saw blade guard for all operations that permit its use.

Kerfing

Kerfing is a special technique that enables a piece of solid wood to be bent in a curve. Rows of parallel cuts are made across the wood on the inside of the curve through half to three-quarters of the wood's thickness, and all the way along the part that is to be curved. The wood can be bent (see Fig. 5)—though it helps the process if you wet it or steam it as well. Use a crosscut or planer blade to make the cuts; a combination blade is too coarse and will give a messy result.

Kerfing reduces the strength of wood sharply, and should not be used for load-bearing frames. It is also only suitable for outside curves—that is, with the saw cuts on the narrower radius; the wood *could* be bent the other way but the surface would probably wrinkle unattractively. But when used correctly, kerfing produces a neat curve that is impossible to make by any other method.

Maintenance of blades

Blades should be kept clean and as sharp as possible. After using a blade, rub a little oil or Vaseline over it to stop it from going rusty. If

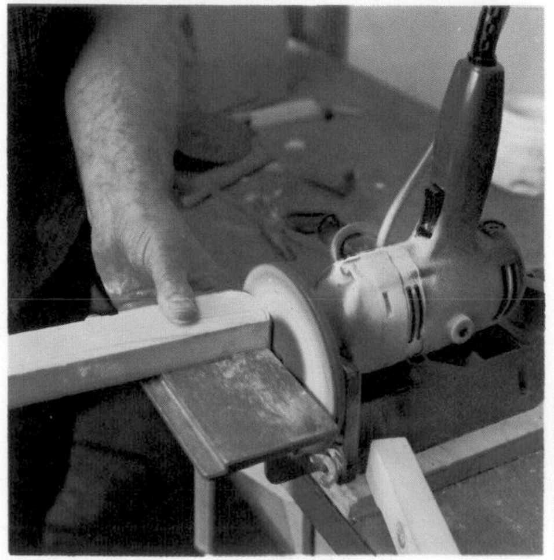

any rust does form, take if off with steel wool. The smoother the surface of the blade is kept, the faster it will cut because there will be less friction between blade and wood.

Circular saw blades are very easy to sharpen yourself. You *can* have them sharpened professionally, and some shops run an exchange service, sharp blades for blunt, but you only need a suitable file and a quarter of an hour to do just as good a job as they can. Simply sharpen the cutting edges of the teeth, retaining their original shape. If you prefer to avoid this work use low cost "throw-away" blades that can be discarded when dull.

Power saws are sometimes fitted with abrasive cutting wheels for sawing masonry block, marble and other materials. When these become worn, they cannot be sharpened, but must be replaced.

The bench grinder

One of the most useful tools in any workshop is the bench grinder. It enables all edge tools to be resharpened in an instant—and sharp tools make for easy work. If you have a grinder attachment for a power drill, you will have to set it up every time you want to sharpen something—which is often. In practice, this means that you will not sharpen things often enough. So a bench grinder, which is not particularly expensive, is a good investment.

Another advantage of an integral bench grinder is that it has two revolving grindstones—one on each end of the motor. Chisel and plane blades are sharpened in two operations; grinding, to get the blade the right shape, and honing, to put an edge on it. Different grinding wheels are needed for each operation, so having both of them on the same machine speeds up work considerably.

Sharpening is done against the front curved edge of the grinding wheel and not against the flat circular face. The wheel revolves so that the front edge moves downward. This keeps sparks and fragments of metal or abrasive from being thrown upward into the eyes (but goggles should *always* be worn). Adjustable tool rests are provided in front of each wheel to hold blades steady while they are sharpened. Cutting wheels come in various grades. For most jobs, a medium wheel for grinding and a very fine one for honing should be all you need.

Fig. 6 *(top left). Sharpening a chisel on a bench grinder. The blade is held steady against the tool rest and moved from side to side to keep the edge straight and even.*
Figs. 7 *and* **8** *(second and third from top). The hand movement for sharpening a twist drill has to be learned. The drill is laid against the wheel with the front of its cutting edge touching (first picture) then turned clockwise and slid forward and up the curve of the wheel (second picture). If you're new at it, buy a drill-grinding attachment.*
Fig. 9 *(bottom). Shaping a curved end on a piece of wood with a bench sander—this one is a power drill attachment.*

NELSON HARGREAVES

The wheels are fastened to their shafts by nuts screwing down on to the threaded ends of the shafts. The wheel on the left has a lefthand thread to stop it from coming undone in use. The wheel on the right has a normal righthand thread.

Sharpening chisel and plane blades

Chisel and plane blades, though completely different in shape and use, are sharpened in exactly the same way. In both types of blade, the preliminary grinding to shape of the edge of the blade should give the ground surface an angle of 25° to the flat face of the blade. Then it should be honed at the slightly greater angle of 30°. The 5° difference saves you from having to hone the whole ground surface. Only the tip is honed (see Fig. 11).

To sharpen a blade, first lay it on the tool rest of the grinder with the point touching the stationary wheel, and measure the angle where the point touches. Move the blade until the angle is 25°, and memorize the position of the blade. Now take the blade away, start the wheels and lay the blade lightly against the coarse wheel. High speed and light pressure are the secret of good grinding. Move a wide blade from side to side across the wheel, so that its whole edge is ground evenly.

Grind on one side only until the blade is properly shaped, when the length of the ground surface should be 2½ times the thickness of the blade. Every few seconds of grinding, remove the blade from the wheel and dip it in cold water to stop it from overheating. An overheated blade "loses its temper" and turns blue. If this happens, grind off the blue part.

The freshly ground surface will be slightly hollow in shape because of the curve of the wheel, but that doesn't matter. The next stage is to hone it.

Find the correct angle of the blade against the stationary wheel as you did before, except that it should be 30° and not 25°. Then start the grinder and lay the sloping side of the blade against the fine wheel—but only for a few seconds. The wheel will turn the edge of the blade over, producing a fine "burr" on the other side. Cool the blade and lay the flat side *flat* on a flat oilstone, and slide it edge-first to remove the burr.

Repeat several times, using very light pressure to produce a razor edge.

Blades can be honed on a flat oilstone to restore the edge bevel several times before they lose their shape and have to be reground.

Sharpening twist drills

Twist drills and high speed drills can also be sharpened on a fine grinding wheel. The angles have to be watched carefully, but otherwise the job is not difficult. Do not cool twist drills in water, because the extra-hard steel might crack. Just try not to overheat them.

There are three important angles that must be maintained on a twist drill. They are marked A, B & C in Fig. 12. Angle A, the angle of the cutting edge to the shaft, varies with the material the drill must cut. Maintain the original angle in grinding, usually 59°. Angle B, the angle of the sloping shoulder of the cutting edge to the horizontal, varies with the size of the drill. This "lip clearance" angle is usually between 12 and 15 degrees.

Fig. 10 *(top right). An adjustable jig for firring. It can be set to a slope of (for example) 1 in 10 by measuring a 1 in. gap 10 in. along from the hinge. The hinge must be inset so that there is no gap when it is shut.*
Fig. 11 *(below left). Sharpening angles for all types of chisel and plane blades.*
Fig. 12 *(below right). Sharpening angles for twist drills—see the text.*

The correct way to sharpen a twist drill is to hold it near the point between the thumb and forefinger of the left hand, gripping it flexibly so that it can be moved about. Rest the left hand comfortably on the tool rest, as shown in Fig. 7, and use the right hand to poke it through the improvised pivot you have made with your left hand until the cutting edge touches the wheel.

The front of the cutting edge should touch the wheel first, at such an angle that its whole length is in contact with the wheel. As soon as it touches, push the shank of the drill down with your right hand so that the cutting edge rises, simultaneously twisting the bit a quarter of a turn clockwise (see Fig. 8). This movement is necessary to achieve the correct curve and angle on each cutting edge. You can practice on an old twist bit until you get it right. Once learned, it is never forgotten.

Sharpen both cutting edges of the drill equally so that the point is in the middle. When it is, check angle C, which should be the same as it was originally. (You can buy drill point gauges for various angles to do this.) To ease drill sharpening buy a drill sharpening attachment for your grinder. The angles are then set by the attachment.

Special wood bits should not be sharpened on a grinding wheel, but with a small flat needle file with medium-fine teeth. Aim only to preserve the original angle of the cutting edges; these bits are so large that sharpening them is a simple job.

Other integral tools

Many other integral tools may be beyond the amateur's budget. They are generally larger and more expensive, and many of them, such as the drill press or orbital sander, can be duplicated satisfactorily by drill-powered accessories.

One that you might come across is a *bench sander*—though even this can be duplicated fairly well with a power drill and a horizontal sanding stand (Fig. 9). It consists of a vertical wheel faced with abrasive paper, and in front of it, a horizontal table equipped with guide fences like those of a bench saw. (Sanding discs are also available for bench saws.)

The purpose of the tool is to sand the surfaces of pieces of wood at an exact angle—like a shooting board, only more accurate and faster. Only one side of the wheel is used, the side which is moving downward as the wheel revolves. This holds the wood down flat on the table for maximum accuracy. The wood is fed past the guide fence, which can be set at 90° or 45° (for really accurate miters) or any angle between.

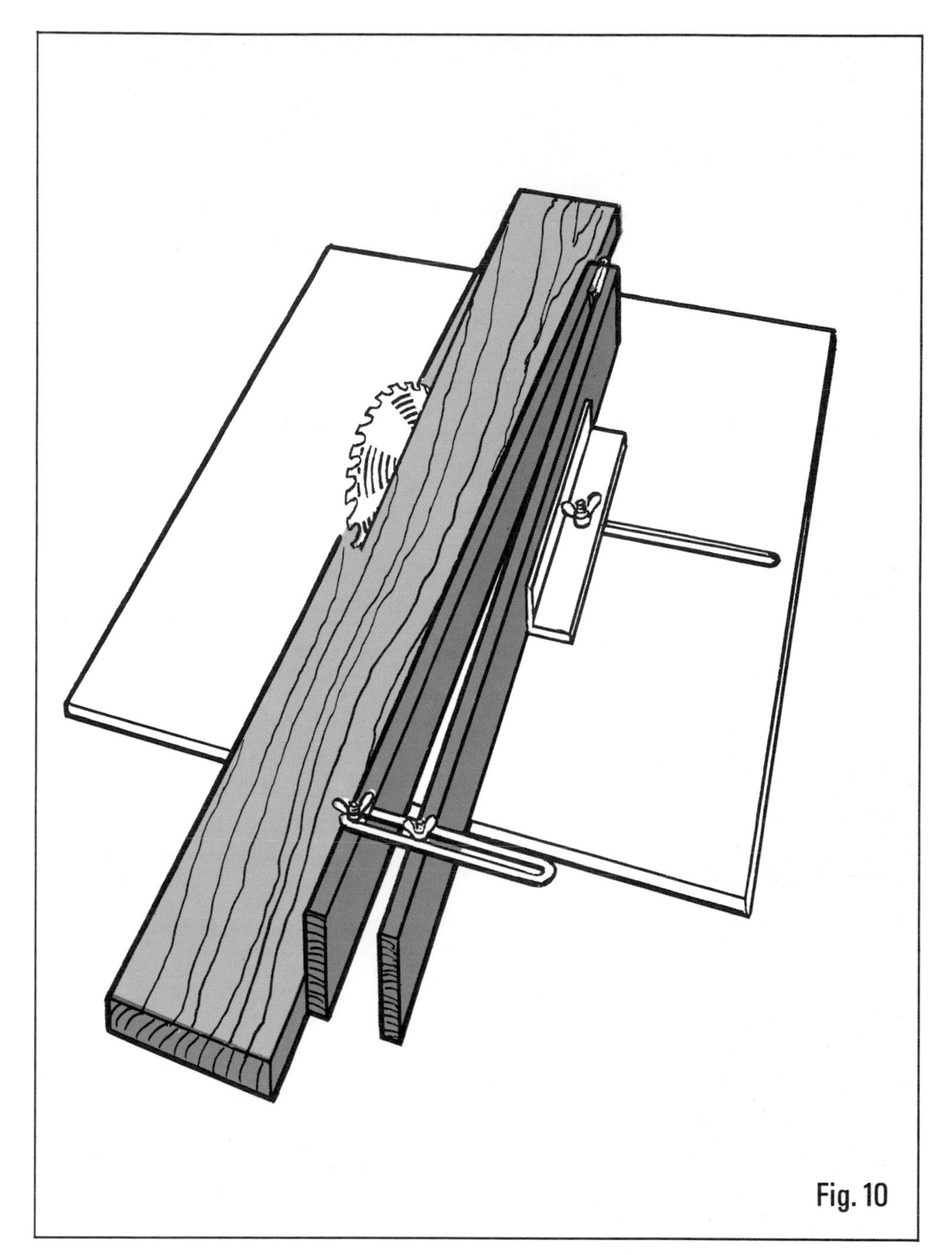
Fig. 10

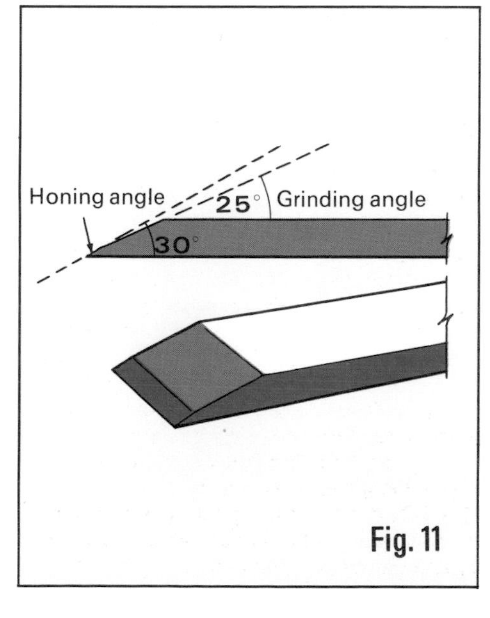

Fig. 11

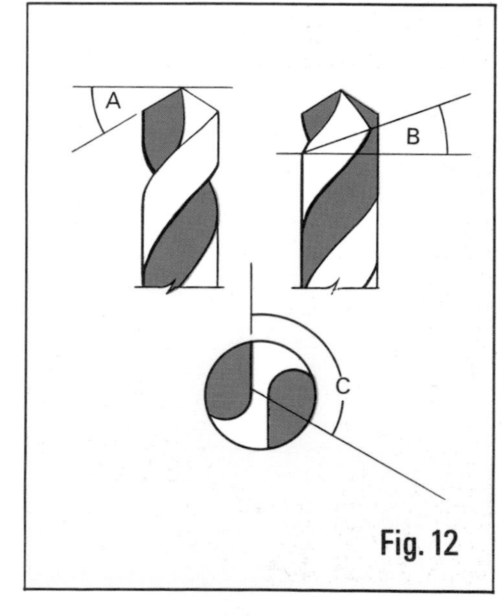

Fig. 12

Sturdy stepladder to build

A stepladder is one of the most useful pieces of equipment in any household. It can be used for painting ceilings, reaching high cupboards, cleaning windows and supporting planks. The ladder described here is simple to build, but as sturdy as any bought stepladder—and of course much cheaper.

The design given here is for a stepladder about 3 ft. tall, a convenient height for general household use. If you want a taller ladder, all you have to do is extend the legs of the design downward (so as to make the ladder wider as well as higher) and put in more steps. If the ladder is longer than about 4 ft. 6 in. you should use heavier lumber for the sides and rear frame, and put a diagonal brace in the rear frame for extra stability. Heavier hinges may also be needed.

This project introduces two variations of ordinary carpentry joints: the *angled dado joint* and the *haunched mortise-and-tenon joint.* It also has an unusual feature: because of the characteristic tapered shape of a stepladder, there are no right angles and no verticals anywhere in the frame. Seen from the front, the sides slope in at 5° to the vertical; seen from the side, they slope at 20°. Achieving these slopes while keeping the frame symmetrical is not nearly as hard as it sounds.

Materials needed

The ladder is made throughout from ¾ in. (nominal 1 in.) thick lumber, which is quite strong enough to support a heavy man if the ladder is kept to the height given here. All the pieces will eventually have their ends cut at an angle, so you will need to cut them out with a more generous allowance for waste than usual. The vertical members of the rear frame, for example, should be cut at least 1 in. over length. A point to watch is that the front vertical members, although 36 in. long, will use about 39 in. of lumber because of their sloping ends—though you can reduce this by cutting them end-to-end out of the same piece of wood.

The lumber required is as follows:

6 ft. 6 in. of 4 in. x ¾ in.
4 ft. 3 in. of 4½ in. x ¾ in.
13 in. of 5⅜ in. x ¾ in.
12½ in. of 3 in. x ¾ in.
11⅜ in. of 2½ in. x ¾ in.
6 ft. 10½ in. of 1⅞ in. x ¾ in.

. . . and a little scrap wood for making wedges for the mortise-and-tenon joints.

These dimensions include the large waste allowances needed for cutting angles, rather than the normal waste allowance for the thickness of saw cuts, etc.

You will also need two 1½ in. "back-flap" hinges, the screws for mounting them (normally ¾ in. No. 8s,) and a few feet of ordinary sash cord. The screws used in construction are 1¾ in. No. 8s; you will need two dozen. Use urea formaldehyde glue (plastic resin), which is very water resistant—stepladders often get wet.

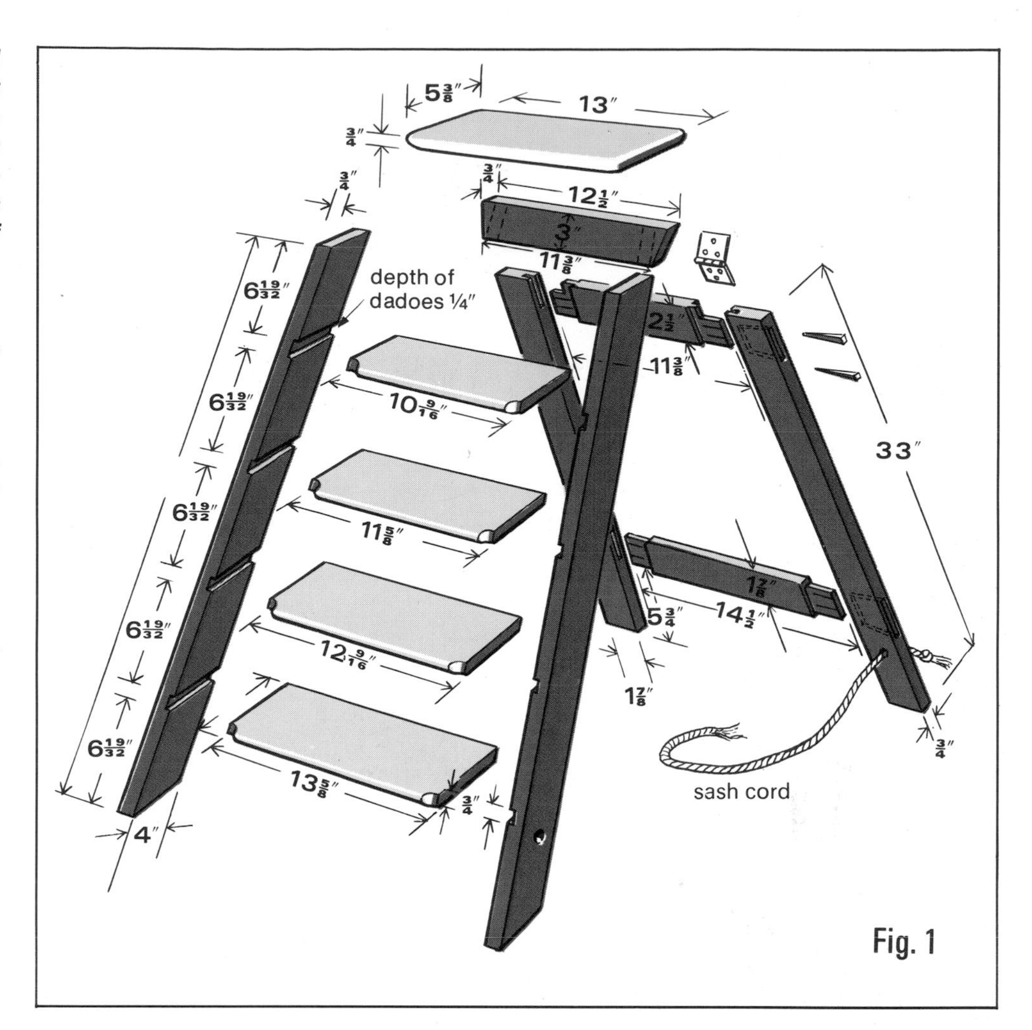

Fig. 1 *(above). An exploded view of the stepladder, showing all measurements.*

Fig. 2 *(below). The shape and dimensions of the haunched mortise-and-tenon joints.*

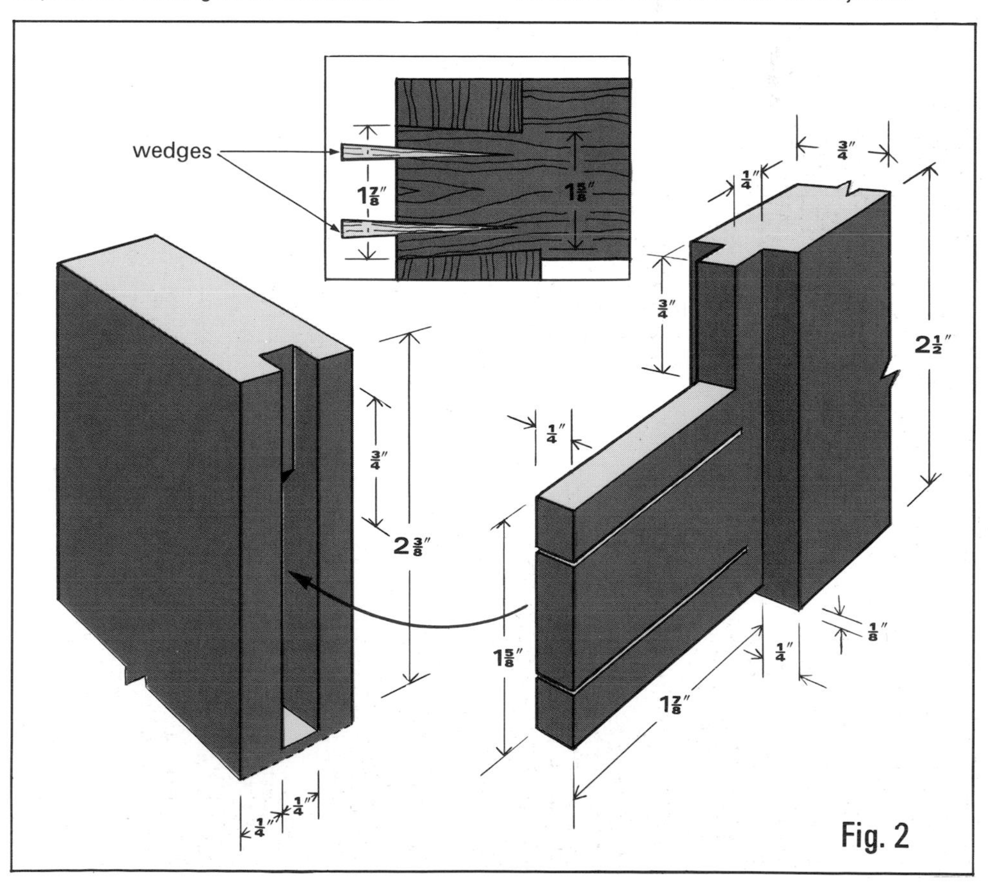

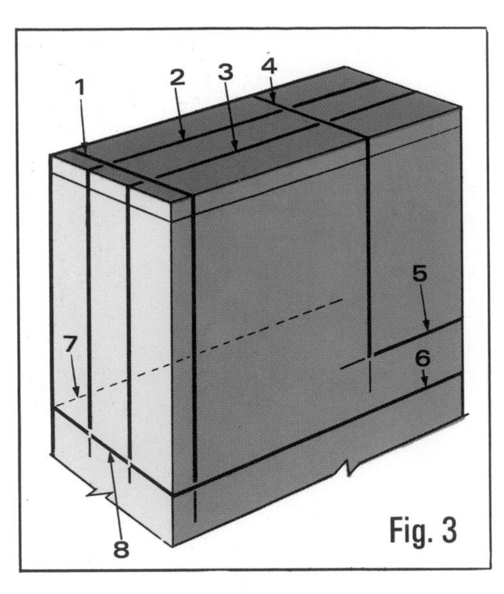

Fig. 3. *When cutting the tenons, make the saw cuts in this order for greatest accuracy.*

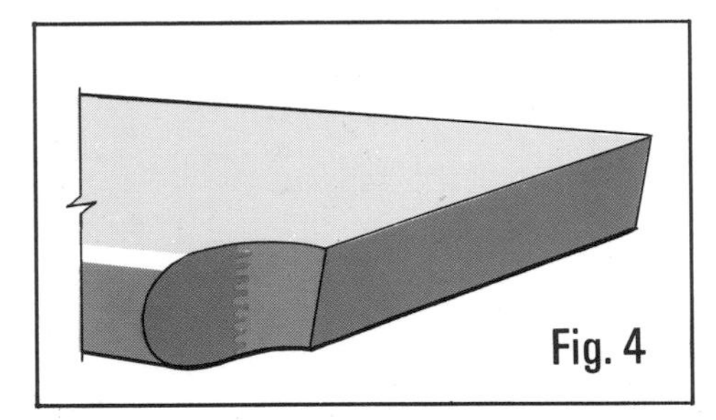

Fig. 4. *A close-up of the end of a tread, showing the shape it should be cut to.*

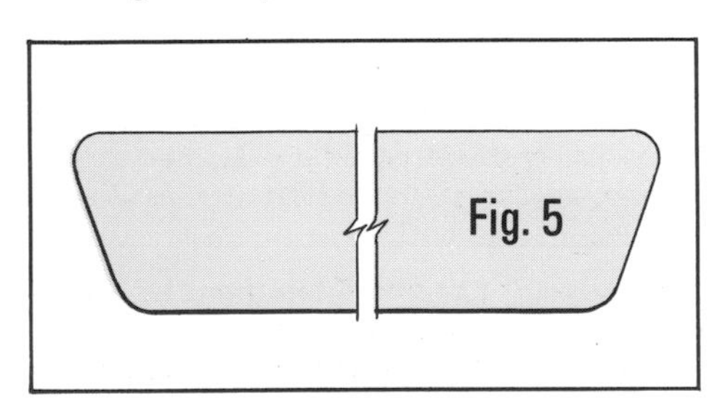

Fig. 5. *The top platform should be planed to this section to make it easier to lift.*

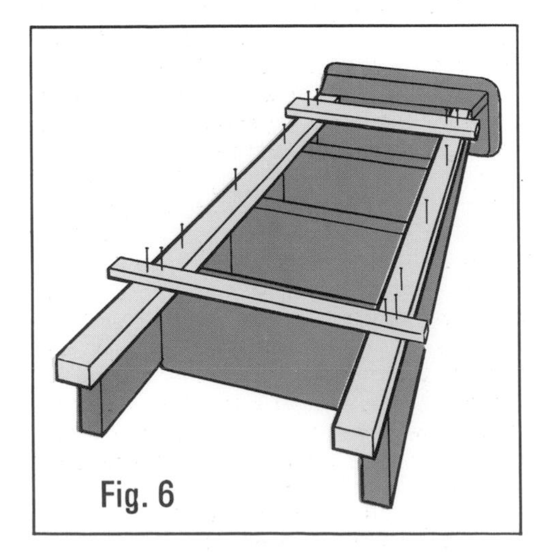

Fig. 6. *The rear frame pieces are nailed to the front frame and marked directly from it.*

The front frame

The front frame, on which the steps are mounted, should be made first. Begin by marking out the vertical side pieces, which are called *stringers*.

The top and bottom of these pieces, and their four dadoes that carry the *treads* or steps, are slanted at an angle of 70° to the sides. This is done quite simply with a bevel gauge set accurately to 70° with a protractor. Use it like a try square, taking care to keep it the right way round. Mark the lines on the edges of the pieces with an ordinary right-angled try square, and use a marking gauge to mark the ends of the dadoes ¼ in. deep.

Cut the stringers to their exact length immediately, since other pieces will be measured against them later on. Saw down the sides of the dadoes, making them a tight fit, and clean them out with a ¾ in. chisel, or a router if you have one.

Now cut out the treads—but cut only the top and bottom ones to the length given; leave the middle two ¼ in.–½ in. over length for the time being. Put the top and bottom treads "dry" into their dadoes. (If you have made the dadoes tight enough, they will probably need to be knocked in.) Clamp up the frame with furniture clamps, or failing those, string twisted tight with pegs. Check the diagonals (the corners are not at right angles, but the diagonals should still match). Now you can scribe the correct length of the middle two treads by laying them against the stringers.

The treads are ½ in. wider, from front to back, than the stringers. Their rear edges are planed off parallel with the rear edge of the stringers, which reduces the extra width slightly. The front corners should be chiseled off at an angle to fair them neatly into the front edges of the stringers, as shown in Fig. 4. This planing and chiseling is best done before the frame is finally assembled, but keep fitting the tread into its housing while you are shaping it, or you might take off too much wood. The front edge of each tread should be planed to a semicircular profile at the same time.

When all the treads are shaped, glue the frame together and brace it with screws passed through the stringers into the ends of the treads—two to each end. Screws do not hold very well in end grain, but glue, screws and dado joints together should give ample strength.

Check the diagonals of the frame and leave it to dry. Meanwhile, cut out the top platform and the top rear piece with its angled ends (the exact angle is not important; it is just for decoration). Plane all the edges of the top platform off at a slight angle, so that the lower side is about ½ in. shorter and narrower each way than the top side (see Fig. 5). Round off the edges and corners. Plane the ends of the top rear piece smooth, but do not round them off.

Glue and screw the top rear piece to the rear edge of the stringers, wide side up, and allow it to project about ⅜ in. above the top of each stringer. Then plane its top edge down at an angle to make it level and parallel with the top of the stringers. Glue and screw the top tread in position, placing the rear two screws so that they pass through the top rear piece—this will give them extra grip. Set the completed front frame aside to dry.

The rear frame

The side members of the rear frame are about 33 in. long, but cut them out 34 in. long to allow for later adjustments. The bottom cross member is the same depth as the side members, but the top one is 2½ in. deep.

Lay the front frame down on its face and place the two rear frame side members on it in the position they will occupy, with their outer edges level with the outer edges of the front stringers.

The top ends of the rear side members will not fit exactly against the lower edge of the top rear piece of the front frame, because they meet it at a slight angle. Saw the top ends so that they do fit, then lightly fasten each rear side member to the front frame with three or four finishing nails. Leave the head of each nail sticking out of the wood, so that they can be removed easily.

Lay the top cross member down on the side members with its top edge directly above the lower edge of the top rear piece (see Fig. 6). Fasten it in place with two finishing nails to each side. Do not drive them right in. Lay the bottom cross member down in the same way, with its top edge level with the top edge of the bottom step, and nail it in place with two finishing nails a side. Then draw pencil lines across each side member along both sides of the two cross members where they cross it. This will give you the angle and position of the mortises.

Remove the nails holding the rear frame to the front frame, but leave the pairs of nails holding the rear frame together. Lift up the rear frame carefully, so as not to disturb it, and mark the cross members where the side members cross them. This marks the position of the tenons.

Dismantle the frame and square all the marks across the edges of the pieces. Then link the edge marks across the other side of the wood, but do not use a try square for this because the lines are not at right angles.

The haunched mortise and tenon

This type of joint is adapted for the top of a frame, where an ordinary mortise-and-tenon joint (see page 39) would not provide the necessary strength. The top of the tenon is cut away so that the mortise can begin below the top of the frame—if it reached to the top it would, of course, weaken the joint seriously.

Full details of the joint are given in Figs. 2 and 3. The upper part of the tenon is not removed completely, but is cut back to leave ¼ in. to make a sort of tongue-and-groove joint. The tongue of the tenon and the slot of the mortise are ¼ in. thick, following the invariable rule that they should be one-third the thickness of the wood.

Mark out both tenons carefully and saw them to shape, making the saw cuts strictly in the order given in Fig. 3 to ensure accuracy. Make two extra saw cuts in the ends of the tenons, reaching almost their full depth. The wedges that hold the joint closed will be in-

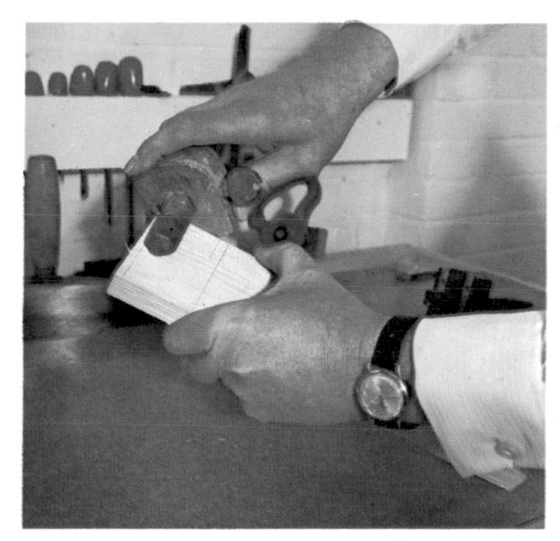

Fig. 7. *To make a haunched mortise-and-tenon joint, first mark out the tenon accurately, marking all the lines before you start cutting.*

Fig. 8. *Next, saw down all the marked lines, tilting the saw blade alternately backward and forward to keep the cut straight.*

Fig. 9. *Don't forget to cut a narrow strip off the lower edge of the tenon; this makes the joint more resistant to distortion.*

Fig. 10. *The mortises are normal in shape except for the narrow groove running up from the top edge to the upper end of the wood.*

Fig. 11. *A well-made joint should be such a tight fit that the pieces have to be knocked together to assemble it.*

Fig. 12. *To complete the joint, hammer small hardwood wedges into the pre-cut slots in the tenons, then plane the projecting ends off level.*

serted into these cuts for maximum strength.

You will have noticed that the lower side of the tenon is cut away too—about ⅛ in. is sawed off it. The normal tenons on the lower cross member are treated in the same way on both sides. The purpose of this is to create a small ledge at the top and bottom of each tenon that will help it to resist being pulled out of shape by sideways strain on the ladder.

The tenons on the lower cross members are conventional except for the ⅛ in. cutback at top and bottom. The shape and dimensions of these tenons are given in Figs. 1 and 2. You can either make saw cuts in them for the wedges or insert the wedges at the top and bottom in the usual way, but the first method is probably a little stronger.

The mortises are not at right angles to the edge of the wood, but their angle is shown by the pencil lines you have already made. The pencil lines are, however, 1⅞ in. apart, and the mortises only 1⅝ in. wide. Place the drill ⅛ in. in from the pencil lines, then drill through the wood in the usual way. Clean out the mortises with a ¼ in. bevel-edged chisel, and use the same tool to cut the ¾ in. long, ¼ in. wide and deep groove from the top mortise up to the top of the side members.

When all the mortises and tenons have been made to fit each other, apply a generous coating of glue to the surfaces to be joined and assemble the frame. Knock wedges into the saw cuts, check the diagonals to ensure that the frame is straight and leave it to dry.

Final assembly

When the glue has set on the rear frame, plane off the projecting ends of the wedges and tenons and lay it on the front frame. Join the two frames together with the back-flap hinges, which should be placed so that the lower half of each hinge lies over the joint between the cross and side members of the rear frame—this will help to hold the joint together.

Now turn the stepladder over on its back, and with a pencil mark the length of the front legs on the projecting ends of the rear legs. Open out the frame and lay a long ruler across both legs at once on the pencil marks. Draw a line across both legs, and then, with a bevel gauge set at 70°, slant the line downward across the edges of the legs—it should, of course, slant the opposite way from the front legs. Link the lines across the back face of the wood with a ruler laid across both legs at once, as you did before. Then saw off the ends of the legs at this compound slant, which should ensure that they rest flat on the ground when the ladder is open.

Drill a hole in each leg a few inches from the bottom for the sash cord to go through—it doesn't matter exactly where, provided it is the same on both sides. Then set the ladder upright and get a friend to open the legs out until you can see that their slanted bottoms are resting flat on the floor. This should happen when the angle between the front and rear frames is about 40°.

Knot the end of a piece of sash cord, pass the other end through the hole in the front frame, then through the hole in the rear frame, and pull it tight without disturbing the setting of the frames. Mark the cord where it emerges from the rear frame, close the ladder up and knot the cord slightly inside the mark, because the knot will be pushed along the cord as the ladder's weight tightens it. Repeat this procedure with the other cord, aiming to get it exactly the same length.

The ladder is now complete. Give it a quick once-over with sandpaper, then varnish it. Ladders should not be painted. Varnish lets you see the condition of the wood.

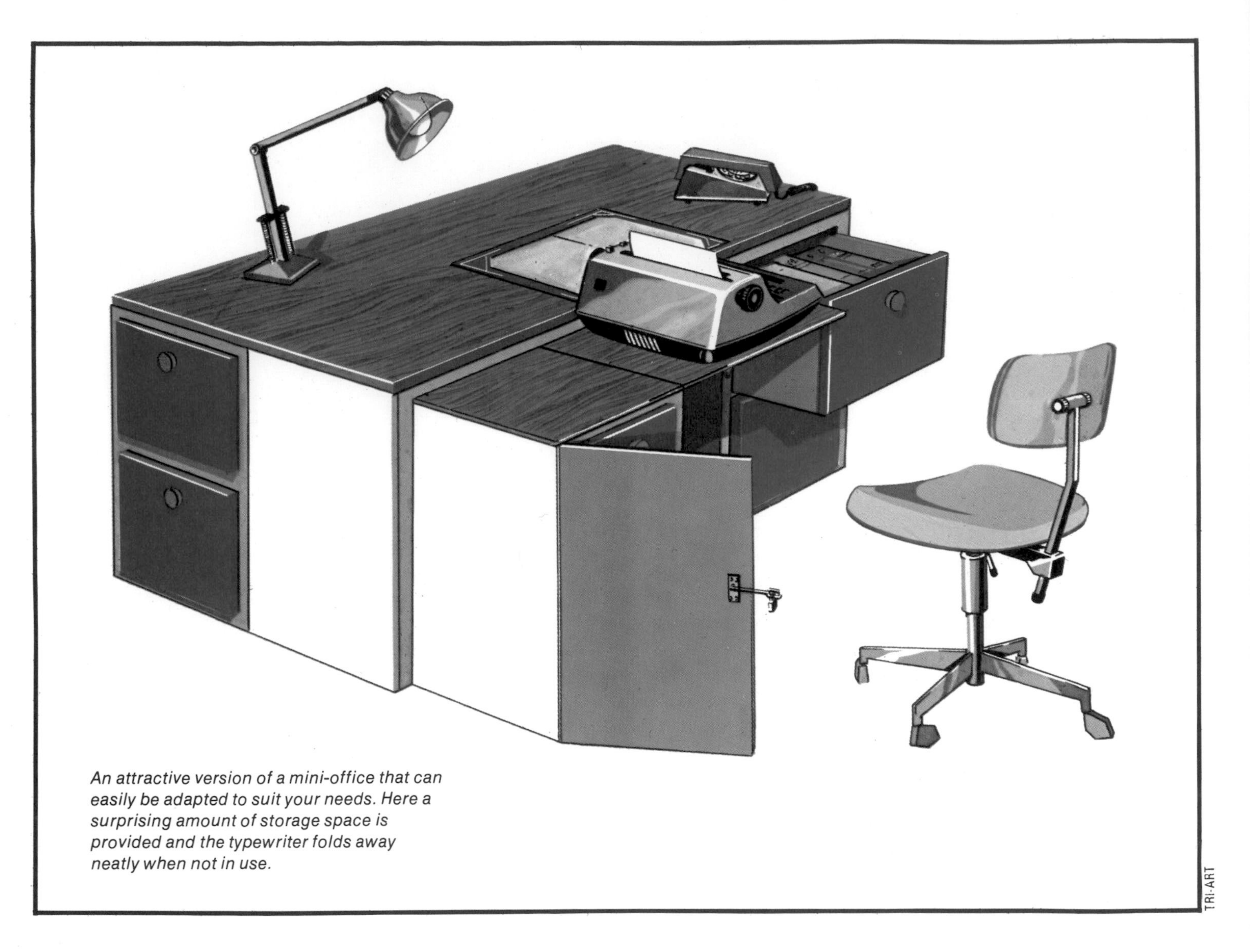

An attractive version of a mini-office that can easily be adapted to suit your needs. Here a surprising amount of storage space is provided and the typewriter folds away neatly when not in use.

Mini-office for the home

If you do work that involves writing or typing at home—whether running a small business, keeping your job records up to date or just coping with household bills—you probably find that your desk, and even the floor around, becomes a welter of loose papers and pencils. What you need is a properly arranged working area—makeshift arrangements never function well. The ideal solution is to build this "mini-office"—a combination of a desk and a huge storage space, which folds up into a compact box when it is not in use.

The mini-office described here is extremely easy to build, and the dimensions given can be altered if its size does not suit you. It has four enormous drawers and a pull-out cabinet with shelves for storing papers that you need to consult quickly. In addition, there is a hinged flap on which a small portable typewriter can be mounted permanently, so that it can be swung up ready for use in a moment. Any or all of these features can, however, be modified to suit your individual requirements.

Design

The mini-office illustrated in Fig. 1 is designed to fit into the right-hand corner of your room; the right side and back are therefore completely flat and free of drawers, shelves and cupboards. To fit it into a left-hand corner simply build a mirror image of the design shown.

As the unit is shown, there is a set of drawers on the right-hand side of the front and a hinged cupboard on the left. When this cupboard is in the closed position, it is the same width as the drawers; this leaves a gap in the center into which a chair can be pushed (a width of approximately 20 in. is adequate for most office-type chairs). If you want to make the whole unit lockable, enlarge the hinged cupboard so that when it is in the closed position, its door can be opened to stretch across the knee-hole of the desk and fasten on to it with a padlock, as shown in Fig. 3. The drawers can be given separate locks of a conventional type. But don't make the cupboard over-large or it will place too great a strain on its hinges.

The cupboard swings open into the working position to give you an additional working surface on your left-hand side. The special feature of this cupboard is the hinged flap fitted to its rear side. This can be lifted and bolted into the main frame, as shown in Fig. 2. It will then be rigid enough to provide a typing surface for a lightweight typewriter. If you bolt the typewriter to the flap, you can leave it in place when you let the flap down and close the cupboard. This arrangement is particularly useful because you can arrange your typing surface to the level that best suits you; although this obviously varies between people and on the height of chair you use, 25 in. or 25½ in. is a good typing surface height, about 2 in. less than the usual height of a desk top.

The back half of the left-hand side is designed to be fitted with two more drawers or shelves, depending on your own requirements.

The dimensions shown in Fig. 1 will give you a desk 54 in. long and 36 in. deep; and since

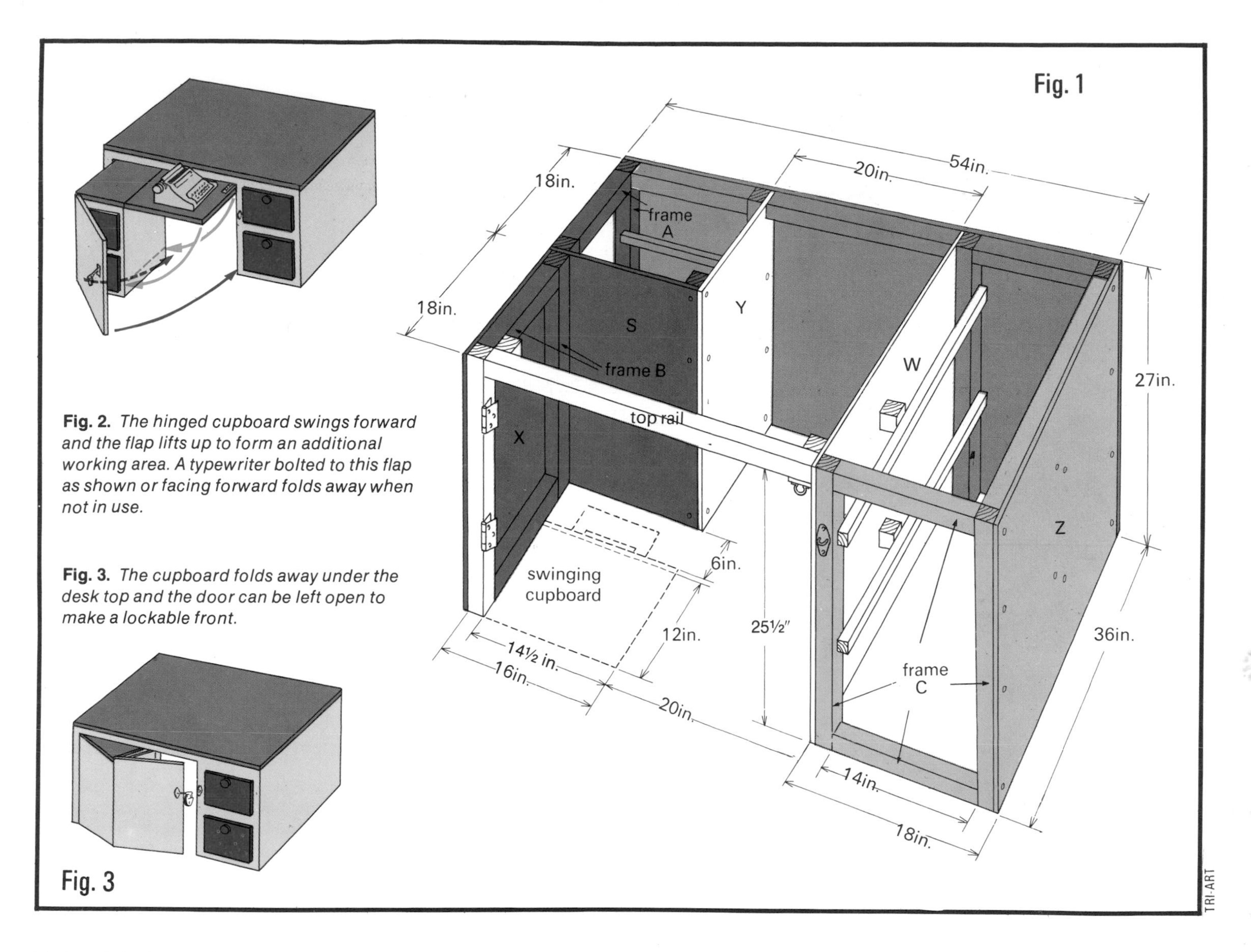

Fig. 2. *The hinged cupboard swings forward and the flap lifts up to form an additional working area. A typewriter bolted to this flap as shown or facing forward folds away when not in use.*

Fig. 3. *The cupboard folds away under the desk top and the door can be left open to make a lockable front.*

Above. *This mini-office with a hinged cupboard and a lockable front is built using an easily constructed framework.*

this is a little deeper than many desk tops, a row of bookshelves made to fit along the back edge of the unit would still leave an adequate working area on the top.

Note that the whole depth of the right-hand side of the unit is available for drawers. You may not need drawers 36 in. deep (which is very deep indeed for a drawer). In this case, you can make the drawers shallower from front to back and use the space that this releases for some other kind of storage space reached through the desk top, for example a locker for stationery or an inset waste-paper basket that can be lifted out to empty it.

If you have full-depth drawers, you should fit them with stops to prevent them from being pulled out the whole way, which would place a great strain on the runners. The best way of doing this is to nail a wood block to the inside of the top (for the top drawer) and to a batten fastened across the inside of the unit (for the lower drawer) *after* the drawers have been inserted, so that the blocks catch on the drawer backs when they have been pulled out a certain distance.

Basic construction

A simple but effective construction system is used for this unit. The top and side panels are made from sheets of plywood which are joined by simply nailing them into lengths of nominal 2 in. x 2 in. or lumber which run the length of the joints. The construction is further braced by three rectangular frames of the same lumber, all of which are of identical size if you keep to the measurements given here.

The *top* gives essential rigidity to the unit and should be cut from a single sheet of ⅝ in. thick particleboard, which is fairly resistant to warping. Don't try to make do with thinner board or a weaker material and don't use two pieces joined together; you won't get the same strength as from a single piece.

The *sides* should be cut from ½ in. plywood. Other materials are unsuitable. Don't use decorative plywood or a board with a plastic-laminated surface, since you will spoil the surface by nailing through it.

All *the frame members* should be made from nominal 2 in. x 2 in. lumber, as this will give plenty of room to nail into from two sides. The frames are assembled simply with butt joints and 4 in. nails driven through pilot holes. The frames do not have to be very rigid, as they are either covered in plywood or braced by adjoining pieces.

All the vertical frame members are the same height, which will speed up the cutting-out process.

Order of construction

Mark and cut all the *vertical members* needed for the frame and make sure that they are all exactly the same length. Then construct the *frame.* Mark and cut the horizontal frame members to size and nail through the uprights into the cross pieces to form the frame rectangles which are marked A, B, and C in Fig. 1. Check that the frame rectangles are square and push them into shape if they aren't. Don't cut any other frame parts yet.

Mark and cut the *particleboard top* and the *plywood side panels* to size. Cut three end pieces the size of piece W (27 in. x 36 in.), checking that the corners are square. Then cut one of the pieces exactly in half to form pieces Y and X. (This can only be done if frames A and B are the same size.) The remaining full-sized pieces are parts W and Z. Mark and cut the back piece to size.

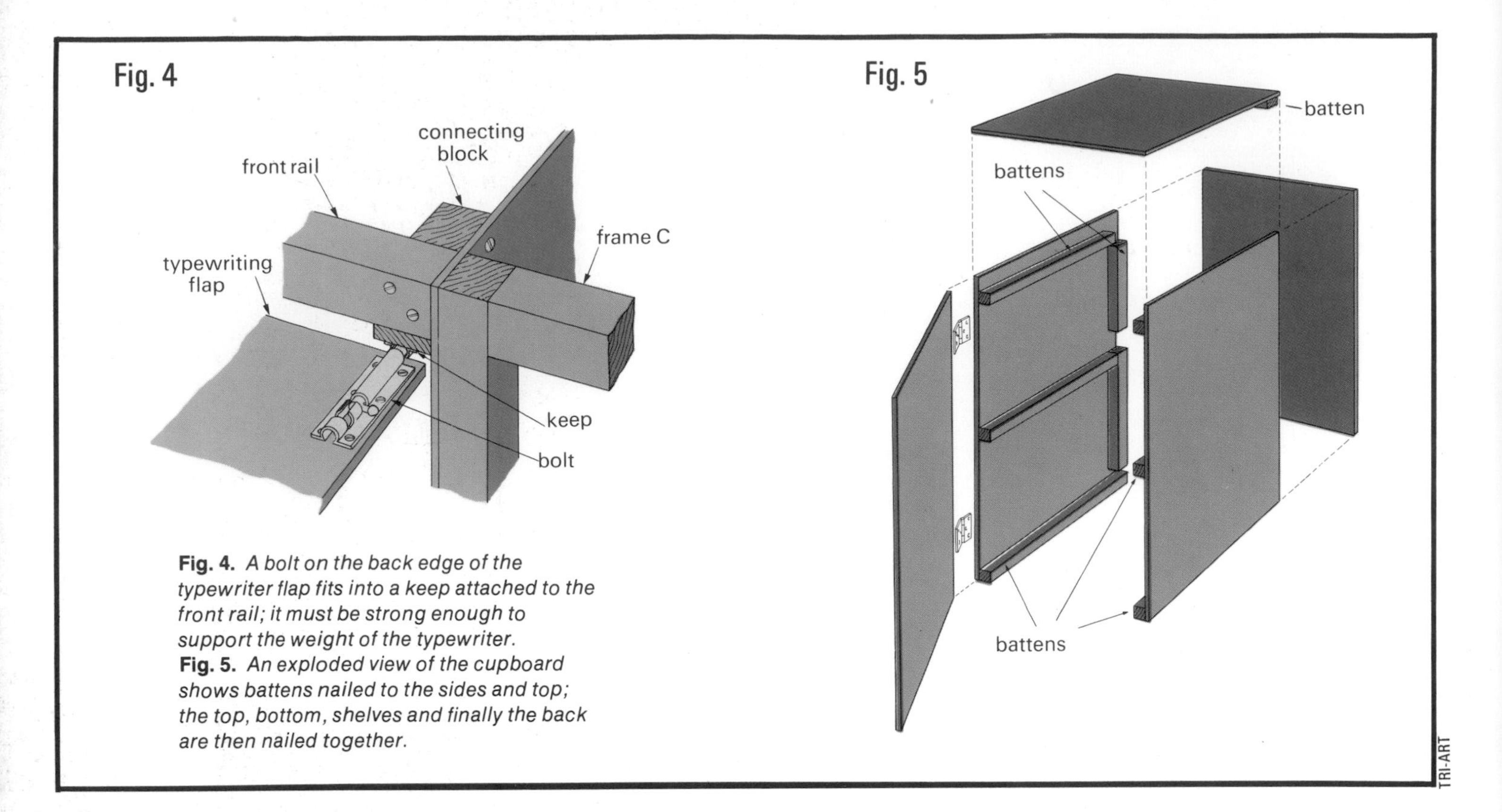

Fig. 4. *A bolt on the back edge of the typewriter flap fits into a keep attached to the front rail; it must be strong enough to support the weight of the typewriter.*
Fig. 5. *An exploded view of the cupboard shows battens nailed to the sides and top; the top, bottom, shelves and finally the back are then nailed together.*

Now partially assemble the parts as follows to give three sections capable of standing upright by themselves. For the *first section,* cover the face of frame rectangle B with plywood piece X, and fasten it on with 1¾ in. nails, which should be used for all ply-to-frame joints. Nail the plywood piece S to the side of frame rectangle B, then nail frame rectangles A and B together to form one end of the unit. Nail a vertical member to the side of piece S that will be joined to piece Y.

For the *second section,* nail the piece of plywood W to the frame rectangle C.

For the *third section,* take the piece of ply that forms the back and nail through the ply into the center verticals (first mark the position of the verticals on both sides of the ply, then place the struts on the floor and nail through the ply into them). Then nail the piece of ply Y to the left-hand center vertical (as seen from the front, but it will be on your right as you nail).

Now that the three sections will stand upright by themselves, the final assembly work can be done. First nail through the back into the two end sections with a few 2 in. nails; you will need a helper to hold the ends in place while you nail into them. Then nail the middle crosspiece W in place.

Cut the front rail, a length of 2 in. x 2 in. lumber, to fit exactly between the inside edge of frame rectangle B and the side of piece W that faces toward it.

From the same size lumber, cut a couple of 2 in. blocks to join the front rail to frame rectangle B and piece W. They will have to be put on with screws to achieve the necessary rigidity; two screws should pass into each side of each block and these should not be in line with each other or the wood may split. Use 2½ in. No. 8 screws to fasten the front rail and frame B to the blocks and to fasten piece W to the right-hand block. Pre-drill the screw holes carefully, preferably using a Screw-mate bit. Screw on the blocks, then screw the front rail to them.

Check the unit for squareness, nail through piece Y into the member fixed to piece S, and add additional nails to all the joints between the pieces. Then cut and fit pieces of 2 in. x 2 in. around the top edges of the unit so that the top can later be screwed down firmly into them.

Shelves and drawers

Fit the drawer slides or shelf supports now, before you fit the top in place and while the inside of the unit is still accessible. For methods of making and fitting drawers refer to CHAPTER 14 and for details of how to mount shelves see CHAPTER 35. It will probably be easiest to fit the drawers on slides of nominal 1 in. x 1 in. lumber as shown in Fig. 1, and hang the drawers by the "three cleat" method; they will then be slightly under width for the space but you can make them false fronts of ¼ in. ply to fit between the frame members. If the drawers are very deep, brace the center of the slides with blocks as shown in Fig. 1.

Fitting the top

The top should be cut to size and nailed—or better screwed—firmly in place at 4 in. intervals all around to keep the unit from twisting. If you cut the top larger than the walls of the unit to give an overhanging edge, make sure that you leave the same amount overhanging all around, unless the unit is to fit flush against a flat wall without a skirting board.

Hinged cupboard

The hinged cupboard is also made from ½ in. plywood and 2 in. x 2 in. battens. Two solid ply sides are joined firmly by the battens to the shelves, top and bottom. The cupboard is held square by a ply back, which should be fastened to battens set all around the inside back edge of the cupboard.

Cut the two sides to size: the height is the same unless the typewriter flap is to be more than 2 in. lower than the desk top, in which case the side that is at the front of the cupboard when it is shut should be higher to bridge the gap between the cupboard and the top rail.

Nail battens to the sides as shown in Fig. 5, then nail the top, the shelves and finally the back to these battens. Then cut and fit the cupboard door, fit a magnetic door catch and hang the cupboard on large, solid hinges, because it will have to take a certain amount of pressure.

Mount the typewriter flap at the same height as the cupboard top on a metal (not plastic) piano hinge and fit a large door bolt to support it as shown in Fig. 4; the *keep* of the bolt (the socket side) may have to be mounted on a block to bring it down to the right level.

Finish

This useful unit can be finished in a variety of ways. If you are going to paint it, punch the nail heads below the surface and fill the nail and screw holes with filler; sand this down well before painting. For a stronger finish you can cover either the top or the entire unit with a plastic laminate, applied with one of the brands of contact adhesive that are available.

Repairing veneers

Much of our finest furniture is veneered, as it has been for centuries. But the veneer that gives it much of its beauty can be damaged by excessive dampness, high heat, as from a too-close fireplace, or by sheer carelessness—like cigarette burns. Fortunately, you can make your own repairs if you work carefully with the right tools and materials.

The base of the mirror stand at the right is an example of the type of veneer damage that can result from heat-drying and blistering, as from too close proximity of a cherry log fire over an extended period. Excessive dampness can produce similar veneer separation, though this type of complete break-away is less likely.

If the veneer is merely rippled

Veneer that has separated from the base wood as the result of dampness, as in unheated storage buildings, is often surprisingly easy to repair. A raised area in a large flat surface, for example, is usually a simple matter of glue separation. Try ironing it back flat, with the iron set at moderate temperature to avoid damaging the finish. Move the iron slowly over the rippled spot, pressing it down, and have a helper follow with a tray of ice cubes over the same spot to chill and harden the glue under the veneer. Older furniture, especially, is likely to have its veneer bonded with animal glue which softens with heat, hardens when chilled. If this method doesn't do the trick, slit the ripples far enough to permit working in liquid glue with a knife blade. Then use heavy weights to flatten the ripples while the glue hardens. Place a polyethylene sheet between the weights and the surface to keep the weights from sticking. This can be peeled and sanded off later. Refinish, if necessary, as described later.

If a patch is necessary

If a section of veneer has actually broken away, and can't be found, you'll have to buy a piece of the same species to make a patch. Cabinetmaker's supply houses and craft and hobby suppliers are the best source. (You'll usually have to buy a larger piece than you require because minimum sizes are based on typical veneer work.) The thickness, however, is usually 1/40 in. on English furniture, 1/28 in. on American furniture. Both are available from many sources.

Cutting the patch

Much damage to veneered furniture is along the edge of a surface. This part is the most

vulnerable to knocks. It is also, fortunately, the easiest part to repair invisibly.

Wherever possible, repair damage to an edge by removing and replacing a straight strip running along the whole edge of the piece, with the grain of the veneer. In this way, the old and new pieces will be joined along the line of their grain, which conceals the joint; and the joint will be a straight line, which makes it very easy to fit. You can leave the outer edges projecting ⅛ in. or so over the edge of the furniture and trim them later, after the glue is dry.

This method is only suitable for veneer with a fairly straight grain. Knotty veneer and veneer damaged at the end of the line of the grain, or in the middle of a surface, are better repaired by the following method.

This involves replacing the damaged area with a "boat-shaped" patch, i.e. an oval one with pointed ends, with the long diameter of the oval pointing along the general direction of the grain. This is the shape that will be least obvious when glued down. The more a joint runs across the grain, the more it will show, so keep the ends of the oval as pointed as possible.

Find out the best shape and size for the patch by putting a piece of tracing paper over the damaged area and drawing "boat" shapes on it until you find one that seems right. Draw a few lines along this shape to give the direction of the grain.

Use a piece of carbon paper to transfer the shape from the tracing paper to the new veneer, pressing very lightly and getting the grain running the right way by means of the lines on the tracing. Then cut the shape out, using a handyman's knife with a brand-new blade. Slope the knife sideways as you cut, so that the cut edge slants; the handle of the blade should point outward at 45°. It is a good idea to cut the shape out slightly too large, as this will give you a chance to move it around over the damage to match the grain more exactly.

When cutting a straight strip of veneer, there is no need to slope the knife. Cut along a steel ruler or straight-edge to keep the line straight.

Take your oval strip of veneer and lay it down over the damaged area. Move it around until the grain matches best, then hold it in place with your left hand and draw round it lightly with a sharp pencil.

Now remove the patch and cut around the mark, just inside the pencil line and holding the knife at the same 45° angle as you used for cutting out the patch. Be sure not to cut out too much of the old veneer; if the patch proves too big you can cut it down, but if it is too small you will have to make a new patch. The slanting joint will hide any small discrepancies, however.

Remove the damaged veneer inside the cut line with a *very sharp* narrow chisel, held bevel (slanted) side down. You should just chisel out the old veneer and the glue under it, without hurting the wood that the veneer is glued to. This is not as hard as it sounds. If your chisel is blunt, sharpen it. Large areas of old glue can be removed with hot water.

Try the patch in the hole and trim it to size if necessary. It is now ready for gluing in.

Gluing in the patch

Patches in veneer need to be glued in very strongly and held in place while the glue dries. A good glue is a white wood glue, but use plastic resin glue for large, difficult patches. The first kind can be attached without clamps if necessary (see below) but it is always best to clamp the patch down if possible.

The patch must be held down firmly. You could use a carpenter's C clamp with a wood block under it and paper in between to avoid sticking the block. The block should cover the whole patch, or the edges will curl. A good improvised clamp for a horizontal surface would be a sheet of glass with three or four bricks carefully laid on top of it and paper underneath.

Apply glue thinly but evenly to the two surfaces to be glued, and press the patch in place. It doesn't matter if a bit of glue oozes out; wipe if off with a damp cloth. Put a sheet of brown paper over the patch to absorb the surplus, then apply the clamp and leave to dry for as long as it needs. The longer you can leave any type of glue, the better.

A patch glued with white glue may be ironed in place with an ordinary electric iron set to "warm." When the adhesive is tacky, cover the patch with brown paper and run the iron back and forth until the glue is dry.

If you don't have an iron or any clamps, you *can* stick the patch on with contact adhesive.

Finishing the patch

When the glue is thoroughly dry, pull off as much of the brown paper as will come off—damping the paper carefully helps. Then wrap some very fine sandpaper around a flat-surfaced cork or wood block and sand off all traces of paper and glue, carrying on until the surface of the patch is level with the veneer around it. Obviously, this will remove some of the polish from the surrounding veneer, but that doesn't matter. Always sand along the grain if possible. Finish the sanding with finishing grade paper (the finest of all), so that the surface is beautifully smooth.

If the surface you are repairing is very light-colored, or the veneer has a coarse, open grain (examples of this type of wood include oak, ash, rosewood and sapele), you should seal it with a proprietary grain filler in the appropriate color. Otherwise enormous amounts of polish will sink into the wood, making it very hard to achieve a good sheen. Rub the filler into the veneer and sand it smooth with finishing grade paper.

Before applying any finish to the patch, make sure it will turn the right color when it is polished. Most finishes darken wood slightly, but not much. As a rough guide, examine the area around the patch where the old finish has been sanded off. If it is not exactly the same color as the patch itself, you will almost certainly need to stain one or the other to match.

You can discover the difference that applying the finish makes to the color of the patch by applying test coats to an offcut of the veneer. For details of how to apply the finish, see below. There is no need to rub down between coats, however.

If you need to stain the patch, buy a good brand of ready-mixed spirit-based wood dye. The *depth* of its color can be controlled, because you can dilute it with alcohol, but the *tone* (i.e., whether it is reddish or yellowish) must match. Experiment to find the right dilution on offcuts of veneer, then paint the diluted mix on to the patch with a brush, being careful not to paint beyond the edge. Water-based wood dye is not suitable because it spreads out.

It is now time to apply a finish or polish the patch and the sanded area around it. "French" polish should be applied undiluted with a rubbing pad made of a small piece of cotton wool wrapped in a lint-free cotton cloth which has been slightly moistened with linseed oil. The purpose of the oil is to stop the cloth from sticking to the sticky polish. If it shows signs of doing this, put a little more oil on the cloth. Ask your paint dealer about ingredients for French polish.

Rub the polish on, not too generously, with small circular motions. Let it dry for 20 minutes, then rub it down very lightly. Apply the next coat, and go on doing this until the sheen matches the surrounding polish. This may take a number of coats, depending on the polish and the wood.

Do not rub down the last coat of polish. Let it dry really thoroughly for two or three days, then complete the job with a good wax.

Some modern veneered furniture has a satin finish. This can be produced with varnish made for the purpose, and sold by paint dealers.

Opposite page. *The veneer on this antique desk lid was badly damaged along one edge, and a strip needed replacing completely.*
1. *The first step: removing the polish around the damaged area with a cabinet scraper.*
2. *Matching a piece of new veneer to the old wood. It should match the area from which the polish has been scraped off.*
3. *A good quick way of seeing whether the new piece will match when it is polished is to wet it—on both sides to prevent warping.*
4. *Removing the damaged strip of old veneer. The chisel is held bevel side up at the edge . . .*
5. *. . . and bevel side down farther in.*
6. *Delicate veneers should be dampened on both sides before cutting and held under a heavy straight edge to avoid splitting them.*
7. *The new strip will take hard knocks, so it is stuck on with plastic resin glue for extra strength.*
8. *Each piece of veneer is cut to length on the spot for maximum accuracy.*
9. *When all the pieces are in place, air bubbles are forced out by rubbing the veneer lightly with a rounded hammer surface. It is then clamped in place for a while.*
10. *If you use white glue, you can iron a patch on. The heavier the iron, the better.*
11. *A short straight-grained patch in an edge, as shown here, should be cut overlength in a wedge shape and gently knocked in sideways to hold it firmly. Trim the ends to length with a very sharp chisel or handyman's knife.*
12. *Burred (that is, knotty) veneer should be joined along the wavy line of its grain.*

1
2
3
4
5
6
7
8
9
10
11
12
NELSON HARGREAVES

Above. *A traditional, well-proportioned, double-ended hipped roof gives a very pleasing finish to this house.*

Hipped roof—English style

The following pages show the construction details of a hipped roof, a roof style used in many parts of the world. Because of the compound angle cuts required, however, you need more skill to build it, but it has advantages in addition to its distinctive appearance. The structure shown in the drawings is that of a home built in Britain, and is somewhat more complex than its American counterparts, but it illustrates basic principles that have long been followed. These principles still apply when you modify the plan to fit the requirements of your local code.

The four sides of a hipped roof (three sides on a duplex house) all slope upward and inward toward the apex of the roof in contrast to a gabled roof where the two sides rise vertically and only the front and back slope. However, the slopes at either end of the roof do not normally meet but lie at either end of the ridge board as in Fig. 1. Here the slope from the eaves on the long sides up to the ridge board are exactly the same as for a gabled roof, and the method of construction is identical. The only difference is that the slope on the long side is continued around the ends of the roof.

The pitch of the roof is usually the same on all sides but the slope on the sides may be different from the slope on the ends. There is no ruling as to what the slope of the roof should be—it can vary for aesthetic reasons, climatic conditions or with the materials used. As stated in CHAPTER 34 an angle between 50 and 60 degrees is usually acceptable.

Main components

The main parts of the hipped roof as shown in Fig. 1 are the same as for the gabled roof as described fully in CHAPTER 34. With the hipped roof the ridge board extends slightly beyond the common rafters to take the hip (corner) rafters. The new components are as follows:

The hip rafters slope upward from the corners of the roof to the ends of the ridge board and they are sometimes birdsmouthed at the top to fit beneath the ridge board; the method of fitting the bottom end is described below. Hip rafters must be one continuous length of timber. As with common rafters, they should never be made from two pieces joined along their length. The size of timber varies as with common rafters, but if you mark timbers of 6 or 7 in. x 1½ in. on your plan the building inspector will correct you if you have over-designed.

Jack rafters slope downward from the hip rafters to meet the eaves, or wall plates, at right angles. These jack rafters are spaced so that they meet the hip rafter from either side at the same point. They should be the same cross section as the common rafters for which sizes are listed in CHAPTER 34.

To calculate the quantities of timber needed count each jack rafter along the long sides of

BARNABY'S PICTURE LIBRARY

Above. *Hipped roofs are especially suitable for bungalows and are easy to maintain as there is no woodwork above the eaves. Here the garage is also covered with a matching hipped roof to make it "belong" to the house.*

the house as another common rafter and discount the jack rafters on the ends. This rough guide works because the shortest jack rafter plus the longest on each hip rafter equal the length of one common rafter, as do the next shortest and next longest and so on. It applies, however, only when the jack rafters are spaced at equal distances from each other and the roof has the same pitch on all sides.

Center section

Cut, mark and fit the wall plates, joists, purlins, collar ties and struts as described for a gabled roof in CHAPTER 34. There are three main differences between a gabled roof and a hipped roof, as described below.

The first compensates for the fact that the hip rafters exert considerable force upon the joint between the wall plates on adjacent walls of the house. Therefore one of the wall plates must be half lapped over the other as indicated in Fig. 3 and both plates should be continued for 3 in. beyond the point where they overlap. This will prevent the wall plates being forced apart which could allow the hip rafter to slip.

The second is that the ridge board must overlap the last pair of common rafters at either end by a little more than the width of the hip rafters, that is about 6 in. When the hip rafter is cut at an angle at the top, this angled end meets the side of the ridge board that extends beyond the last pair of common rafters.

The third difference is that you should cut the purlins longer than you have drawn them on your plan and fit these overlength purlins in position. Then when you fit the purlins at either end you can cut all the purlins to the correct length and angle at the same time.

End sections

When your center section is completed this far check that it is square and rigid. Then fit the following pieces to turn the angle of the roof around the corner.

Hip rafters

The hip rafters set the angle of slope for the ends of the roof and support the jack rafters, so it is essential that you fit them properly. The instructions below give the right way to do it.

Take a length of timber for each hip rafter and cut the bottom end at an angle so that when in place the lower end of the timber will meet flush with the top horizontal surface of the wall plates (note that this is not the final cut). Then cut it slightly longer than the distance from the top of the wall plate at the corner of the house to the point where the ridge board and the end common rafter meet. Then hold the timber in place over and parallel with the position it eventually has to take, with the lower end resting on the wall plate in order to mark out the joints at either end.

There are three cuts to make to fit the hip rafter to the ridge board and end common rafter (see Fig. 2): a short horizontal cut to make a rabbet under the ridge board, a vertical angled cut to fit the hip rafter to the ridge board, and another vertical angled cut to fit the hip rafter to the end common rafter. (On some roofs, where the hip rafter does not meet the common rafter, this third cut is unnecessary).

With a pencil, mark on the underneath and side of the hip rafter the horizontal plane of the bottom ridge board. Then, from the marked horizontal line, draw vertical lines parallel with the side of the ridge board on the side of the hip rafter. These lines should be an inch or so from the point where the marked horizontal line meets the inside edge of the hip rafter to provide a rabbet when the joint is cut. Also mark lines on the sides, top and bottom of the hip rafter that are parallel with the side of the common rafter which it meets. Then cut out the slightly rabbeted joint shown in Fig. 2. Check that the top end of the rafter fits and then mark and cut the lower end.

The foot of the hip rafter exerts considerable pressure upon the wall plate at the corner of the house, so use the method shown in Fig. 3 to spread the weight as much as possible over the two adjacent walls. Here a length of wood, an *angle tie,* is notched and nailed into the two adjacent wall plates to form a right angled triangle. Another piece of wood, a *dragging beam,* stretches between the middle of the angle tie and the joint between the wall plates. This beam supports the foot of the rafter and spreads the weight evenly onto both adjacent walls.

Cut the *angle tie* out of 7 in. x 3 in. timber and make a hole in the middle to take the end of the dragging beam. Nail the angle tie in place. Then cut and nail a batten across the joint between adjacent wall plates so that it is parallel with the angle tie.

The *dragging beam* is cut from the same size timber as the angle tie and rabbeted on either side at one end to pass through the hole already made in the angle tie. Then cut a groove in the underside at the other end of the dragging beam so that it fits over the strip of batten already nailed across the joint between the wall plates. Make two saw cuts in the top of the dragging beam forming an oblique tenon as shown in Fig. 4 to take the foot of the hip rafter. Then fit and nail the dragging beam in place between the angle tie and wall plates.

Place the hip rafter in position so that it fits tightly against the ridge board and end common rafter but with the foot against the side of the dragging beam; mark the outline of the oblique tenon cut in the dragging beam on the foot of the rafter and cut to shape. Now check to see whether this hip rafter fits the spaces for the other hip rafters; if so mark the other rafters from the first (allowing for left and right), but if not repeat the measurement by the direct method. Then nail the hip rafters firmly in position.

Jack rafters

Mark the positions of the jack rafters on the side wall plates so that they are spaced at equal distances from each other. This ensures that the two longest jack rafters on the end walls fall the same distance either side of the ridge board, but it is a point worth checking. Mark two lines on the wall plate for each rafter with a cross in between indicating clearly where the foot of the rafter is to be. Mark the top ends of the jack rafters on the hip rafters in the same way. Then rest a length of timber in place at a number of these points stretching from the wall plate to hip rafter and check that it lies at right angles to the wall plate.

Mark and cut the jack rafters to length individually. Lay a length of timber over the wall plate and hip rafter and mark it both for length and the angle of the birdsmouth. Then cut the first piece to length. Check to see whether the angles at the ends of the rafter fit all the way along the hip rafters; they should. If so, cut one end of the next rafter from the first and then hold the timber in place while you mark it to length. Cut and nail all the jack rafters in position.

Purlins

You have already fitted the purlins on the longer sides of the roof in position so that they extend beyond the hip rafters. You must now cut the ends of these purlins to length at the correct angle and the purlins for the ends of the roof to size so that they meet in a miter.

To cut the ends of the purlins already fixed in position, take a saw and lay it against the side of the hip rafter that faces the end of the roof. Then cut through the ends of the side purlins at the same angle as the side of the hip rafter.

To find the angle for the ends of the purlins take a piece of timber the size of the purlins and lay it on the outside of the end of the roof so that it overlaps the hip rafter at the level of the side purlin. Take a straight edge and continue the line of the side of the hip rafter (the side that you previously lay the saw against) around the side of the piece of timber. Cut to these lines to give you a template for marking the end purlin. Measure the length of the purlin either from your plan or from the roof itself, and cut the ends to the angle of the template. Then nail the purlins in place at either end of the house—nail at the ends first and then to all the jack rafters in between.

With the framework of your hipped roof completed you have met all the differences between the construction of a gabled and a hipped roof. There is a considerable amount of work still to be done such as covering the roof with felt, nailing the battens in place and laying the tiles in place but this is all described fully in CHAPTER 34.

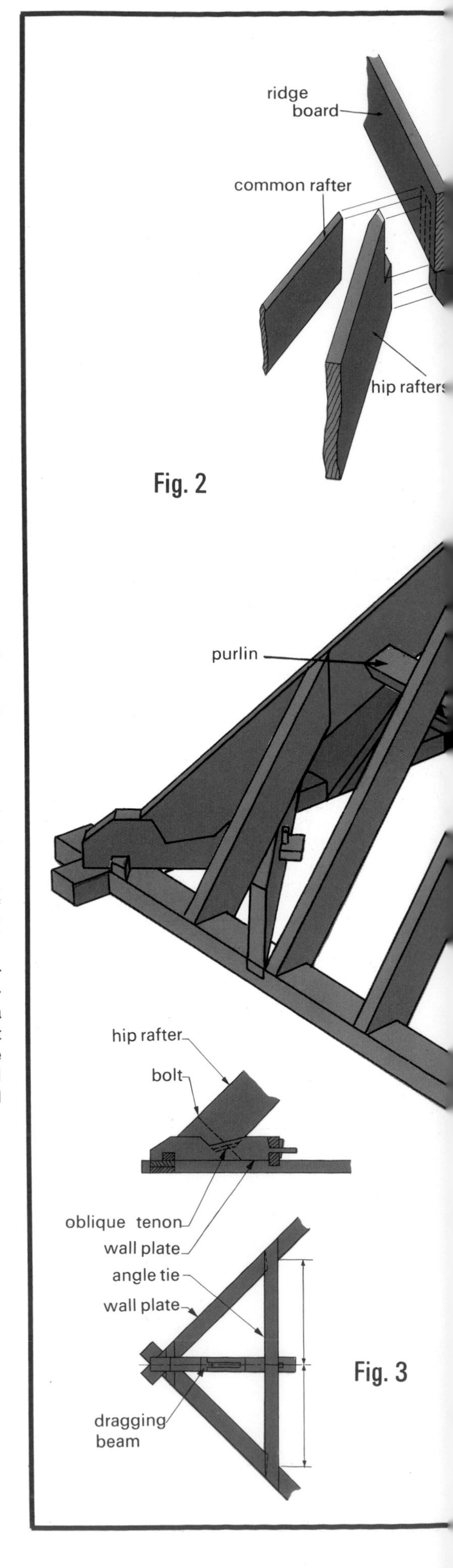

Fig. 1. *The framework for the center section of a hipped roof is the same as for a gabled roof. Here the hipped end of the roof slopes up from the wall plate (bottom left) to the ridge board (top right) in contrast to the vertical end of a gabled roof (for details of a gabled roof see* CHAPTER *34).*
Fig. 2. *The common rafters meet the ridge board at right angles, but the hip rafters are birdsmouthed as shown to slip under and meet the ridge board at an angle.*
Fig. 3. *A plan (bottom) and side view (top) of the construction in* **Fig. 4** *which shows how the force exerted by the foot of the hip rafter is spread over both wall plates. The dragging beam which supports the hip rafter stretches from the corner of the roof to the angle tie fixed to the wall plates.*

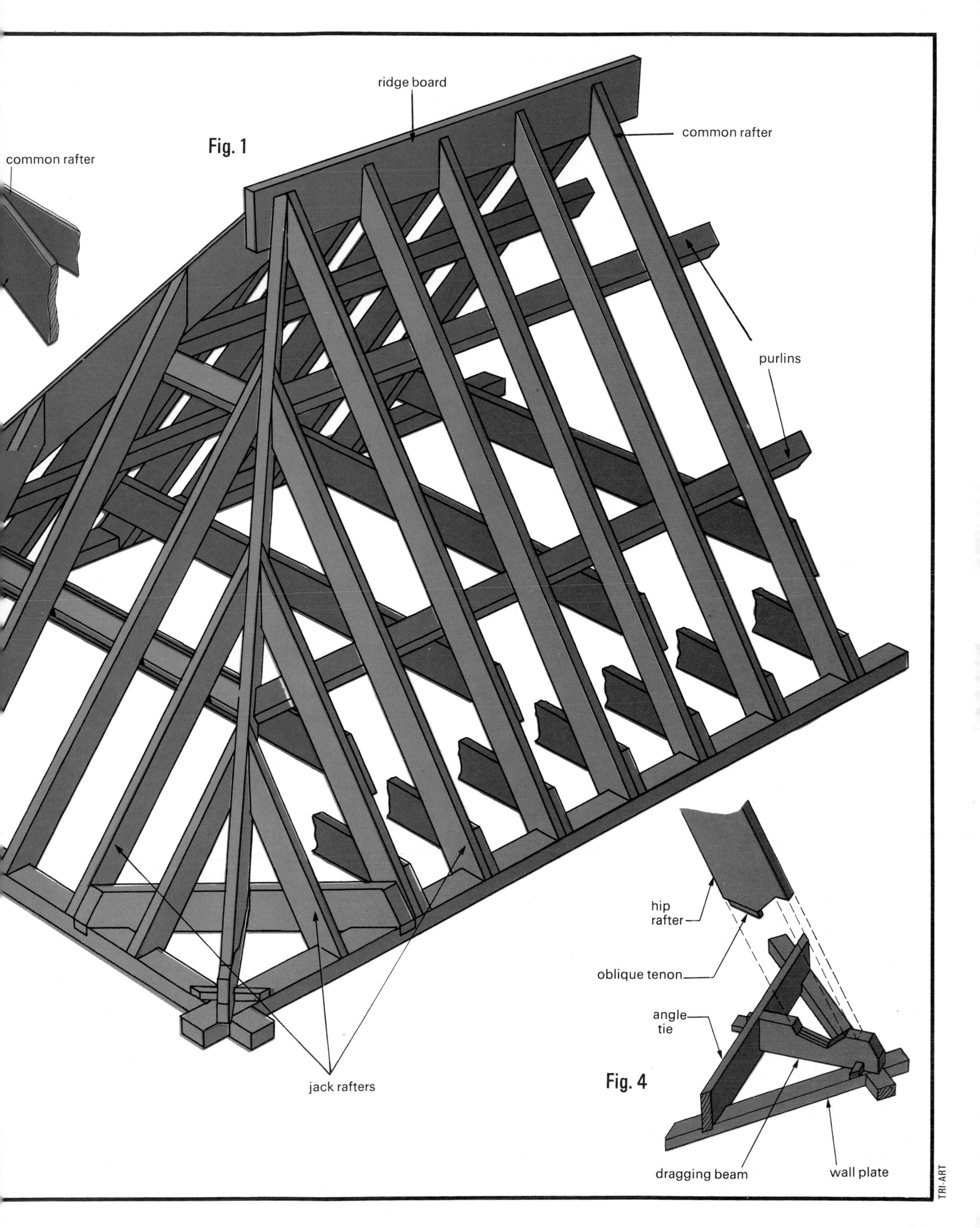
ridge board
Fig. 1
common rafter
common rafter
purlins
jack rafters
hip rafter
oblique tenon
angle tie
Fig. 4
dragging beam
wall plate
TRI-ART

Hipped roof—with an angle

The techniques you use in building a hipped roof can be applied to other special roof construction situations. And the form of the hipped roof may help solve other problems. The illustrations that follow show how hipped roof design may simplify additions to your house, major or minor. The methods used for the basic hipped roof are followed in the right angle addition shown—all in British style, described in British terms.

You can roof over an interesting L-shaped house, for example, or erect a new extension at right angles to the main roof if the original line of the house cannot be extended. These techniques can be adapted to other situations such as building a sloping roof over a bay window.

The extension that is built at right angles to the main part of the roof, can itself be hipped or gable ended. But the joint between the extension and the main part of the roof must necessarily follow the line of a hipped roof (see Fig. 1). In order to build two parts of the roof at right angles to each other you must first familiarize yourself with the techniques described for building a gabled roof (CHAPTER 34 tells you how to build a gabled roof) and a hipped roof (see CHAPTER 45).

The few differences between a roof which turns through 90 degrees and a simple rectangular roof are the positioning of the valley rafter (which is essentially the same as the hip rafter reversed), the method of providing a watertight joint between the two slopes of the roof directly above the valley rafter, the method of joining the ridge boards and the order of construction. These are discussed in the same order below.

Hips and valleys

The two sections of the roof meet to form an internal corner between the outside of the walls of 90 degrees and an external corner of 270 degrees (see Fig. 1).

The external corner where the two sections of the roof meet is the same as for the corner of a normal hipped roof and identical construction techniques are used.

At the internal corner the sides of the roof slope upward and ''inward'' to the point where the two ridge boards meet. The sides of the roof meet above the valley rafter which is in many ways similar to the hip rafter. This valley rafter slopes from the joint between the wall plates in the internal corner up to the joint between the boards—and if seen in a plan view continues the line of the hip rafter on the external corner.

The valley rafter differs from the hip rafter in two ways.

First, the foot of the valley rafter pushes both the wall plates and walls together and not apart as with a hip rafter on an external corner.

Fig. 1 (right). *This L-shaped roof structure combines a gabled and a hipped roof; the major feature is in the joint between the two sections at the hip and valley rafter.*

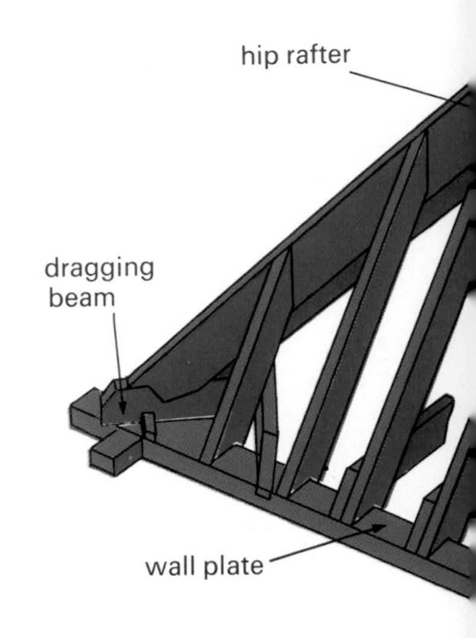

Left. *Here the valley, that is the joint between the two sections of the roof, has been made watertight with lead flashing before tiling the roof on either side.*

Therefore, the valley rafter can be birdsmouthed directly over the joint between the wall plates and does not have to be set in a dragging beam as with a hip rafter. The vertical face of the valley rafter at the birdsmouth does not have to be cut to fit around the right angle formed by the wall plates but can be a straight line cut which butts up against the corner formed by the wall plates.

Second, the jack rafters which run from the valley rafter up to the ridge board carry the weight to that section of the roof and transfer it to the valley rafter. The valley rafter should therefore be braced as much as possible at intermediate points along its length. Struts should run from the valley rafter down to the joists or ideally to runner joists so that the weight is distributed over a wider area (see Fig. 1).

Valley gutters

The joint between the two slopes of the roof in the valley can be made watertight in two ways.

The first method is to use special tiles which

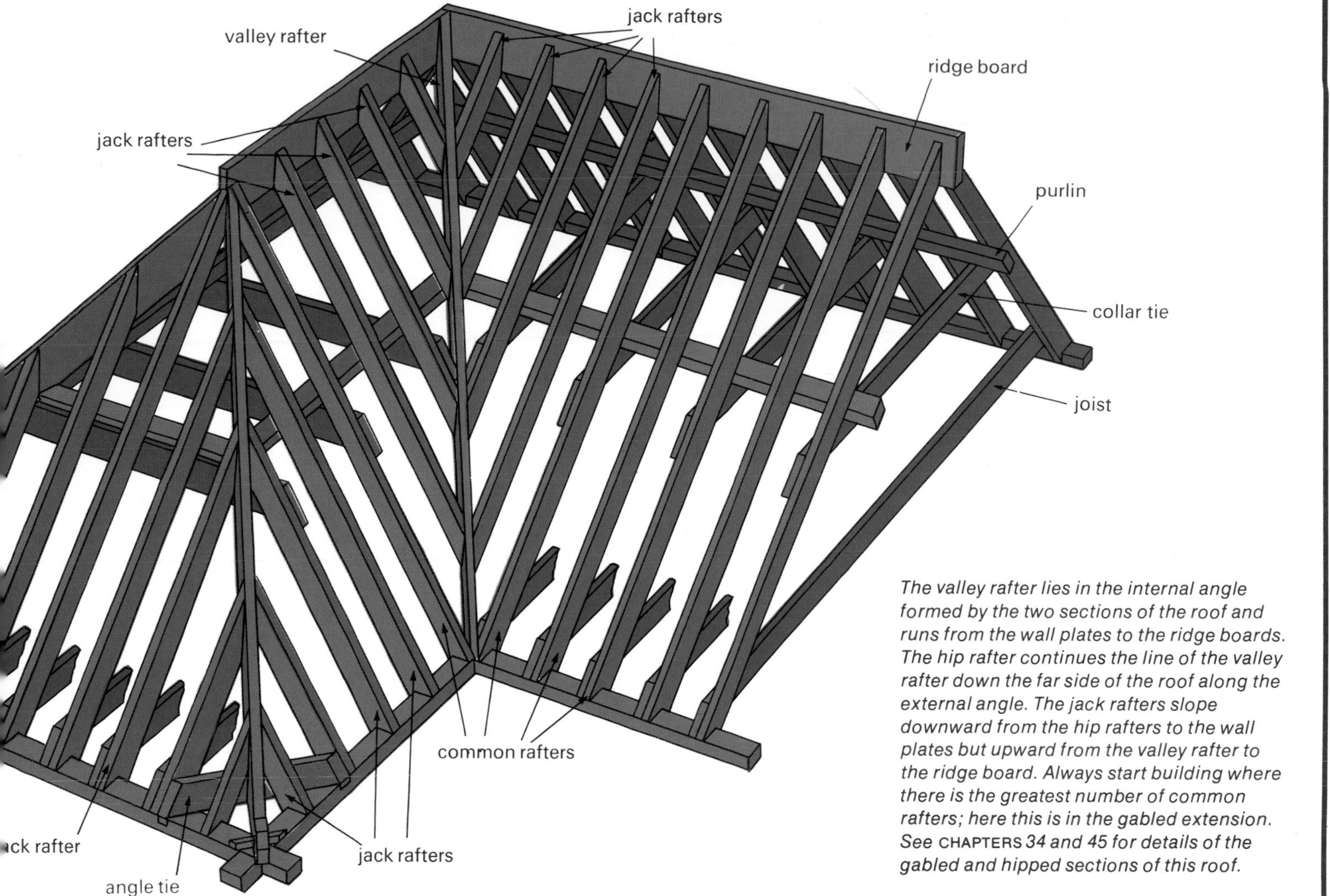

The valley rafter lies in the internal angle formed by the two sections of the roof and runs from the wall plates to the ridge boards. The hip rafter continues the line of the valley rafter down the far side of the roof along the external angle. The jack rafters slope downward from the hip rafters to the wall plates but upward from the valley rafter to the ridge board. Always start building where there is the greatest number of common rafters; here this is in the gabled extension. See CHAPTERS *34 and 45 for details of the gabled and hipped sections of this roof.*

can be purchased for this joint just as they can for the joints over the hip rafter. But since they are made in only a limited number of angles check that you can obtain some locally which are the same angle as the slope of your roof.

In the second method, lead flashing can be run down the joint between the roofs along the line of the valley rafter producing a watertight gutter. In this method the roof tiles stop short of the joint (see Fig. 2).

In an open gutter the space left between the edges of the tiles on either slope is about 8 in. A valley board 1 in. thick and slightly narrower than the gap between the tile edges is nailed in place over the valley rafter and between the jack rafters so that it runs the length of the valley rafter and lies a couple of inches below the edges of tiles. Tilting fillets set either side of the valley board support the lower edge of the bottom row of tiles. The entire gutter, from just above the tilting fillet on either side of the valley board is covered with lead flashing from the ridge board down to the wall plates. Any debris, leaves, moss and so on, that collects in an open gutter is relatively easy to clean out.

In a secret (hidden) gutter there is only about a 1 in. gap between the edges of the tiles so that the gutter is barely noticeable, but the gutter itself is correspondingly smaller than the open gutter. There is no valley board providing a wide flat base to the gutter; instead the tilting fillets are nailed in place either side of the valley rafter leaving only a relatively narrow V-shaped gutter between. This is then covered with lead flashing as previously described. If the roof is not boarded all over, but only battened ready for tiling, a board (called lier board) must be fitted either side of the joint between the two slopes to provide a firm base for the lead flashing.

Joint between the ridge boards

Various joints can be used at the ends of the ridge boards. A simple butt joint, however, is normally sufficient since the hip, valley and common rafters all meet at this point and make a strong joint.

Order of construction

Start by fixing the wall plates around the walls and setting the joists in place as described in CHAPTER 34. Remember, though, to half lap the wall plates in the corners as described in CHAPTER 45 to give extra strength in the corners. The wall plates should also be overlapped in the internal corner where the valley rafter meets the wall plate.

Then mark on the joists the positions of the common rafters as described fully in CHAPTER 34. Mark also the line of the ridge boards and the points where they end. One end of each ridge board is above the wall plate with a gabled roof or at the intersection of the hipped rafters with a hipped roof. The other end of each ridge board is at the point where they meet—but remember that with a butt joint one ridge board ends where it meets the second but the second continues to the far side of the first.

The construction of a roof which turns through 90 degrees is really the construction of two roofs: first build the main framework of one of the roofs and then build the main framework for the other roof while using the framework of the first as partial support.

The best place to start building is that part of the roof where the greatest number of full length common rafters run from the wall plates up to a ridge board. This is not necessarily the main part of the roof; for instance if the main part of the roof is hipped at both ends while the extension is gabled the largest number of full length common rafters may fall in the extension.

Now build the framework for the first section of the roof. Follow the instructions given in CHAPTER 34 for a gabled roof or read CHAPTER 45 for a hipped roof. Once the common rafters and ridge board are in position check that the framework is square and add collar ties and struts to make the framework rigid.

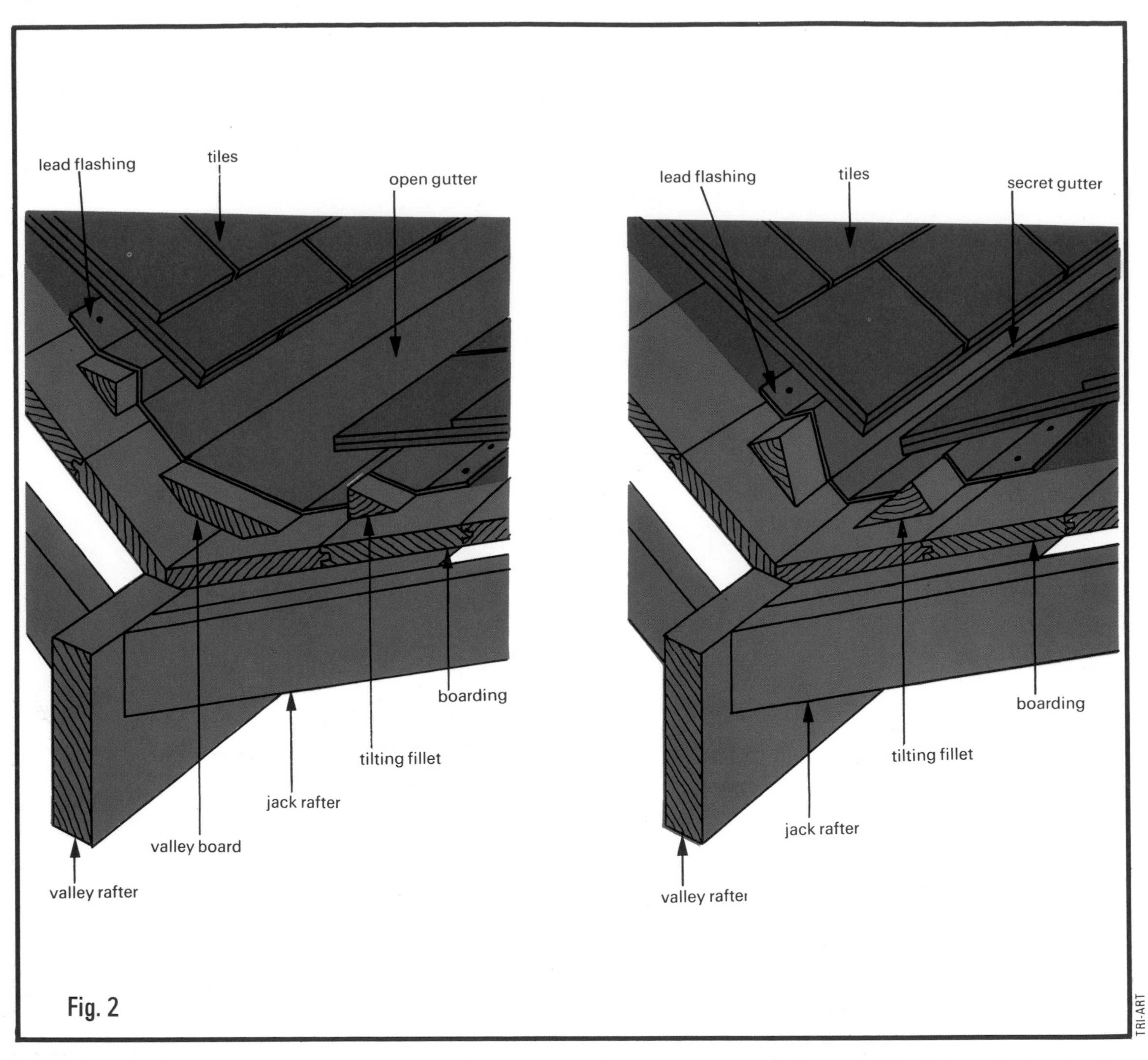

Fig. 2. *The construction of two types of gutter over the valley rafter. The bottom of the tiles are supported by a couple of tilting fillets. In an* open gutter *(left) there is about 8 in. between the tiles, so the tilting fillets are spaced some distance apart and a valley board fitted. In a* secret gutter *(right) there is only a 1 in. gap between the edges of the tiles and the tilting fillets are set at either side of the valley rafter.*

Then erect the framework for the second section of the roof so that the ends of the ridge boards for both sections meet. Place the pair of common rafters at the end of the second part of the roof in position and brace them with temporary supports. Place the end of the ridge board in the birdsmouth in the common rafters as described fully in CHAPTER 34. Position the pair of common rafters nearest the intersection of the two roof sections flat on the joists, place the other end of the ridge board in the birdsmouth and raise the rafter into position. Check that the ends of one of the ridge boards meets the side of the other, and then nail this butt joint firmly in position. Fix the other common rafters and hip rafters in position.

Hold the valley rafter above and parallel with its eventual position and cut as for the hip rafter. Make sure you allow enough to cut a birdsmouth in the bottom end to fit over the wall plates. Nail the rafter in place and fit extra support struts in position underneath.

The jack rafters are then cut and fitted between the hip rafter and wall plates, and between the valley rafter and ridge board.

Once the framework is completed all that has to be done is to fit tiling battens and roofing felt and tile the roof as previously described. If you are using lead flashing to make the joint between the sides of the roof in the valley watertight, fit the valley boards, tilting fillets and flashing in position before you start tiling.

Valley over bays

In a number of older terraced houses (row houses) a sloping gabled roof projects from the middle of the main roof out over a bay as in Fig. 3. Very often the ridge board on this extension is considerably lower than the ridge board of the main roof. This type of sloping roof is largely decorative and not intended to house another room in the loft. Therefore its construction is fairly simple.

The main roof structure is built in the normal way with common rafters running from the wall plates to the ridge board where the extension is to be fitted. A couple of valley boards (in place of rafters) are then nailed to the common rafters so that they slope upward from the

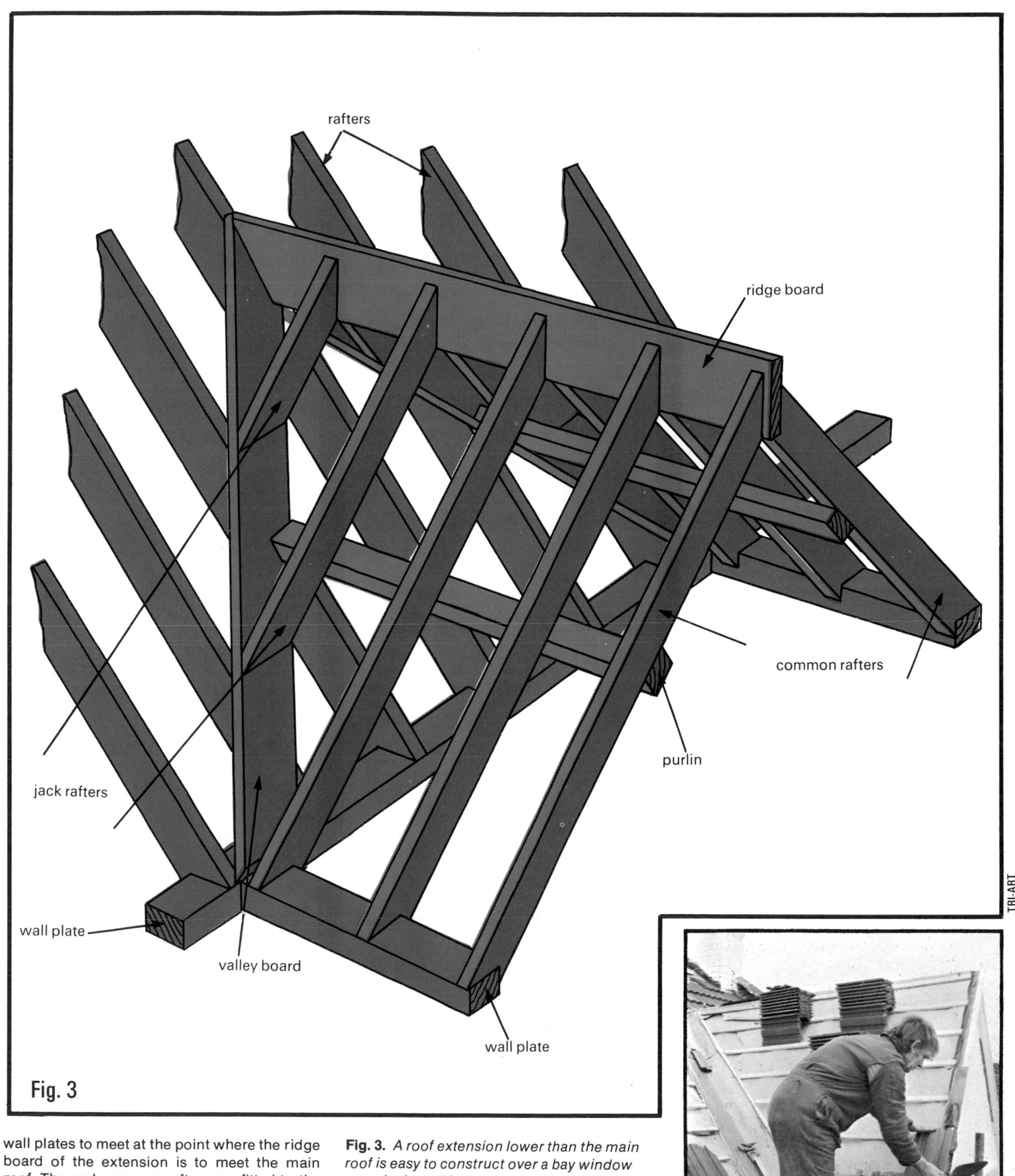

wall plates to meet at the point where the ridge board of the extension is to meet the main roof. The end common rafters are fitted to the end of the bay window (as in Fig. 3) and the ridge board is fitted and nailed to the valley boards where they meet. The jack rafters are then cut and fitted to run between the ridge board and valley board. When tiled, this type of sloping roof over a bay provides a pleasing break to an otherwise plain roof line.

Fig. 3. *A roof extension lower than the main roof is easy to construct over a bay window or an L-shaped house extension. Two valley boards are nailed to the common rafters of the main roof so that they run from the wall plates to the top of the roof extension. The method of construction is then the same as for a gabled roof.* **Right.** *An L-shaped roof with a gabled end to the extension. The gutter is completed and the roof battened ready for tiling.*

Index